T0310688

Big Data

Storage, Sharing, and Security

OTHER BOOKS BY FEI HU

Associate Professor
Department of Electrical and Computer Engineering
The University of Alabama

Cognitive Radio Networks
with Yang Xiao
ISBN 978-1-4200-6420-9

Wireless Sensor Networks: Principles and Practice
with Xiaojun Cao
ISBN 978-1-4200-9215-8

Socio-Technical Networks: Science and Engineering Design
with Ali Mostashari and Jiang Xie
ISBN 978-1-4398-0980-8

**Intelligent Sensor Networks: The Integration of Sensor Networks,
Signal Processing and Machine Learning**
with Qi Hao
ISBN 978-1-4398-9281-7

Network Innovation through OpenFlow and SDN: Principles and Design
ISBN 978-1-4665-7209-6

Cyber-Physical Systems: Integrated Computing and Engineering Design
ISBN 978-1-4665-7700-8

**Multimedia over Cognitive Radio Networks: Algorithms, Protocols,
and Experiments**
with Sunil Kumar
ISBN 978-1-4822-1485-7

**Wireless Network Performance Enhancement via Directional Antennas:
Models, Protocols, and Systems**
with John D. Matyjas and Sunil Kumar
ISBN 978-1-4987-0753-4

**Security and Privacy in Internet of Things (IoTs): Models, Algorithms,
and Implementations**
ISBN 978-1-4987-2318-3

Spectrum Sharing in Wireless Networks: Fairness, Efficiency, and Security
with John D. Matyjas and Sunil Kumar
ISBN 978-1-4987-2635-1

Big Data: Storage, Sharing, and Security
ISBN 978-1-4987-3486-8

Opportunities in 5G Networks: A Research and Development Perspective
ISBN 978-1-4987-3954-2

Big Data

Storage, Sharing, and Security

Edited by Fei Hu

CRC Press
Taylor & Francis Group
Boca Raton London New York

CRC Press is an imprint of the
Taylor & Francis Group, an **informa** business

AN AUERBACH BOOK

CRC Press
Taylor & Francis Group
6000 Broken Sound Parkway NW, Suite 300
Boca Raton, FL 33487-2742

© 2016 by Taylor & Francis Group, LLC
CRC Press is an imprint of Taylor & Francis Group, an Informa business

No claim to original U.S. Government works

Printed on acid-free paper
Version Date: 20160226

International Standard Book Number-13: 978-1-4987-3486-8 (Hardback)

This book contains information obtained from authentic and highly regarded sources. Reasonable efforts have been made to publish reliable data and information, but the author and publisher cannot assume responsibility for the validity of all materials or the consequences of their use. The authors and publishers have attempted to trace the copyright holders of all material reproduced in this publication and apologize to copyright holders if permission to publish in this form has not been obtained. If any copyright material has not been acknowledged please write and let us know so we may rectify in any future reprint.

Except as permitted under U.S. Copyright Law, no part of this book may be reprinted, reproduced, transmitted, or utilized in any form by any electronic, mechanical, or other means, now known or hereafter invented, including photocopying, microfilming, and recording, or in any information storage or retrieval system, without written permission from the publishers.

For permission to photocopy or use material electronically from this work, please access www.copyright.com (http://www.copyright.com/) or contact the Copyright Clearance Center, Inc. (CCC), 222 Rosewood Drive, Danvers, MA 01923, 978-750-8400. CCC is a not-for-profit organization that provides licenses and registration for a variety of users. For organizations that have been granted a photocopy license by the CCC, a separate system of payment has been arranged.

Trademark Notice: Product or corporate names may be trademarks or registered trademarks, and are used only for identification and explanation without intent to infringe.

Visit the Taylor & Francis Web site at
http://www.taylorandfrancis.com

and the CRC Press Web site at
http://www.crcpress.com

For Gloria, Edwin & Edward (twins)......

Contents

Preface

Big Data is one of the hottest topics today because of the large-scale data generation and distribution in computing products. It is tightly integrated with other cutting-edge networking technologies, including cloud computing, social networks, Internet of things, and sensor networks. Characteristics of Big Data may be summarized as four Vs, that is, volume (great volume), variety (various modalities), velocity (rapid generation), and value (huge value but very low density). Many countries are paying high attention to this area. As an example, in the United States in March 2012, the Obama Administration announced a US$200 million investment to launch the "Big Data Research and Development Plan," which was a second major scientific and technological development initiative after the "Information Highway" initiative in 1993.

Because Big Data is a relatively new field, there are many challenging issues to be addressed today: (1) *Storage*—How do we aggregate heterogeneous types of data from numerous sources, and then use fast database management technology to store the Big Data? (2) *Sharing*—How do we use cloud computing to share the Big Data among large groups of people? (3) *Security*—How do we protect the privacy of Big Data during the network sharing? This book will cover the above 3S designs, through the detailed description of the concepts and implementations.

This book is unlike any other similar books. Because Big Data is such a new field, there are very few books covering its implementation. Although a few similar books are already published, they are mostly about the basic concepts and society impacts. They are thus not suitable for R&D people. Instead, this book will discuss Big Data management from an R&D perspective.

Targeted Audiences: (1) Industry—company engineers can use this book as a reference for the design of Big Data processing and protection. There are many practical design principles covered in the chapters. (2) Academia—researchers can gain much knowledge on the latest research topics in this area. Graduate students can resolve many issues by reading the chapters. They will gain a good understanding of the status and trend of Big Data management.

Book Architecture: The book consists of two sections:

- *Section I. Big Data management*: In this section we cover the following important topics:

    - *Spatial management*: In many applications and scientific studies, there is a growing need to manage spatial entities and their topological, geometric, or geographic properties. Analyzing such large amounts of spatial data to derive values and guide decision making has become essential to business success and scientific progress.

- *Data transfer*: A content delivery network with large data centers located around the world requires Big Data transfer for data migration, updates, and backups. As cloud computing becomes common, the capacity of the data centers and both the intranetwork and internetwork of those data centers increase.

- *Data processing*: Dealing with "Big Data" problems requires a radical change in the philosophy of the organization of information processing. Primarily, the Big Data approach has to modify the underlying computational model to manage uncertainty in the access to information items in a huge nebulous environment.

- *Section II. Big Data Security*: Security is a critical aspect after Big Data is integrated with cloud computing. We will provide technical details on the following aspects:

  - *Security*: To achieve a secure, available, and reliable Big Data cloud-based service, we not only present the state-of-the-art of Big Data cloud-based services, but also a novel architecture to manage reliability, availability, and performance for accessing Big Data services running on the cloud.

  - *Privacy*: We will examine privacy issues in the context of Big Data and potential data mining of that data. Issues are analyzed based on the emerging unique characterizations associated with Big Data: the Big Data Lake, "thing" data, the quantified self, repurposed data, and the generation of knowledge from unstructured communication data, that is, Twitter Tweets. Each of those sets of emerging issues is analyzed in detail for their potential impact on privacy.

  - *Accountability*: Accountability of user data access on a specific application helps in monitoring, controlling, and assessing data usage by the user for the application. Data loss is the main source of leaking information that may possibly compromise the privacy of individual and/or organization. Therefore, the naive question is, "how can data leakages be controlled and detected?" The simple answer to this would be audit logs and effective measures of data usage.

The chapters have detailed technical descriptions of the models, algorithms, and implementations of Big Data management and security aspects. There are also accurate descriptions on the state-of-the-art and future development trends of Big Data applications. Each chapter also includes references for readers' further studies.

Thank you for reading this book. We believe that it will help you with the scientific research and engineering design of Big Data systems. We welcome your feedback.

Fei Hu
University of Alabama, Tuscaloosa, Alabama

Editor

Dr. Fei Hu is currently a professor in the Department of Electrical and Computer Engineering at the University of Alabama, Tuscaloosa, Alabama. He earned his PhD degrees at Tongji University (Shanghai, China) in the field of signal processing (in 1999), and at Clarkson University (New York) in electrical and computer engineering (in 2002). He has published over 200 journal/conference papers and books. Dr. Hu's research has been supported by the U.S. National Science Foundation, Cisco, Sprint, and other sources. His research expertise can be summarized as *3S: Security, Signals, Sensors*: (1) *Security*—This deals with overcoming different cyber attacks in a complex wireless or wired network. His current research is focused on cyber-physical system security and medical security issues. (2) *Signals*—This mainly refers to *intelligent signal processing*, that is, using machine learning algorithms to process sensing signals in a smart way to extract patterns (i.e., pattern recognition). (3) *Sensors*—This includes microsensor design and wireless sensor networking issues.

Contributors

Emad Abd-Elrahman
RST Department
Telecom Sudparis
Evry, France

Ablimit Aji
Analytics Lab
Database Systems
Hewlett Packard Labs
Palo Alto, California

Usamah AlGemili
Department of Computer Science
George Washington University
Washington, DC

Adi Alhudhaif
Department of Computer Science
Prince Sattam bin Abdulaziz University
Al-Kharj, Saudi Arabia

Faisal Alsaby
Department of Computer Science
George Washington University
Washington, DC

Nadia Bennani
INSA-Lyon
LIRIS Department
University of Lyon
Lyon, France

Simon Y. Berkoich
COMStar Computing Technology Institute
and
Department of Computer Science
George Washington University
Washington, DC

Nevil Brownlee
Department of Computer Science
University of Auckland
Auckland, New Zealand

Lionel Brunie
INSA-Lyon
LIRIS Department
University of Lyon
Lyon, France

Thomas Cerqueus
INSA-Lyon
LIRIS Department
University of Lyon
Lyon, France

Ernesto Damiani
Department of Computer Technology
University of Milan
Milan, Italy

Manik Lal Das
Dhirubhai Ambani Institute
    of Information and Communication
    Technology
Gujarat, India

Vijay Gadepally
MIT Lincoln Laboratory
Lexington, Massachusetts

Ibrahim A. Gomaa
Computer and Systems Department
National Telecommunication Institute
Cairo, Egypt

Fouad Amine Guenane
ENST
Telecom ParisTech
Paris, France

Benjamin Habegger
INSA-Lyon
LIRIS Department
University of Lyon
Lyon, France

Ariel Hamlin
MIT Lincoln Laboratory
Lexington, Massachusetts

Omar Hasan
INSA-Lyon
LIRIS Department
University of Lyon
Lyon, France

Yuh-Jong Hu
Department of Computer Science
National Chengchi University Taipei
Taipei, Taiwan

Rasheed Hussain
Department of Computer Science
and Engineering
Hanyang University
Ansan, South Korea

Jeremy Kepner
MIT Lincoln Laboratory
Lexington, Massachusetts

Donghyun Kim
Department of Mathematics and Physics
North Carolina Central University
Durham, North Carolina

Harald Kosch
Department of Informatics and Mathematics
University of Passau
Passau, Germany

Duoduo Liao
COMStar Computing Technology Institute
Washington, DC

Dongxi Liu
CSIRO
Clayton South Victoria, Australia

Wen-Yu Liu
Department of Computer Science
National Chengchi University Taipei
Taipei, Taiwan

Jianguo Lu
School of Computer Science
University of Windsor
Ontario, Canada

Aniket Mahanti
Department of Computer Science
University of Auckland
Auckland, New Zealand

Michele Nogueira
Department of Informatics
NR2—Federal University of Parana
Curitiba, Brazil

Heekuck Oh
Department of Computer Science
and Engineering
Hanyang University
Ansan, South Korea

Daniel E. O'Leary
University of Southern California
Los Angeles, California

Albert Reuther
MIT Lincoln Laboratory
Lexington, Massachusetts

Jun Sakuma
Department of Computer Science
University of Tsukuba
Tsukuba, Japan

Nabil Schear
MIT Lincoln Laboratory
Lexington, Massachusetts

Ahmed Serhrouchni
ENST
Telecom ParisTech
Paris, France

Emily Shen
MIT Lincoln Laboratory
Lexington, Massachusetts

Junggab Son
Department of Mathematics and Physics
North Carolina Central University
Durham, North Carolina

Mayank Varia
Boston University
Boston, Massachusetts

Dong Wang
Department of Computer Science
and Engineering
University of Notre Dame
Notre Dame, Indiana

Fusheng Wang
Department of Biomedical Informatics
and
Department of Computer Science
Stony Brook University
Stony Brook, New York

Shenlu Wang
School of Computer Science and Engineering
University of New South Wales
Sydney, Australia

Yan Wang
School of Information
Central University
    of Finance and Economics
Beijing, China

J. Gerard Wolff
CognitionResearch.org
Menai Bridge, United Kingdom

Sophia Yakoubov
MIT Lincoln Laboratory
Lexington, Massachusetts

Maryam Yammahi
Department of Computer Science
George Washington University
Washington, DC

and

College of Information
    Technology
United Arab Emirates University
Al Ain, United Arab Emirates

Arkady Yerukhimovich
MIT Lincoln Laboratory
Lexington, Massachusetts

Se-young Yu
Department of Computer Science
University of Auckland
Auckland, New Zealand

John Zic
CSIRO
Clayton South Victoria, Australia

BIG DATA MANAGEMENT: STORAGE, SHARING, AND PROCESSING

I

Chapter 1

Challenges and Approaches in Spatial Big Data Management

Ablimit Aji

Fusheng Wang

CONTENTS

1.1 Introduction

Advancements in computer technology and the rapid growth of the Internet have brought many changes to society. More recently, the Big Data paradigm has disrupted many industries ranging from agriculture to retail business, and fundamentally changed how businesses operate and make decisions at large. The rise of Big Data can be attributed to two main reasons:

3

First, high volumes of data generated and collected from *devices*. The rapid improvement of high-resolution data acquisition technologies and sensor networks have enabled us to capture large amounts of data at an unprecedented scale and rate. For example, the GeoEye-1 satellite has the highest resolution of any commercial imaging system and is able to collect images with a ground resolution of 0.41 m in the panchromatic or black and white mode [1]; the Sloan Digital Sky Survey (SDSS), with a rate of about 200 GB per night, has amassed more than 140 TB of information [5]; and the modern medical imaging scanners can capture the micro-anatomical tissue details at the billion pixel resolution [13].

Second, traces of *human* activity and crowd-sourcing efforts facilitated by the Internet. The proliferation of cost-effective and ubiquitous positioning technologies, mobile devices, and sensors have enabled us to collect massive amounts of spatial information of human and wildlife activity. For example, FourthSquare—a popular local search and discovery service—allow users to *check-in* at more than 60 million venues, and so far has more than 6 billion check-ins [2]. Driven by the business potential, more and more businesses are providing services that are *location-aware*. At the same time, the Internet has made remote collaboration so easy that, now, a crowd can even generate a free mapping of the world autonomously. OpenStreetMap [3] is a large collaborative mapping project, which is generated by users around the globe, and it has more than two million registered users as of this writing.

In many applications and scientific studies, there is a growing need to manage spatial entities and their topological, geometric, or geographic properties. Analyzing such large amounts of spatial data to derive values and guide decision making have become essential to business success and scientific progress. For example, location-based social networks (LBSNs) are utilizing large amounts of user location information to provide geo-marketing and recommendation services. Social scientists are relying on such data to study dynamics of social systems and understand human behavior. Epidemiologists are combining such spatial data with public health data to study the patterns of disease outbreak and spread. In all those domains, spatial Big Data analytics infrastructure is a key enabler.

Over the last decade, the Big Data technology stack and the software ecosystem has evolved to cope with most common use cases. However, modern data-intensive spatial applications require a different approach to be able to handle unique requirements of spatial Big Data.

In the rest of this chapter, first we provide examples of data-intensive spatial applications, and discuss the unique challenges that are common to them. Then, we present major research efforts, data-intensive computing techniques, and software systems that are intended to address these challenges. Lastly, we conclude the chapter with a discussion on future outlook of this area.

1.2   Big Spatial Data and Applications

The rapid growth of spatial data is driven not only by conventional applications, but also by emerging scientific applications and large internet services that have become data-intensive and compute-intensive.

1.2.1   *Spatial analytics for derived scientific data*

With the rapid improvement of data acquisition technologies such as high-resolution tissue slide scanners and remote sensing instruments, it has become more efficient to capture extremely large spatial data to support scientific research. For example, digital pathology imaging has become an emerging field in the past decade, where examination of high-resolution images of tissue specimens enables novel, more effective ways of screening for disease, classifying disease states, understanding its progression, and evaluating the efficacy of therapeutic strategies. In clinical environment, medical professionals have been relying on the manual judgment from pathologists—a process inherently subject to human bias—to diagnose, and understand the disease condition.

Today, *in silico* pathology image analysis offers a means of rapidly carrying out quantitative, reproducible measurements of micro-anatomical features in high-resolution pathology images and large image datasets. Medical professionals and researchers can use computer algorithms to calculate the distribution of certain cell types, and perform associative analysis with other data such as patient genetic composition and clinical treatment.

Figure 1.1 shows a protocol for *in silico* pathology image analysis pipeline. From left to the right, the sub-figures represent: glass slides, high-resolution image scanning, whole slide images, and automated image analysis. The first three steps are data acquisition processes that are mostly done in a pathology laboratory environment, and the final step is where the computerized analysis is performed. In the image analysis step, regions of micro-anatomical objects (millions per image) such as nuclei and cells are computed through image segmentation algorithms, represented with their boundaries, and image features are extracted from these objects. Exploring the results of such analysis involves complex queries such as spatial cross-matching, overlay of multiple sets of spatial objects, spatial proximity computations between objects, and queries for global spatial pattern discovery. These queries often involve billions of spatial objects and heavy geometric computations.

Scientific simulation also generates large amounts of spatial data. Scientists often use models to simulate natural phenomena, and analyze the simulation process and data. For example, earth science uses simulation models to help predict the ground motion during earthquakes. Ground motion is modeled with an octree-based hexahedral mesh, using soil density as input. Simulation tools calculate the propagation of seismic waves through the Earth by approximating the solution to the wave equation at each mesh node. During each time step, for each node in the mesh, the simulator calculates the node velocity in spatial directions, and records those information to the primary storage. The simulation result is a spatio temporal earthquake data set describing the ground velocity response [6]. As the scale of the experiment increases, the resulting dataset also increases, and scientists often struggle to query and manage such large amounts of spatio temporal data in an efficient and cost-effective manner.

1.2.2   GIS and social media applications

Volunteered geographic information (VGI) further enriched global information system (GIS) world with massive amounts of user-generated geographical and social data. VGI is a special case of the larger Internet phenomenon—user-generated content—in the GIS domain. Everyday Internet users can provide, modify, and share geographical data using interactive online

Figure 1.1: Derived spatial data in pathology image analysis.

services such as OpenStreetMap [3], Wikimapia, GoogleMap, GoogleEarth, and Microsoft's Virtual Earth. The spatial information needs to be constantly analyzed and corroborated to track changes, and understand the current status. Most often, a spatial database system is used to perform such analysis.

Recently, the explosive growth of social media applications contributed massive amounts of user-generated geographic information in the form of tweets, status updates, check-ins, Waze, and traffic reports. Furthermore, if such geospatial information is not available, automated geo tagging/coding tools can infer and assign an approximate location to those contents. Analysis of such large amounts of data has implications for many applications—both commercial and academic. In [11] authors have used the geospatial information to investigate the relationship between the geographic location of protestors attending demonstrations in the 2013 Vinegar protests in Brazil and the geographic location of users that tweeted the protests. Another example is location-based targeted advertising [24] and recommendation [18]. Those online services and GIS systems are backed by conventional spatial database systems that are optimized for different application requirements.

1.3 Challenges and Requirements

Modern data-intensive spatial analytics applications are different from conventional applications in several aspects. They involve the following:

- *Large volumes of multidimensional data*: Conventional warehousing applications deal with data generated from business transactions. As a result, the underlying data (such as numbers and strings) tend to be relatively *simple* and *flat*. However, this is not the case for the spatial applications which deal with massive amounts of geometry shapes and spatial objects. For example, a typical whole slide pathology contains more than 20 billion pixels, millions of objects, and 100 million derived image features. A single study may involve thousands of images analyzed with dozens of algorithms— with varying parameters—to generate many different result sets to be compared and consolidated, at the scale of tens of terabytes. A moderate-size healthcare operation can routinely generate thousands of whole slide images per day, leading to petabytes of analytical results per year. A single 3D pathology image could come from a thousand slices and take 1 TB storage, containing several millions to 10 millions of derived 3D surface objects.

- *High computation complexity*: Most spatial queries involve multidimensional geometric computations that are often compute-intensive. While spatial filtering through minimum bounding rectangles (MBRs) can be accelerated through spatial access methods, spatial refinements such as polygon intersection verification are highly expensive operations. For example, spatial join queries such as spatial cross-matching or spatial overlay can be very expensive to process. This is mainly due to the polynomial complexity of many geometric computation methods. Such compute-intensive geometric computation, combined with the large volumes of Big Data requires a high-performance solution.

- *Complex spatial queries*: Spatial queries are complex to express in current spatial data analytics systems. Most scientific researchers and spatial application developers are often interested in running queries that involve complex spatial relationships such as nearest neighbor query, and spatial pattern queries. Such queries are not well supported in current spatial database systems. Frequently, users are forced to write database user-defined functions to be able to perform the required operations. SQL—structured query language—has gained tremendous momentum in the relational database field and become the de facto standard for querying the data. While most spatial queries can be expressed in SQL, due to the structural differences in the programming model, efficient SQL-based spatial queries are often hard to write and requires considerable optimization efforts.

A major requirement for the spatial analytics systems is fast query response. Scientific research or analytics in general, is an iterative and exploratory process in which large amounts of data can be generated quickly for the initial prototyping and validation. This requires a scalable architecture that can query spatial data on a large scale. Another requirement is to support queries on a cost-effective architecture such as commodity clusters or cloud environments. Meanwhile, scientific researchers and application developers often prefer expressive query languages over programming API, without worrying about how the queries are translated, optimized, and executed. With the rapid improvement of instrument resolutions, the increased accuracy of data analysis methods, and the massive scale of observed data, complex spatial queries have become increasingly compute-intensive and data-intensive.

1.4 Spatial Big Data Systems and Techniques

Two mainstream approaches for large-scale data analysis are parallel database systems [15] and MapReduce-based systems [14]. Both approaches share certain common design elements: they both employ a shared-nothing architecture [25], and deployed on a cluster of independent nodes via a high-speed interconnecting network; both achieve parallelism by partitioning the data and processing the query in parallel on each partition.

However, parallel database approach has major limitations on managing and querying spatial data at massive scale. Parallel database management systems (DBMSs) tend to reduce the I/O bottleneck through partitioning of data on multiple parallel disks and are not optimized for computational-intensive operations such as spatial and geometric computations. Partitioned parallel DBMS architecture often lacks effective spatial partitioning to balance data and task loads across database partitions. While it is possible to induce a spatial partitioning, fixed grid tiling, for example, and map such partitioning to one dimensional attribute distribution key, such an approach fails to handle boundary objects for accurate query processing. Scaling out spatial queries through a parallel database infrastructure is possible while being costly, and such approach is explored in [27,28]. More recently, Spark [31] has emerged as a new data processing framework for handling iterative and interactive workloads.

Due to both computational intensity and data intensity of spatial workloads, large-scale parallelization often holds the key to achieving high-performance spatial queries. As the cloud-based cluster computing technology gets mature and economically scalable, MapReduce-based systems offer an alternative solution for data and compute-intensive spatial analytics at large scale. Meanwhile, parallel processing of queries rely on effective data partitioning to scale. Considering that spatial workloads are often compute-intensive [7,22], how to utilize hardware accelerators for query co-processing is a very promising technique as modern computer systems are embracing heterogeneous architecture that combines graphics processing unit (GPU) and CPU [12]. In the rest of the chapter, we elaborate each of these techniques in greater detail, and summarize state-of-the-art approaches and systems.

1.4.1 MapReduce-based spatial query processing

MapReduce is a very scalable parallel processing framework that is designed to process flat unstructured data. However, it is not particularly well suited to process multidimensional spatial objects, and several systems have emerged over the past few years to fill this gap. Well-known systems and prototypes include HadoopGIS [8–10], SpatialHadoop [16,17], Parallel Secondo [20,21], and GIS tools for Hadoop [4,30]. These systems are based on the open source implementation of MapReduce—Hadoop, and provides similar analytics functionality. However, they differ in implementation details and architecture: HadoopGIS and SpatialHadoop are pure MapReduce-based query evaluation systems; Parallel Secondo is a hybrid system that combines a database engine with MapReduce; and GIS tools for Hadoop is a functional extension of Hive [26] with user-defined functions.

MapReduce relies on the partitioning of data to process them in parallel, and it is the key for a high-performance system. In the context of large-scale spatial data analytics, an intuitive approach is to partition the dataset based on the spatial attribute, and assign spatial objects to partitioned regions (or tiles). Consequently, generated tiles form a parallelization unit that can be processed independently in parallel. A MapReduce-based spatial query processing system takes advantage of such partitioning to achieve high performance. Algorithm 1.1 illustrates a general design framework for such systems, and all the above-mentioned systems follow this framework while implementation details may vary.

Algorithm 1.1: Typical workflow of spatial query processing on MapReduce

---

1 A. Data/space partitioning
2 B. Data storage of partitioned data on HDFS
3 **for** *tile* **in** *input_collection* **do**
4   Indexing building for objects in the tile
5   Tile based spatial querying processing

6 E. Boundary object handling
7 G. Data aggregation
8 H. Result storage on HDFS

Initially the dataset is spatially partitioned to generate tiles as shown in step A. In step B, spatial objects are assigned unique tile identifiers (UIDs), merged, and stored into Hadoop Distributed File System (HDFS). Step C is for pre-processing queries, which could be queries that do global index-based filtering. Step D does tile-based spatial query processing independently and parallelized across a large number of cluster nodes. Step E provides handling of boundary objects that arise from the partitioning. Step F is for post-query processing, and step G performs data aggregation. Finally, the query results are persisted to HDFS, which can be input to the next query operator.

Following such framework, spatial queries such as spatial join query, spatial range query, and nearest neighbor query can be implemented efficiently. Reference implementations are provided in HadoopGIS, SpatialHadoop, and Parallel Secondo.

1.4.2 Effective spatial data partitioning

Spatial data partitioning is an essential initial step to define, generate, and represent partitioned data. Effective data partitioning is critical for task parallelization, load balancing, and directly affects system performance. Generally a space-oriented partitioning can be applied to generate data partitions, and the concept is illustrated in Figure 1.2 in which the spatial data is partitioned into uniform grids.

However, there are several problems with this approach: (1) As spatial objects (e.g., polygons and polylines) are extent, regular grid-based spatial partitioning would undesirably

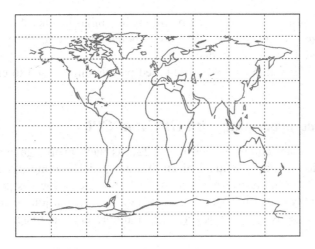

Figure 1.2: An example of spatial data partitioning.

produce objects spanning multiple cell grids, which need to be replicated and post-processed. If such objects account for a considerable fraction of the dataset, the overall query performance would suffer from such boundary handling overhead. (2) Fixed grid partitioning is skew-averse, whereas data in most real-world spatial applications are inherently highly skewed. In such case, it is very likely that parallel processing nodes assigned to process those dense regions will become the *stragglers*, and the overall query processing efficiency will suffer [23].

The boundary problem can be addressed in two different ways—*multi-assignment, single-join* (MASJ) and *single-assignment, multi-join* (SAMJ) [19,32]. In MASJ approach, each boundary object is replicated to each tile that overlaps with the object. During the query processing phase, each partition is processed only once without considering the boundary objects. Then a *de-duplication* step is initiated to remove the redundancies that resulted from the replication. However, in SAMJ approach, each boundary object is only assigned to one tile. Therefore, during the query processing phase, each tile is processed multiple times to account for the boundary objects.

Both approaches introduce extra query processing overhead. In MASJ, the replication of boundary objects incurs extra storage cost and computation cost. In SAMJ, however, only extra computation cost is incurred by processing the same partition multiple times. Hadoop-GIS and SpatialHadoop takes the MASJ approach and modify, query processing steps account for replicated objects. However, depending on the application requirement, such design choice can be re-evaluated and modified to achieve better performance.

The data skew problem can be mitigated through skew-aware partitioning approaches that can create balanced partitions. HadoopGIS uses a multi-step approach named *SATO* which can partition a geospatial dataset into *balanced regions* while *minimizing the number of boundary objects*. SATO represents the four main steps in this framework for spatial data partitioning: **S**ample, **A**nalyze, **T**ear, and **O**ptimize. First, a small fraction of the dataset is sampled to identify overall global data distribution with potential dense regions. The sampled data is analyzed with a *partition analyzer* that produces a coarse partition scheme in which each partition region is expected to contain roughly equal amounts of spatial objects. Later, these coarse partition regions are further processed with a partitioning component that *tears* the regions into more granular partitions that are much less skewed. Finally, generated partition meta-information is aggregated to produce multilevel partition indexes and additional partition statistics that are used to *optimize* spatial queries.

SpatialHadoop also creates balanced partitions in similar manner. Specifically, an appropriate partition size is estimated, and rectangular boundaries are generated according to such partition parameter. Then, a MapReduce job is initiated to create spatial partitions that corresponds to such configuration. An $R+$-Tree-based partitioning is used to ensure the partition size constraint.

1.4.3 Query co-processing with GPU and CPU

Most spatial queries are compute-intensive [7,22], as they involve geometric computations on complex multidimensional spatial objects. While spatial filtering through MBRs can be accelerated through spatial access methods, spatial refinements such as polygon intersection verification are highly expensive operations. For example, spatial join queries such as spatial cross-matching or overlaying multiple sets of spatial objects on an image or map can be very expensive to process.

GPUs have been successfully utilized in numerous applications that require high-performance computation. Mainstream general purpose GPUs come with hundreds of cores, and can run thousands of threads in parallel. Compared to the multi-core computer systems (dozens of cores), GPUs can scale to large number of threads in a cost-effective manner. In the coming years, such heterogeneous parallel architecture will become dominant, and software systems must fully exploit such heterogeneity to deliver performance growth [12].

In most cases, spatial algorithms are designed for executing on the CPUs, and the branch-intensive nature of CPU-based algorithms require the algorithm to be rewritten for GPUs for satisfactory performance. For such need, PixelBox is proposed in [29]. PixelBox is an algorithm specifically designed for accelerating cross-matching queries on the GPUs. It first transforms the vector-based geometry representation into raster representation using a pixelization method, and performs operations on such representations in parallel. The pixelization method reduces the geometry calculation problem into simple pixel position checking problem, and it is very suitable for execution on GPUs. Since testing the position of one pixel is totally independent of another, it can parallelize the computation by having multiple threads process the pixels in parallel. Furthermore, since the position of different pixels are computed against the same pair of polygons, the operations performed by different threads follow the single instruction multiple data (SIMD) fashion, a parallel computation model that GPUs are designed for. Experimental results [29] suggest that PixelBox achieves almost an orders of magnitude speedup compared to the CPU implementation, and significantly reduces the cost of computation.

1.4.3.1 Task assignment

One critical issue for GPU-based parallelization is task assignment. For example, in a recent approach that combines MapReduce and GPU-based query processing [7], tasks arrive in the form of data partitions along with spatial query operation on the data. Given a partition, the query optimizer has to decide which device should be assigned to execute the task. Such decision is not simple, and it depends on how much speedup can be obtained by assigning it to CPU or GPU. If we schedule a small task on GPU, we may not only get very little speedup, the opportunity cost of executing some other high speedup task on GPU can be high.

In such cases, a predictive modeling approach can offer a reasonable solution. Similar to the speculative execution model in Hadoop, a fraction of data (10%, for example) is used for performance profiling and model training. Then, regression or machine learning approaches are used to derive the performance model, and corresponding model parameters. Later during the runtime, the derived model is used to predict the potential speedup factor for current task, and tasks are scheduled to execute on GPU if the speedup factor is higher than certain threshold.

1.4.3.2   Effects of task granularity

Data need to be shipped to the device memory to be executed on the GPU device. Such data transfer incurs certain I/O cost. While the memory bandwidth between GPU and CPU is much higher compared to the bandwidth between memory and the disk, it should be minimized to achieve optimal performance. To achieve optimal speedup, the compute-to-transfer ratio needs to be high for GPU applications. Therefore, applications need to adjust the partition granularity to fully utilize system resources. While larger partitioning is ideal for achieving higher speedup on GPU, it causes data skew which is detrimental for MapReduce system performance. At the same time, a very small partition is not a good candidate for hardware acceleration.

1.5   Discussion and Conclusion

In this chapter, we have discussed several representative spatial Big Data applications, common challenges in this domain, and potential solutions toward those challenges. Spatial Big Data from various application domains share many similar requirements with enterprise Big Data, but has its own unique characteristics—spatial data are multidimensional, spatial queries are complex, and spatial query processing comes with high computational complexity. As the volume of data grow continuously, we need efficient Big Data systems and data management techniques to be able to cope with such challenges. MapReduce-based massively parallel query processing systems offer a scalable, yet cost-effective solution for processing large amounts of spatial data. While relying on such framework, effective data partitioning techniques can be critical for facilitating massive parallelism, and achieving satisfactory query performance. Meanwhile, as multi-core computer architecture and programming techniques become mature, hardware-accelerated query co-processing on GPUs can further improve query performance for large-scale spatial analytics tasks.

Acknowledgments

Fusheng Wang acknowledges that this material is based on work supported in part by NSF CAREER award IIS 1350885, NSF ACI 1443054, and by the National Cancer Institute under grant No. 1U24CA180924-01A1.

References

1. Satellite imagery. https://en.wikipedia.org/wiki/Satellite_imagery.

2. Foursquare Labs, Inc. https://foursquare.com/about.

3. OpenStreetMap: A map of the world, free to use. http://www.openstreetmap.org.

4. GIS tools for Hadoop, Big Data spatial analytics. http://esri.github.io/gis-tools-for-hadoop.

5. York DG, Adelman J, Anderson Jr. JE, Anderson SF, Annis J, Bahcall NA, Bakken JA et al. The sloan digital sky survey: Technical summary. *The Astronomical Journal*, 120(3):1579, 2000.

6. Anastasia Ailamaki, Verena Kantere, and Debabrata Dash. Managing scientific data. *Commun. ACM*, 53(6):68–78, 2010.

7. Ablimit Aji, Teodoro George, and Fusheng Wang. Haggis: Turbocharge a mapreduce based spatial data warehousing system with gpu engine. In *Proceedings of the 3rd ACM SIGSPATIAL International Workshop on Analytics for Big Geospatial Data*, pp. 15–20, Dallas, TX, 2014.

8. Ablimit Aji, Xiling Sun, Hoang Vo, Qioaling Liu, Rubao Lee, Xiaodong Zhang, Joel Saltz et al. Demonstration of hadoop-gis: A spatial data warehousing system over mapreduce. In *SIGSPATIAL/GIS*, pp. 518–521. ACM, New York, 2013.

9. Ablimit Aji, Fusheng Wang, and Joel H. Saltz. Towards building a high performance spatial query system for large scale medical imaging data. In *SIGSPATIAL/GIS*, pp. 309–318. ACM, New York, 2012.

10. Ablimit Aji, Fusheng Wang, Hoang Vo, Rubao Lee, Qiaoling Liu, Xiaodong Zhang, and Joel Saltz. Hadoop-GIS: A high performance spatial data warehousing system over MapReduce. *Proc. VLDB Endow.*, 6(11):1009–1020, August 2013.

11. Marco Bastos, Raquel Recuero, and Gabriela Zago. Taking tweets to the streets: A spatial analysis of the vinegar protests in Brazil. *First Monday*, 19(3), 2014. http://firstmonday.org/ojs/index.php/fm/article/view/5227/3843.

12. Shekhar Borkar and Andrew A Chien. The future of microprocessors. *Commun. ACM*, 54(5):67–77, 2011.

13. Lee AD Cooper, Alexis B Carter, Alton B Farris, Fusheng Wang, Jun Kong, David A Gutman, Patrick Widener et al. Digital pathology: Data-intensive frontier in medical imaging. *Proc. IEEE*, 100(4):991–1003, 2012.

14. Jeffrey Dean and Sanjay Ghemawat. Mapreduce: Simplified data processing on large clusters. *Commun. ACM*, 51(1):107–113, 2008.

15. David DeWitt and Jim Gray. Parallel database systems: The future of high performance database systems. *Commun. ACM*, 35(6):85–98, 1992.

16. Ahmed Eldawy and Mohamed F Mokbel. A demonstration of spatialhadoop: An efficient mapreduce framework for spatial data. *Proc. VLDB Endow.*, 6(12):1230–1233, 2013.

17. Ahmed Eldawy and Mohamed F Mokbel. Spatialhadoop: A mapreduce framework for spatial data. In *Proceedings of the IEEE International Conference on Data Engineering*. IEEE, Seol, Korea, 2015.

18. Justin J Levandoski, Mohamed Sarwat, Ahmed Eldawy, and Mohamed F Mokbel. Lars: A location-aware recommender system. In *IEEE 28th International Conference on Data Engineering*, pp. 450–461. IEEE, Arlington, VA, 2012.

19. Ming-Ling Lo and Chinya V Ravishankar. Spatial hash-joins. In *ACM SIGMOD Record*, pp. 247–258. ACM, Montreal, Canada, 1996.

20. Jiamin Lu and Ralf H Guting. Parallel secondo: Practical and efficient mobility data processing in the cloud. In *IEEE International Conference on Big Data*, pp. 107–125. IEEE, Silicon Valley, CA, 2013.

21. Jiamin Lu and Ralf Hartmut Guting. Parallel secondo: A practical system for large-scale processing of moving objects. In *IEEE 30th International Conference on Data Engineering*, pp. 1190–1193. IEEE, Chicago, IL, 2014.

22. Bogdan Simion, Suprio Ray, and Angela D Brown. Surveying the landscape: An in-depth analysis of spatial database workloads. In *SIGSPATIAL*, pp. 376–385. ACM, New York, 2012.

23. Benjamin Sowell, Marcos V Salles, Tuan Cao, Alan Demers, and Johannes Gehrke. An experimental analysis of iterated spatial joins in main memory. *Proc. VLDB Endow.*, 6(14):1882–1893, 2013.

24. Jack Steenstra, Alexander Gantman, Kirk Taylor, and Liren Chen. Location based service (lbs) system and method for targeted advertising, March 23, 2006. US Patent App. 10/931,309.

25. Michael Stonebraker. The case for shared nothing. *IEEE Database Eng. Bull.*, 9(1):4–9, 1986.

26. Ashish Thusoo, Joydeep S Sarma, Namit Jain, Zheng Shao, Prasad Chakka, Suresh Anthony, Hao Liu et al. Hive: A warehousing solution over a map-reduce framework. *Proc. VLDB Endow.*, 2(2):1626–1629, August 2009.

27. Fusheng Wang, Jun Kong, Lee Cooper, Tony Pan, Tahsin Kurc, Wenjin Chen, Ashish Sharma et al. A data model and database for high-resolution pathology analytical image informatics. *J. Pathol. Inform.*, 2(1):32, 2011.

28. Fusheng Wang, Jun Kong, Jingjing Gao, Lee Cooper, Tahsin M Kurc, Zhengwen Zhou, David Adler et al. A high-performance spatial database based approach for pathology imaging algorithm evaluation. *J. Pathol. Inform.*, 4:5, 2013.

29. Kaibo Wang, Yin Huai, Rubao Lee, Fusheng Wang, Xiaodong Zhang, and Joel H Saltz. Accelerating pathology image data cross-comparison on CPU-GPU hybrid systems. *Proc. VLDB Endow.*, 5(11):1543–1554, 2012.

30. Randall T Whitman, Michael B Park, Sarah M Ambrose, and Erik G Hoel. Spatial indexing and analytics on hadoop. In *Proceedings of the 22nd ACM SIGSPATIAL International Conference on Advances in Geographic Information Systems*, pp. 73–82. ACM, New York, 2014.

31. Matei Zaharia, Mosharaf Chowdhury, Michael J Franklin, Scott Shenker, and Ion Stoica. Spark: Cluster computing with working sets. In *Proceedings of the 2nd USENIX Conference on Hot Topics in Cloud Computing*, p. 10. USENIX Association, Berkeley, CA, 2010.

32. Xiaofang Zhou, David J Abel, and David Truffet. Data partitioning for parallel spatial join processing. *GeoInformatica*, 2(2):175–204, 1998.

Chapter 2

Storage and Database Management for Big Data*

Vijay Gadepally

Jeremy Kepner

Albert Reuther

CONTENTS

*This work is sponsored by the Assistant Secretary of Defense for Research and Engineering under Air Force Contract #FA8721-05-C-0002. Opinions, interpretations, recommendations and conclusions are those of the authors and are not necessarily endorsed by the United States Government.

2.1 Introduction

The ability to collect and analyze large amounts of data is a growing problem within the scientific community. The growing gap between data and users calls for innovative tools that address the challenges faced by big data volume, velocity, and variety. While there has been great progress in the world of database technologies in the past few years, there are still many fundamental considerations that must be made by scientists. For example, which of the seemingly infinite technologies are the best to use for my problem? Answers to such questions require a careful understanding of the technology field in addition to the types of problems that are being solved. This chapter aims to address many of the pressing questions faced by individuals interested in using storage or database technologies to solve their big data problems.

Storage and database management is a vast field with many decades of results from very talented scientists and researchers. There are numerous books, courses, and articles dedicated to the study. This chapter attempts to highlight some of these developments as they relate to the equally vast field of big data. However, it would be unfair to say that this chapter provides a comprehensive analysis of the field—such a study would require many volumes. It is our hope that this chapter can be used as a launching pad for researchers interested in the study. Where possible, we highlight important studies that can be pursued for further reading.

In Section 2.2, we discuss the big data challenge as it relates to storage and database engines. The chapter goes on to discuss database utility compared to large parallel storage arrays. Then, the chapter discusses the history of database management systems with special emphasis on current and upcoming database technology trends. In order to provide readers with a deeper understanding of these technologies, the chapter will provides a deep dive into two canonical open source database technologies: Apache Accumulo [1], which is based on the popular Google BigTable design, and a NewSQL array database called SciDB [59]. Finally, we will provide insight into technology selection and walk readers through a case study which highlights the use of various database technologies to solve a medical big data problem.

2.2 Big Data Challenge

Working with big data is prone to a variety of challenges. Very often, these challenges are referred to as the three Vs of big data: Volume, Velocity and Variety [45]. Most recently, there has been a new emergent challenge (perhaps a fourth V): Veracity. These combined challenges constitute a large reason why big data is so difficult to work with.

Big data volume stresses the storage, memory, and computational capacity of a computing system and often requires access to a computing cloud. The National Institute of Science and Technology (NIST) defines cloud computing to be "a model for enabling ubiquitous, convenient, on-demand network access to a shared pool of configurable computing resources ... that can be rapidly provisioned and released with minimal management effort or service provider interaction" [47]. Within this definition, there are different cloud models that satisfy different problem characteristics and choosing the right cloud model is problem specific. Currently, there are four multibillion dollar ecosystems that dominate the cloud-computing landscape: enterprise clouds, big data clouds, Structured Query Language (SQL) database clouds, and supercomputing clouds. Each cloud ecosystem has its own hardware, software, conferences, and business markets. The broad nature of business big data challenges makes it unlikely that one cloud ecosystem can meet its needs, and solutions are likely to require the tools and techniques from more than one cloud ecosystem. For this reason, at the Massachusetts Institute of Technology (MIT) Lincoln Laboratory, we developed the MIT SuperCloud architecture [51] that enables the prototyping of four common computing ecosystems on a shared hardware platform as depicted in Figure 2.1. The velocity of big data stresses the rate at which data can be absorbed and meaningful answers produced. Very often, the velocity challenge is mitigated through high-performance databases, file systems, and/or processing. Big data variety may present the largest challenge and greatest opportunities. The promise of big data is the ability to correlate diverse and heterogeneous data to form new insights. A new fourth V [26], veracity, challenges our ability to perform computation on data while preserving privacy.

As a simple example of the scale of data and how it has changed in the recent past, consider the social media analysis developed by [24]. In 2011, Facebook had approximately 700,000 pieces of content per minute; Twitter had approximately 100,000 tweets per minute; and YouTube had approximately 48 hours of video per minute. By 2015, just 4 years later, Facebook had 2.5 million pieces of content per minute; Twitter had approximately 277,000 tweets per minute; and YouTube had approximately 72 hours of new video per minute. This increase in data generated can be roughly approximated to be 350 MB/min for Facebook, 50 MB/min for Twitter, and 24–48 GB/min for YouTube! In terms of the sheer volume of data, IDC estimates that from the year 2005 to the year 2020, there will an increase in the amount of data generated from 130 EB to 40,000 EB [30].

One of the greatest big data challenges is in determining the ideal storage engine for a large dataset. Databases and file systems provide access to vast amounts of data but differ at a fundamental level. File system storage engines are designed to provide access to a potentially large subset of the full dataset. Database engines are designed to index and provide access to a smaller, but well defined, subset of data. Before looking at particular storage and database

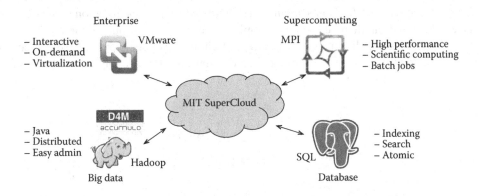

Figure 2.1: The MIT SuperCloud infrastructure allows multiple cloud environments to be launched on the same hardware and software platform in order to address big data volume.

engines, it is important to take a look at where these systems fall within the larger big data system.

2.3 Systems Engineering for Big Data

Systems engineering studies the development of complex systems. Given the many challenges of big data as described in Section 2.2, systems engineering has a great deal of applicability to developing a big data system. One convenient way to visualize a big data system is as a pipeline. In fact, most big data systems consist of different steps which are connected to each other to form a pipeline (sometimes, they may not be explicitly separated though that is the function they are performing). Figure 2.2 shows a notional pipeline for big data processing.

First, raw data is often collected from sensors or other such sources. These raw files often come in a variety of formats such as comma-separated values (CSVs), JavaScript Object Notation (JSON) [21], or other proprietary sensor formats. Most often, this raw data is collected by the system and placed into files that replicate the formatting of the original sensor. Retrieval of raw data may be done by different interfaces such as cURL (http://curl.haxx.se/) or other messaging paradigms such as publish/subscribe. The aforementioned formats and retrieval interfaces are by no means exhaustive but highlight some of the popular tools being used.

Once the raw data is on the target system, the next step in the pipeline is to parse these files into a more readable format or to remove components that are not required for the end-analytic. Often, this step involves removing remnants of the original data collection step such as unique identifiers that are no longer needed for further processing. The parsed files are often kept on a serial or parallel file system and can be used directly for analytics by scanning files. For example, a simple word count analytic can be done by using the Linux grep command on the parsed files, or more complex analytics can be performed by using a parallel processing framework such as Hadoop MapReduce or the Message Passing Interface (MPI). As an example of an analytic which works best directly with the file system, dimensional analysis [27] performs aggregate statistics on the full dataset and is much more efficient working directly from a high-performance parallel file system.

For other analytics (especially those that wish to access only a small portion of the entire dataset), it is convenient to ingest this data into a suitable database. An example of such an analytic is given in [28], which performs an analysis on the popularity of particular entities in a database. This example takes only a small, random piece of the dataset (the counts of words is much smaller than the full dataset) and is well suited for database usage. Once data is in the database or on the file system, a user can write queries or scans depending on their use case to produce results that can then be used for complex analytics such as topic modeling.

Each step of the pipeline involves a variety of choices and decisions. These choices may depend on hardware, software, or other factors. Many of these choices will also make a

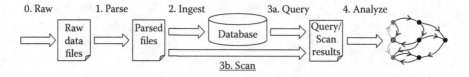

Figure 2.2: A standard big data pipeline consists of five steps to go from raw data to useful analytics.

difference to the later parts of the pipeline and it is important to make informed decisions. Some of the choices that one may have at each step include the following:

- *Step 0*: Size of individual raw data files, output format

- *Step 1*: Parsed data contents, data representation, parser design

- *Step 2*: Size of database, number of parallel processors, pre-processing

- *Step 3*: Scan or query for data, use of parallel processing

- *Step 4*: Visualization tools, algorithms

For the remainder of this chapter, we will focus on some of the decisions in steps two and three of the pipeline. By the end of the chapter, we hope that readers will have an understanding of different storage and database engines, the right time to use technology, and how these pieces can come together.

2.4 Disks and File Systems

One of the most common ways to store a large quantity of data is through the use of traditional storage media such as hard drives. There are many storage options that must be carefully considered that depend upon various parameters such as total data volume and desired read and write rates. In the pipeline of Figure 2.2, the storage engine plays an important part of steps two and three.

In order to deal with many challenges such as preserving data through failures, the past decades have seen the development of many technologies such as RAID (redundant array of independent disks) [17], NFS (network file system), HDFS (Hadoop Distributed File System) [11], and Lustre [67]. These technologies aim to abstract the physical hardware away from application developers in order to provide an interface for an operating system to keep track of a large number of files while allowing support for data failure, high-speed seeks, and fast writes. In this section, we will focus on two leading technologies, Lustre and HDFS.

2.4.1 Serial memory and storage

The most prevalent form of data storage is provided by an individual's laptop or desktop system. Within these systems, there are different levels of memory and storage that trade off speed with cost calculated as bytes per dollar. The fastest memory provided by a system (apart from the relatively low capacity system cache) is the main memory or random access memory (RAM). This volatile memory provides relatively high speed (10s of GB/s in 2015) and is often used to store data up to hundreds of gigabytes in 2015. When the data size is larger than the main memory, other forms of storage are used. Within serial storage technologies, some of the most common are traditional spinning magnetic disc hard drives and solid-state drives (solid-state drives may be designed to use volatile RAM or nonvolatile flash technology). The capacity of these technologies can be in the 10s of TB each and can support transfer rates anywhere from approximately 100 MB/s to GB/s in 2015.

2.4.2   Parallel storage: Lustre

Lustre is designed to meet the highest bandwidth file requirements on the largest systems in the world [12] and is used for a variety of scientific workloads [49]. The open source Lustre parallel file system presents itself as a standard POSIX, general-purpose file system and is mounted by client computers running the Lustre client software. Files stored in Lustre contain two components—metadata and object data. Metadata consists of the fields associated with each file such as i-node, filename, file permissions, and timestamps. Object data consists of the binary data stored in the file. File metadata is stored in the Lustre metadata server (MDS). Object data is stored in object storage servers (OSSes) shown in Figure 2.3. When a client requests data from a file, it first contacts the MDS, which returns pointers to the appropriate objects in the OSSes. This movement of information is transparent to the user and handled fully by the Lustre client. To an application, Lustre operations appear as standard file system operations and require no modification of application code.

A typical Lustre installation might have many OSSes. In turn, each OSS can have a large number of drives that are often formatted in a RAID configuration (often RAID6) to allow for the failure of any two drives in an OSS. The many drives in an OSS allows data to be read in parallel at high bandwidth. File objects are striped across multiple OSSes to further increase parallel performance. The above redundancy is designed to give Lustre high availability while avoiding a single point of failure. Data loss can only occur if three drives fail in the same OSS prior to any one of the failures being corrected. For Lustre, the typical storage penalty to provide this redundancy is approximately 35%. Thus, a system with 6 PB of raw storage will provide 4 PB of data capacity to its users.

Lustre is designed to deliver high read and write performance to many simultaneous large files. Lustre systems offer very high bandwidth access to data. For a typical Lustre configuration, this bandwidth may be approximately 12 GB/s in 2015 [2]. This is achieved by the clients having a direct connection to the OSSes via a well-designed high-speed network. This connection is brokered by the MDS. The peak bandwidth of Lustre is determined by the aggregate network bandwidth to the client systems, the bisection bandwidth of the network switch, the aggregate network connection to the OSSes, and the aggregate bandwidth of all

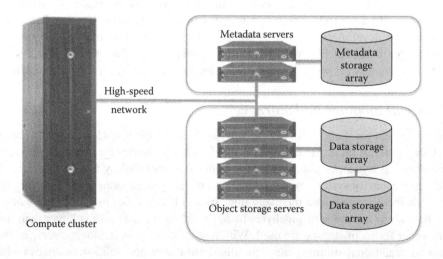

Figure 2.3: A Lustre installation consists of metadata servers and object storage servers. These are connected to a compute cluster via a high-speed interconnect such as at 10 GB Ethernet or Infiniband.

the disks [42]. Like most file systems, Lustre is designed for sequential read access and not random lookups of data (unlike a database). To find a particular data value in Lustre requires, on average, scanning through half the file system. For a typical system with approximately 12 GB/s of maximum bandwidth and 4 PB of user storage, this may require approximately 4 days.

2.4.3   Parallel storage: HDFS

Hadoop is a fault-tolerant, distributed file system and distributed computation system. An important component of the Hadoop ecosystem is the supporting file system called the HDFS that enables MapReduce [22] style jobs. HDFS is modeled after the Google File System (GFS) [33] and is a scalable distributed file system for large, distributed, and data-intensive applications. GFS and HDFS provide fault tolerance while running on inexpensive off-the-shelf hardware, and deliver high aggregate performance to a large number of clients. The Hadoop distributed computation system uses the MapReduce parallel programming model for distributing computation onto the data nodes.

The foundational assumptions of HDFS are that its hardware and applications have the following properties [11]: high rates of hardware failures, special purpose applications, large datasets, write-once-read-many data, and read-dominated applications. HDFS is designed for an important, but highly specialized class of applications for a specific class of hardware. In HDFS, applications primarily employ a co-design model whereby the HDFS is accessed via specific calls associated with the Hadoop API.

A file stored in HDFS is broken into two pieces: metadata and data blocks as shown in Figure 2.4. Similar to the Lustre file system, metadata consists of fields such as the filename, creation date, and the number of replicas of a particular piece of data. Data blocks consist of the binary data stored in the file. File metadata is stored in an HDFS name node. Block data is stored on data nodes. HDFS is designed to store very large files that will be broken up into multiple data blocks. In addition, HDFS is designed to support fault-tolerance in massive distributed data centers. Each block has a specified number of replicas that are distributed across different data nodes. The most common HDFS replication policy is to store three copies of each data block in a location-aware manner so that one replica is on a node in the local rack, the second replica on a node in a different rack, and the third replica on another node in the same different rack [3]. With such a policy, the data will be protected from node and rack failure.

The storage penalty for a triple replication policy is 66%. Thus, a system with 6 PB of raw storage will provide 2 PB of data capacity to its users with triple replication. Data loss can only occur if three drives fail prior to any one of the failures being corrected. Hadoop is written in Java and is installed in a special Hadoop user account that runs various Hadoop daemon processes to provide services to connecting clients. Hadoop applications contain special

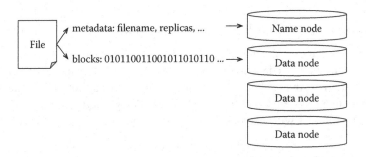

Figure 2.4: Hadoop splits a file into metadata and replicates it in data blocks.

application program interface (API) calls to access the HDFS services. A typical Hadoop application using the MapReduce programming model will distribute an application over the file system so that each application is exclusively reading blocks that are local to the node on which it is running. A well-written Hadoop application can achieve very high performance if the blocks of the files are well distributed across the data nodes. Hadoop applications use the same hardware for storage and computation. The bandwidth achieved out of HDFS is highly dependent upon the computation to communication ratio of the Hadoop application. For a well-designed Hadoop application, this aggregate bandwidth may be as high as 100 GB/s for a typical HDFS setup. Like most other file systems, HDFS is designed for sequential data access and no random access of data.

2.5   Database Management Systems

Relational or SQL databases [20,62] have been the de facto interface to databases since the 1980s and are the bedrock of electronic transactions around the world. For example, most financial transactions in the world make use of technologies such as Oracle or dBase. With the great rise in quantity of unstructured data and analytics based on the statistical properties of datasets, NoSQL (Not Only SQL) database stores such as the Google BigTable [19] have been developed. These databases are capable of processing the large heterogeneous data collected from the Internet and other sensor platforms. One style of NoSQL databases that have become used for applications that require support for high velocity data ingest and relatively simple cell-level queries are key-value stores.

As a result, the majority of the volume of data on the Internet is now analyzed using key-value stores such as Amazon Dynamo [23], Cassandra [44], and HBase [32]. Key-value stores and other NoSQL databases compromise on data consistency in order to provide higher performance. In response to this challenge, the relational database community has developed a new class of relational databases (often referred to as NewSQL) such as SciDB [16], H-Store [37], and VoltDB [64] to provide the features of relational databases while also scaling to very large datasets. Very often, these newSQL databases make use of a different datamodel [16] or advances in hardware architectures. For example, MemSQL [56] is a distributed in-memory database that provides high-performance, atomicity, consistency, isolation, and durability (ACID)-compliant relational database management. Another example, BlueDBM [36], provides high-performance data access through flash storage and field programmable gate arrays (FPGA).

In this section, we provide an overview of database management systems, the different generations of databases, and a deep dive into two newer technologies: a key-value store—Apache Accumulo and an array database—SciDB.

2.5.1   *Database management systems and features*

A database is a collection of data and all of the supporting data structures. The software interface between users and a database is known as the database management system. Database management systems provide the most visible view into a dataset. There are many popular database management systems such as MySQL [4], PostgreSQL [63], and Oracle [5]. Most commonly, users interact with database management systems for a variety of reasons, which are listed as follows:

1. To define data, schema, and ontologies

2. To update/modify data in the database

3. To retrieve or query data

4. To perform database administration or modify parameters such as security settings

5. More recently, to perform analytics on the data within the database

Databases are used to support data collection, indexing, and retrieval through transactions. A database transaction refers to the collection of steps involved in performing a single task [31]. For example, a single financial transaction such as *credit $100 towards the account of John Doe* may involve a series of steps such as locating the account information for John Doe, determining the current account value, adding $100 to the account, and ensuring that this new value is seen by any other transaction in the future. Different databases provide different guarantees on what happens during a transaction.

Relational databases provide ACID guarantees. Atomicity provides the guarantee that database transactions either occur fully or completely fail. This property is useful to ensure that parts of a transaction do not occur successfully if other parts fail, which may lead to an unknown state. The second guarantee, consistency, is important to ensure that all parts of the database see the same data. This guarantee is important to ensure that when different clients perform transactions and query the database, they see the same results. For example, in a financial transaction, a bank account may be debited before further transactions can occur. Without consistency, parts of the database may see different amounts of money (not a great database property!). Isolation in a database refers to a mechanism of concurrency control in a database. In many databases, there may be numerous transactions occurring at the same time. Isolation ensures that these transactions are isolated from other concurrent transactions. Finally, database durability is the property that when a transaction has completed, it is persisted even if the database has a system failure. Nonrelational databases such as NoSQL databases often provide a relaxed version of ACID guarantees referred to as BASE guarantees in order to support a distributed architecture or performance. This stands for Basically Available, Soft State, Eventual Consistency guarantees [50]. As opposed to the ACID guarantees of relational databases, nonrelational databases do not provide strict guarantees on the consistency of each transaction but instead provide a looser guarantee that *eventually* one will have consistency in the database. For many applications, this may be an acceptable guarantee.

For these reasons, financial transactions employ relational databases that have the strong ACID guarantees on transactions. More recent trends that make use of the vast quantity of data retrieval from the Internet can be done via nonrelational databases such as Google BigTable [19], which are responsible for fast access to information. For instance, calculating statistics on large datasets are not as susceptible to small eventual changes to the data.

While many aspects of learning how to use a database can be taught through books or guides such as this, there is an artistic aspect to their usage as well. More practice and experience with databases will help overcome common issues, improved performance tuning, and help with improved database management system stability. Prior to using a database, it is important to understand the choices available, properties of the data, and key requirements.

2.5.2 History of open source databases and parallel processing

Databases and parallel processing have developed together over the past few decades. Parallel processing is the ability to take a given program and split it across multiple processors in order to reduce computation time or resource availability for the application. Very often, advances in parallel processing are directly used for the computational piece of databases such as sorting and indexing datasets.

Open source databases have been around since the mid-1990s. Some of the first relational databases were based on the design of the Ingres database [62] originally developed at UC Berkeley. During the same time period, there were many parallel processing or high-performance computing paradigms [38,53] that were being developed by industry and academia. The first few (popular) open source databases that were created were PostgreSQL and MySQL. The earliest forms of parallel cluster processing take their root in the early 1990s with the wide proliferation of *nix-based operating systems and parallel processing schedulers such as Grid Engine. For about 10 years, until the mid-2000s, these technologies continued to mature, and developers saw the need for greater adoption of distributed computing and databases.

Based on a series of papers from Google in the mid-2000s, the MapReduce computing paradigm was created which gained wide acceptance through the open source Apache Hadoop soon after. These technologies, combined with the seminal Google BigTable [19] paper helped spark the NoSQL movement in databases. Not long after this, numerous technologies such as GraphLab [46], Neo4j [68], and Giraph [9] were developed to apply parallel processing to large unstructured graphs such as those being collected and stored in NewSQL databases. Since the year 2010, there has been renewed interest in developing technologies that offer high performance along with some of the ACID guarantees of relational databases (which will be discussed in Section 2.5.3). This requirement has driven the development of a new generation of relational databases often called NewSQL. In the parallel processing world, users are looking for better ways to deal with streaming data or machine learning and graph algorithms than the Hadoop framework offered and are developing new technologies such as Apache Storm [65] and Spark [69]. Of course, the worlds of parallel processing and databases will continue to evolve, and it will be interesting to see what lies ahead! A brief informational graphic of the history of parallel cluster processing and databases is provided in Figure 2.5.

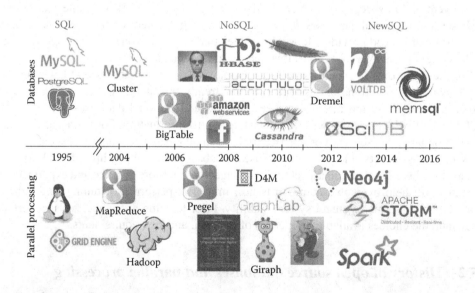

Figure 2.5: An incomplete history of open source databases and parallel cluster computing technologies.

2.5.3 CAP theorem

The CAP theorem is a seminal theorem [13] used to specify what guarantees can be provided by a distributed database. The CAP theorem states that no distributed database can simultaneously provide strong guarantees on the consistency, availability, and partition tolerance of a database. This is often stated as the two-out-of-three rule, though in reality it is more of a loose guarantee rather than losing the guarantee completely. In practice partition tolerance is an important aspect of NoSQL distributed databases; the two-out-of-three rule of the CAP theorem implies that most such databases fall into a consistency-partition tolerance or availability-partition tolerance style. Traditional relational databases are examples of choosing consistency and availability as the two-out-of-three CAP theorem guarantees.

Unlike the definition of consistency in ACID (which refers to a single node view of data in a database), consistency in the CAP theorem refers to the property that all nodes in a distributed database see the same data and provide a client with the same results regardless of which server is responding to a client request. Very often, a strong consistency guarantee is enforced by placing locks on a table, column, row or cell until all parts of the transaction are performed. While this property is very useful to ensure that all queries subsequent to the completion of the transaction see the same value, locking can hinder performance and availability guarantees. For example, in the case of a partition, enforcing consistency implies that certain nodes will not be able to respond to requests until all nodes have a consistent view of the data; thus compromising the availability of these nodes. A NoSQL database that prioritizes consistency over availability is Google BigTable [19] (though it may still have relaxed consistency between data replicas spread across database instances).

Database availability is a property in a distributed database which implies that every transaction must receive a response about whether the transaction has succeeded or failed. Databases that provide strong availability guarantees typically prioritize nodes responding to requests over maintaining a consistent system-wide view of the data. This availability, however, often comes at the cost of consistency. For example, in the event of a database partition, in order to maintain availability, certain parts of a distributed database may provide different results until a synchronization occurs. An example of a NoSQL database that provides a strong availability guarantee is Amazon Dynamo [23], which provides a consistency model often called *eventual consistency* [66].

Consider the example transaction given in Figure 2.6. In this example, the transaction is to update the count of the word *Apple* in *Doc*1 to be 5 from an application that computes a word count in documents. In a relational database, this transaction can occur within a single transaction that locks the row (*Doc*1) and performs the update before relinquishing the lock. In a highly available distributed database that supports eventual consistency, this update may be performed in parallel and eventually combined to show the correct count of the word Apple in *Doc*1 to be 5. For a short period of time, until this consistency is achieved, different nodes may provide a different response when queried about the count of the word Apple in *Doc*1.

The final aspect of the CAP theorem is database partition tolerance. Database partition tolerance is a database property that allows a database to function even after system failures. This property is often a fundamental requirement for large-scale distributed databases. In the event of failure of a piece of a distributed database, a well-designed database will handle the failure and move pieces of data to working components. This property is usually guaranteed by most of NoSQL databases. Traditional relational databases do not rely on distributed networks which are prone to disruption, thus avoiding the need for partition tolerance.

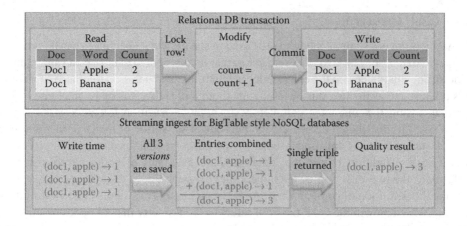

Figure 2.6: Relational update transaction compared with nonrelational database update transaction.

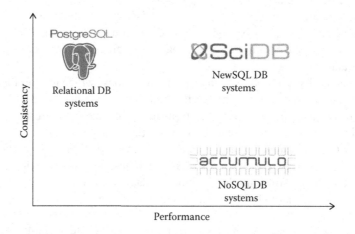

Figure 2.7: Notional guide to implication of the CAP theorem for database design. Traditional relational databases provide high consistency, NoSQL databases provide high performance at the cost of consistency, and NewSQL databases attempt to bridge the gap.

In recent years, there has been some controversy [14,34,57] surrounding the use of the CAP theorem as a fundamental rule in the design of modern databases. Most often, the CAP theorem is used to imply that one can have an all or nothing of two of the three aspects. However, it has been shown in [15] that careful partition and availability optimization may be able to achieve a database that provides a version of all three guarantees. While the CAP theorem can be used for high-level understanding of tradeoffs and design of current technologies, it is certainly possible to design databases that provide versions of guarantees on all three properties through different data models or hardware. In Figure 2.7, we provide a notional guide to the CAP theorem and database classes and also show an example technology for each of these database classes.

2.5.4 Relational databases

Relational databases such as MySQL, PostgreSQL, and Oracle form the bedrock of database technologies today. They are by far the most widely used and accessed databases. We interact with these databases daily: everywhere financial transactions, medical records, and purchases are made. From the CAP theorem, relational databases provide strong consistency and availability; however, they do not support partition tolerance. In order to avoid issues with partition tolerance in distributed databases, relational databases are often vertically scalable. Vertical scalability refers to systems that scale by improving existing software or hardware. For example, vertically scaling a relational database involved improving the resources of a single node (more memory, faster processor, faster disk drive, etc.). Thus, relational databases often run on high-end, expensive nodes and are often limited by the resources of a single node. This is in contrast to nonrelational database that are designed to support horizontal scalability. Scaling a database horizontally involves adding more nodes to the system. Most often, these nodes can be inexpensive commercial off-the-shelf systems (COTS) that are easy to add as resource requirements change.

Relational databases provide ACID guarantees and are used extensively in practice. Relational databases are called *relational* because of the underlying data model. A relational database is a collection of tables that are connected to each other via relations expressed as keys. The specification of tables and relations in a database is referred to as the schema. Schema design requires thorough knowledge of the dataset. Consider a very simple example of a relational database that maintains a record of purchases made by customers as depicted in Figure 2.8. The main purchase table can be used to track purchases. This table is related to a customer table via the customer ID key. The purchase table is also connected to a product table via the product ID key. Using a combination of these tables, one can query the database for information such as *who purchased a banana on March 22, 2010?*. Many databases may contain tens to hundreds of tables and require careful thought during the design.

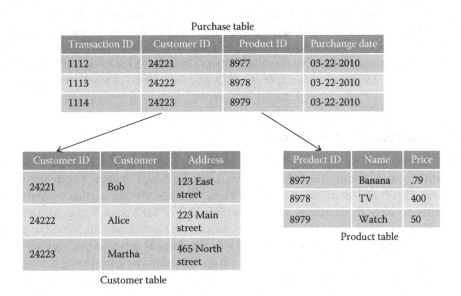

Figure 2.8: A simple relational database that contains information about purchases made. The database consists of three tables: a purchase table, a customer table, and a product table.

2.5.5 NoSQL databases

Since the mid-2000s and the Google BigTable paper, there has been a rise in popularity of NoSQL databases. NoSQL databases support many of the large-scale computing activities with which we interact regularly such as web searches, document indexing, large-scale machine learning, and graph algorithms. NoSQL databases support horizontal scaling: you can increase the performance through the addition of nodes. This allows for scaling through the addition of inexpensive COTS as opposed to expensive hardware upgrades required for vertical scaling. NoSQL databases often need to relax some of the consistency or availability guarantees of relational databases in order to take advantage of strong partition tolerance guarantees. In order to keep up with rising data volumes, organizations such as Google looked for ways to incorporate inexpensive off-the-shelf systems for scaling their hardware. However, incorporating such systems requires the use of networks which can be unreliable. Thus, partition tolerance to network disruptions became an important design criteria. In keeping with the CAP theorem, either consistency or availability must be relaxed to provide partition tolerance in a distributed database.

At a transaction level, NoSQL databases provide BASE guarantees. These guarantees may not be suitable for many applications where strong consistency or availability is required. However, for a variety of *big data* applications, BASE guarantees are sufficient for the purpose. For example, recall the example transaction described in Figure 2.6. If the end analytic (step 5 in the big data pipeline) is an approximate algorithm to look for trends in word count, the exact count of the word Apple in *Doc*1 may not be as important as the fact that the word Apple exists in the document. In this case, BASE guarantees may be sufficient for the application. Of course, before choosing a technology to use for an application, it is important to be aware of all design constraints and the impact of technology choice on the final analytic requirements.

NoSQL database use a variety of data models and typically do not have a pre-defined schema. This allows developers the flexibility to specify a schema that leverages the database capabilities while supporting the desired analytics. Further, the lack of a well-defined schema allows dynamic schemas that can be modified as data properties change. Certain graph databases [8] use data structures based on graphs. In such databases, an implicit schema is often generated based on the graph representation of the data in the database. Key-value databases [35] take a given dataset and organize them as a list of keys and values. Document stores such as those described in [6] may use a schema based on a JSON or XML representation of a dataset. Figure 2.9 shows an example of a JSON-based schema applied to the dataset shown in Figure 2.8.

2.5.6 New relational databases

The most recent trend in database design is often referred to as NewSQL databases. Given the controversy surrounding the CAP theorem, such databases attempt to provide a version of all three distributed database properties. These databases were created to approach the performance of NoSQL databases while providing the ACID transaction guarantees of traditional relational databases [58]. In order to provide this combination, NewSQL databases often employ different hardware or data models than traditional database management systems. NewSQL databases may also make use of careful optimizations on partitioning and availability in order to provide a version of all three aspects of the CAP theorem. NewSQL databases may be considered as an alternative to both SQL and NoSQL style databases [43]. Most NewSQL databases provide support for the SQL.

NewSQL databases, while showing great promise, are a relatively new technology area. In the market now are databases designed for sensor processing [25], high-speed online transaction processing (OLTP) [56], and streaming data and analytics [18].

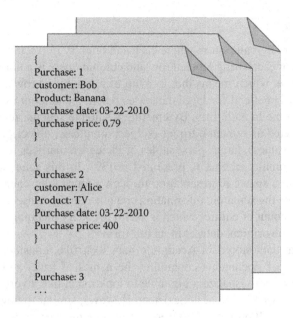

Figure 2.9: An example of using a JSON-based data model for the purchase table of Figure 2.8.

	Relational Databases	NoSQL	NewSQL
Examples	MySQL, PostgreSQL, Oracle	HBase, Cassandra, Accumulo	SciDB, VoltDB, MemSQL
Schema	Typed columns with relational keys	Schema-less	Strongly typed structure of attributes
Architecture	Single-node or sharded	Distributed, scalable	Distributed, scalable
Guarantees	ACID transactions	Eventually consistent	ACID transactions (most)
Access	SQL, indexing, joins, and query planning	Low-level API (scans and filtering)	Custom API, JDBC, bindings to popular languages

Figure 2.10: A simple guide to differentiate between SQL, NoSQL, and NewSQL style databases.

A quick guide to the major differences between SQL, NoSQL, and NewSQL style database is provided in Figure 2.10. Later in Section 2.5.8, we will provide a deeper look at a relatively new array-based NewSQL database called SciDB.

2.5.7 Deep dive into NoSQL technology

Apache Accumulo is an open source database used for high-performance ingest and retrieval [54]. Accumulo is based on the Google BigTable. Apache Accumulo is a suitable technology for environments where large quantities of text data need to be indexed and inserted into a database at a high rate. In this deep dive, we will discuss the Accumulo data model, design, and performance.

2.5.7.1 Data model

Accumulo is a tabular key-value store where each cell or entry in Accumulo is a key (or tuple) mapped to a value. For each string labeled row and column, there is a unique value. Accumulo is a row store database, which means that look-up of a row by its row ID can occur quickly. This property is very beneficial for large databases where the amount of data stored does not significantly increase the lookup time. By creating a suitable schema for a given dataset, this model can be interpreted as semantic triples (subject, predicate, object), documents (mapping of text documents to values), large sparse tables or incidence matrices for graphs. One widely adopted Apache Accumulo schema is presented in [39]. In this schema, the original dense dataset it converted to a sparse representation through a four table schema. Two of the tables are used to encapsulate the semantic information contained in the dense dataset. A degree table is used to represent a count of entities which is useful for query planning and an additional raw table is used to store the original dataset in its full form.

To visualize what data stored in Accumulo may look like, consider Figure 2.11, which is a schema for a set of documents containing the names of fruits and vegetables. From a logical perspective, we can visualize a big table to look much like a very large spreadsheet in which many of the values are null or nonexistent. However, similar to how sparse matrices are stored [55], Accumulo will only store nonzero entries in a manner similar to what is shown in the sparse tuples part of the figure. The row key (or row ID) refers to the document name or number, and the column key represents the name of the fruit or vegetable. The value in such a situation may represent the count or the number of times a particular fruit or vegetable was mentioned in a document. For example, in the figure, the fruit banana was mentioned in *doc1* five times and one time in *doc2*.

In reality, the Accumulo data model is more complicated. Instead of just a three tuple (row, column, and value), each cell is actually a seven tuple where the column is broken into three parts, and there is an additional field for a timestamp as seen in Figure 2.12.

Sparse tuples:

(doc1, apple)	→ 1
(doc1, banana)	→ 5
(doc1, carrot)	→ 2
(doc2, banana)	→ 1
(doc2, daikon)	→ 2
(doc3, carrot)	→ 4
(doc3, eggplant)	→ 1

Logical table

Row ID	apple	banana	carrot	daikon	eggplant
doc1	1	5	2		
doc2		1		2	
doc3			4		1

Figure 2.11: The logical table (right) is stored as a series of tuples (left) in a key-value store database such as Accumulo.

	Key				Value
Row ID	Column			Timestamp	
	Family	Qualifier	Visibility		

Figure 2.12: An Accumulo entry consists of a seven tuple. Each entry is organized lexicographically based on the row key.

The column family is often used to define classes within the dataset. The column qualifier is used to define the unique aspect of the tuple. The visibility field is used to specify cell-level visibility for controlling who can access this data item. The value field is often used to store aspects of the dataset that must be stored but not searched upon. Given the Accumulo properties of fast row lookup, a good schema will usually store the semantic information of the dataset into the rows and columns of the table.

2.5.7.2 Design

Each key-value combination or cell in Accumulo is stored lexicographically and a number of cells are stored in tables. Tables can be arbitrarily large. Contrary to the design of relational tables, very often an entire dataset is put into a single table. In order to provide high-performance distributed access, portions of a table are persisted on disk as tablets which are in turn stored in tablet servers as shown in Figure 2.13.

Since a single table can often be very large in a distributed database, each table is split into a number of tablets. A table in Accumulo is defined to be a map of key-value pairs (or cells) with a global order among the keys. A tablet is a row range within a table that is stored on a logical or physical computational element such as a server. A tablet server is the mechanism that hosts tablets and provides functionality such as resource management, scheduling, and hosting remote procedure calls.

2.5.7.3 Performance

Accumulo supports high-performance data ingest and queries. A test performed at MIT Lincoln Laboratory was able to demonstrate a database insert rate of nearly 115 million entries (key-value tuples) per second on an Apache Accumulo instance running on over 200 servers simultaneously [41]. The results from this experiment are shown in Figure 2.14.

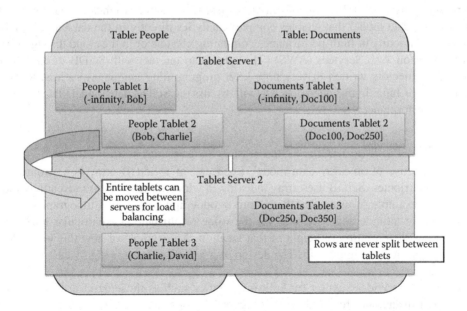

Figure 2.13: Accumulo tables are split into tablet which are hosted on tablet servers. Tablets associated with different tables may exist on the same tablet servers.

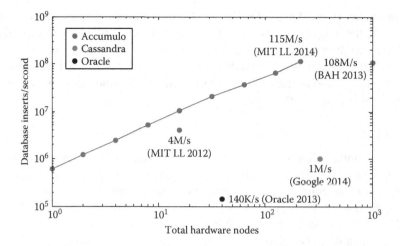

Figure 2.14: Demonstration of high-performance data ingest in Accumulo. The *y*-axis is a measure of the number of seven tuples shown in Figure 2.12 that Accumulo can insert per second.

2.5.8 Deep dive into NewSQL technology

NewSQL databases provide a new view into the design of databases. By shunning the popular CAP theorem, this style of database technology has the ability to combine the advantages of both SQL and NoSQL database for high-performance databases that provide ACID guarantees of single node relational databases. One technology in this category is an array-based database called SciDB. This database is designed for datasets and applications which can be represented as arrays or matrices. For example, SciDB is an ideal choice for signals or imagery which can be represented as a 1D or 2D array of values, respectively. SciDB is a massive parallel database with an array data model that enables complex analytics directly in the database. SciDB can be run on commodity or high-performance computing clusters or in the cloud through services such as Amazon Web Services (AWS) [7]. A user can interact with SciDB through a series of SciDB connectors written in very high-level programming languages such as R, Python, MATLAB®, or Julia. In this deep dive, we will discuss the SciDB data model, database design, and tested performance.

2.5.8.1 Data model

SciDB makes use of an array data model. Each cell of a SciDB array is a strongly typed structure of attributes. SciDB uses array indexing in which dimensions are essentially indices. Consider an example of using SciDB to store a topographic map. In a topographic map, the dimensions are latitude and longitude. As the value or attribute at a particular location, one usually stores the elevation or height of that location. An example of how this would look in SciDB is presented in Figure 2.15. The design of a SciDB schema is highly customized to an application.

SciDB also provides support for built-in analytics such as those described in [60]. Using SciDBs built-in analytics for the dataset of Figure 2.15, one can find the latitude/longitude pairs which satisfy certain criteria such as an elevation threshold without moving the full dataset back to the client in order to perform the analysis.

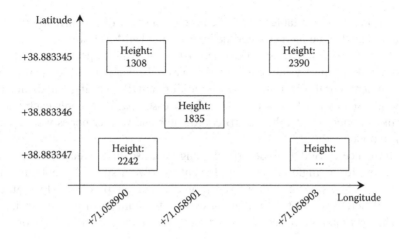

Figure 2.15: A notional example of a SciDB array for storing the height or elevation at a particular latitude and longitude.

2.5.8.2 Design

SciDB is deployed on a cluster of servers. Each server in the cluster has access to local processing, memory, and storage capabilities. These servers can be in the cloud through services such as Amazon Elastic Compute Cloud (EC2) or hosted on high-performance computing systems or even commodity clusters. Standard Ethernet connections are used to move data between servers. Each physical server hosts a SciDB instance (or instances) that is responsible for local storage and processing.

When a client or external application requests a connection with SciDB, it creates a connection with one of the instances running on a physical server in the cluster. A coordinator instance is responsible for communicating query results to the connecting client and all other instances participate in the execution of the query and data storage. The other instances in the cluster are referred to as worker instances which work with the coordinator node.

2.5.8.3 Performance

SciDB benchmarking was performed at MIT Lincoln Laboratory. The data used for benchmarking was generated using a random graph generator from the Graph500 benchmark [48]. The Graph500 scalable data generator can efficiently generate power-law graphs that represent common graphs such as those generated by social media. The number of vertices and edges in the graph are set using a positive integer called the SCALE parameter. Given a SCALE parameter, the number of vertices, N, and the number of edges, M, are then computed as $N = 2^{\text{SCALE}}$ and $M = 8N$.

SciDB is a highly scalable database and is capable of connecting with multiple clients at once. In order to test the scalability of SciDB, the test was performed using the parallel MATLAB tool, pMATLAB [10], in addition to the Dynamic Distributed Dimensional Data Model (D4M) [40] to insert data from multiple clients simultaneously. In order to overcome a SciDB bottleneck that applies a table lock when data is being written to a particular table, D4M can create multiple tables based on the total number of processes that are being used to ingest data. For example, if there are four ingestors (four systems simultaneously writing data to SciDB), D4M creates four tables into which each ingestor will concurrently insert data. The resulting tables can then be merged after the ingest using D4M if desired.

For this benchmark, SciDB was launched using the MIT SuperCloud [51] architecture through a database hosting system that launches a SciDB cluster in a high-performance computing environment. For the purpose of benchmarking SciDB on a single node, instances were launched on a system with dual Intel Xeon E5 processors with 16 cores each and 64 GB of RAM. For the reported results, the SciDB coordinator and worker nodes were located on the same physical node.

The performance of SciDB is described using weak scaling. Weak scaling is a measure of the time taken for a single processing element to solve a specific problem and measures the performance when scaling with a fixed problem size per processor. In Figure 2.16, the performance of SciDB for a SCALE that varies with the number of processors into SciDB is presented. The maximum performance (insert rate) was observed at 10 processors.

2.6 How to Choose the Right Technology

One of the guiding principles in technology selection is that *one size does not fit all* as mentioned in [61]. In the data storage world, this can translate to *there is no single technology that will solve all of your problems*. For a given problem mapped to the pipeline of Figure 2.2, a typical solution may have a combination of file systems, SQL, NoSQL, and NewSQL databases and different parallel processing strategies. In fact, a good big data solution will make use of technologies that best satisfy the overall application goal. Choosing the right storage or database engine can be a challenging problem and often requires careful thought and a deep

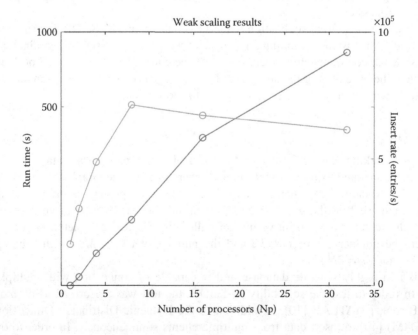

Figure 2.16: Weak scaling of D4M-SciDB insert performance for problem size that varies with number of processors.

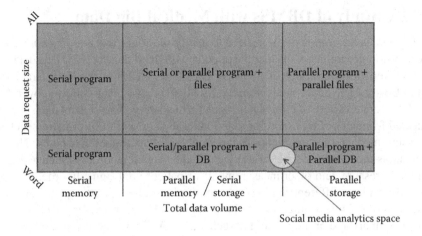

Figure 2.17: Choosing the right combination of technologies is a factor of many parameters such as total data volume and data request size. An example decision point for common social media analytics is shown in the small circle.

understanding of the problem being solved. Sometimes, one may go in looking for a database solution only to find that a database is not the right option! In Figure 2.17, we describe how various technologies can be selected based on the property of data size and data request size. Of course, this is a simple guide to technology selection and there may be many other factors that come into consideration when making a decision in the real world.

As described in Figure 2.17, when choosing a storage or database engine, one must take into account the data request size (as a percentage of the entire corpus), and the total data volume. For a small amount of data volume (10s of GB in 2015) and any request size, the most efficient solution is to make use of a systems onboard memory for the storage and retrieval of information. For a larger dataset (100s of GB to 10 TB in 2015), the most efficient solution for small requests (less than 5% of the entire dataset) is to use a database. If your request size is larger than approximately 5% of the entire dataset, it is often faster and more efficient to write a parallel program that operates on data in the file system. For much larger datasets (10 TB to many PBs in 2015), one will need to make use of parallel storage technologies such as HDFS or Lustre or parallel databases such as Accumulo or SciDB. Again, if the total data request size is greater than 5% of the entire dataset, it may be more efficient to work directly on the files.

Databases are designed to pull small chunks of information out (finding a needle in a haystack) and not for sequential access (the forte of distributed file systems). Much of the overhead incurred in using a database is for extensive indexing. Very often, this may be dictated by the application at hand, other components in a pipeline, and/or the desired interface and language support. For example, many relational databases support the Java Database Connector, which is a convenient Java API for accessing relational databases. Performance characteristics and data properties may also dictate database choice. For example, for a rapidly changing data schema, key-value store NoSQL databases may be a suitable choice. For applications that require high performance and ACID compliance, NewSQL databases are a suitable option. The important thing to keep in mind is that technologies continue to evolve, and it is important to re-evaluate technology choices as requirements change or are not being satisfied.

2.7 Case Study of DBMSs with Medical Big Data

Medical big data is a common example used to justify the adage that *one size does not fit all* for database and storage engines. Consider the popular MIMIC II dataset [52]. This dataset consists of data collected from a variety of intensive care units (ICU) at the Beth Israel Deaconess Hospital in Boston. The data contained in the MIMIC II dataset was collected over 7 years and contains data from a variety of clinical and waveform sources. The clinical dataset contains the data collected from tens of thousands of individuals and consists of information such as patient demographics, medications, interventions, and text-based doctor or nurse notes. The waveform dataset contains thousands of time series physiological signal recordings such as ECG signals, arterial blood pressure, and other measurements of patient vital signs. In order to support data extraction from these different datasets, one option would be to attempt to organize all the information into a single storage or database engine. However, existing technologies would prove to be cumbersome or inefficient for such a task. A distributed file system would provide inefficient random access to data (e.g., looking up details of a patient). Similarly, a relational database would be inefficient for searching and operating on the waveform signals.

The next solution is to store and index each of the individual components into a storage or database engine that is the most efficient for a given data modality. In such a system, individual components of the entire dataset could be stored in the storage or database engine that best supports the types of queries and analytics one wishes to perform. In this system, one may place the clinical dataset in a relational database such as MySQL. The text notes may go into a NoSQL database such as Apache Accumulo and the waveform data may go into an array-based NewSQL database such as SciDB. At MIT Lincoln Laboratory, we developed a prototype of such a system using these technologies [29].

The prototype developed supports cross-database analytics such as "tell me about what happens to heart rate variance of patients who have taken a particular medication." Naturally, such a query needs information from the clinical data contained in MySQL database, the patient database contained in Accumulo, and the waveform data contained in SciDB. The sample query provided is then broken up into three distinct queries where (1) tell me which patients have taken a particular medication goes to MySQL, (2) tell me which of these patients have heart beat waveforms goes to Accumulo, and (3) show me what happened to these patients heart rate variance goes to the waveform database. At each of these sub-queries, associative arrays are

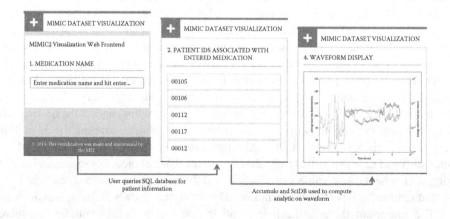

Figure 2.18: Screenshots of MIMIC II visualization that uses a combination of SQL, NoSQL, and NewSQL style databases in concert for a single analytic.

generated that can be used to move the results of one query to the next database engine. In Figure 2.18, we show the web front end that uses the D4M to implement the query described above.

This example and many more highlight the importance of technology selection. By breaking up the problem into smaller pieces, choosing the right technology was much easier. Recall the pipeline of Figure 2.2. Filling in the details for steps 2 and 3 often requires careful consideration.

2.8 Conclusions

The world of storage and database engines is vast, and there are many competing technologies out there. The technologies we discussed in this chapter discuss methods to address three of the four Vs of big data—volume (storage and database engines), velocity (distributed NoSQL and NewSQL databases), and variety (database schemas and data models). One V that we did not discuss is big data veracity—or privacy-preserving technologies. For a thorough understanding of this topic, we recommend you read Chapter 10.

Acknowledgments

The authors thank the LLGrid team at MIT Lincoln Laboratory for their support in setting up the computational environment used to test the performance of Apache Accumulo and SciDB. The authors also thank the following individuals: Lauren Edwards, Dylan Hutchison, Scott Sawyer, Julie Mullen, and Chansup Byun.

References

1. The Apache Software Foundation. Apache Accumulo. https://accumulo.apache.org/.

2. Es7k: World fastest entry-level lustre appliance. DDN Product Brochure. http://www.ddn.com/products/es7k-fastest-entry-level-lustre-appliance/.

3. HDFS architecture guide. http://hadoop.apache.org/docs/r1.2.1/hdfs_design.html.

4. Mysql website. https://www.mysql.com/.

5. Oracle database 12c. https://www.oracle.com/database/index.html.

6. Veronika Abramova and Jorge Bernardino. Nosql databases: Mongodb vs cassandra. In *Proceedings of the International C* Conference on Computer Science and Software Engineering*, pp. 14–22. ACM, New York, 2013.

7. Amazon, Inc. Amazon EC2. http://aws.amazon.com/ec2/.

8. Renzo Angles and Claudio Gutierrez. Survey of graph database models. *ACM Computing Surveys (CSUR)*, 40(1):1, 2008.

9. Ching Avery. Giraph: Large-scale graph processing infrastructure on hadoop. In *Proceedings of the Hadoop Summit*, Santa Clara, CA, 2011.

10. Nadya T Bliss and Jeremy Kepner. pMATLAB parallel MATLAB library. *International Journal of High Performance Computing Applications*, 21(3):336–359, 2007.

11. Dhruba Borthakur. Hdfs architecture guide. HADOOP APACHE PROJECT, http://hadoop. apache.org/common/docs/current/hdfs design. pdf, 2008.

12. Peter J Braam. The lustre storage architecture, 2004.

13. Eric Brewer. A certain freedom: Thoughts on the cap theorem. In *Proceedings of the 29th ACM SIGACT-SIGOPS Symposium on Principles of Distributed Computing*, pp. 335–335. ACM, New York, 2010.

14. Eric Brewer. Cap twelve years later: How the "rules" have changed. *Computer*, 45(2): 23–29, 2012.

15. Eric Brewer. Pushing the cap: Strategies for consistency and availability. *Computer*, 45(2):23–29, 2012.

16. Paul G Brown. Overview of scidb: Large scale array storage, processing and analysis. In *Proceedings of the ACM SIGMOD International Conference on Management of Data*, pp. 963–968. ACM, New York, 2010.

17. Daniel Carteau. Three interconnected raid disk controller data processing system architecture, December 11, 2001. US Patent 6,330,642.

18. Ugur Cetintemel, Jiang Du, Tim Kraska, Samuel Madden, David Maier, John Meehan, Andrew Pavlo et al. S-store: A streaming newsql system for big velocity applications. *Proceedings of the VLDB Endowment*, 7(13):1633–1636, 2014.

19. Fay Chang, Jeffrey Dean, Sanjay Ghemawat, Wilson C Hsieh, Deborah A Wallach, Mike Burrows, Tushar Chandra et al. Bigtable: A distributed storage system for structured data. *ACM Transactions on Computer Systems (TOCS)*, 26(2):4, 2008.

20. Edgar F Codd. A relational model of data for large shared data banks. *Communications of the ACM*, 13(6):377–387, 1970.

21. Douglas Crockford. The application/JSON media type for javascript object notation (JSON). 2006.

22. Jeffrey Dean and Sanjay Ghemawat. Mapreduce: Simplified data processing on large clusters. *Communications of the ACM*, 51(1):107–113, 2008.

23. Giuseppe DeCandia, Deniz Hastorun, Madan Jampani, Gunavardhan Kakulapati, Avinash Lakshman, Alex Pilchin, Swaminathan Sivasubramanian et al. Dynamo: Amazon's highly available key-value store. In *ACM SIGOPS Operating Systems Review*, volume 41, pp. 205–220. ACM, New York, 2007.

24. Domo. https://www.domo.com/learn.

25. Aaron Elmore, Jennie Duggan, Michael Stonebraker, Magda Balazinska, Ugur Cetintemel, Vijay Gadepally, Jeff Heer et al. A demonstration of the bigdawg polystore system. *Proceedings of the VLDB Endowment*, 8(12):1908–1911, 2015.

26. Vijay Gadepally, Braden Hancock, Benjamin Kaiser, Jeremy Kepner, Pete Michaleas, Mayank Varia, and Arkady Yerukhimovich. Improving the veracity of homeland security big data through computing on masked data. In *IEEE Technologies for Homeland Security*, Waltham, MA, 2015.

27. Vijay Gadepally and Jeremy Kepner. Big data dimensional analysis. In *IEEE High Performance Extremem Computing Conference*, 2014.

28. Vijay Gadepally and Jeremy Kepner. Using a power law distribution to describe big data. In *IEEE High Performance Extremem Computing Conference*, 2015.

29. Vijay Gadepally, Sherwin Wu, Jeremy Kepner, and Sam Madden. Mimicviz: Enabling visualization of medical big data. In *New England Database Summit*, 2015.

30. John Gantz and David Reinsel. The digital universe in 2020: Big data, bigger digital shadows and biggest growth in the far east. In *IDC iView*, 2012.

31. Hector Garcia-Molina. Using semantic knowledge for transaction processing in a distributed database. *ACM Transactions on Database Systems (TODS)*, 8(2):186–213, 1983.

32. Lars George. *HBase: The Definitive Guide*. O'Reilly Media, Cambridge, MA, 2011.

33. Sanjay Ghemawat, Howard Gobioff, and Shun-Tak Leung. The google file system. In *ACM SIGOPS Operating Systems Review*, volume 37, pp. 29–43. ACM, New York, 2003.

34. Seth Gilbert and Nancy A Lynch. Perspectives on the cap theorem. *Computer*, 45(2): 30–36, 2012.

35. Jing Han, E Haihong, Guan Le, and Jian Du. Survey on nosql database. In *6th International Conference on Pervasive Computing and Applications*, pp. 363–366. IEEE, New York, 2011.

36. Sang-Woo Jun, Ming Liu, Sungjin Lee, Jamey Hicks, John Ankcorn, Myron King, Shuotao Xu et al. Bluedbm: An appliance for big data analytics. In *Proceedings of the 42nd Annual International Symposium on Computer Architecture*, pp. 1–13. ACM, New York, 2015.

37. Robert Kallman, Hideaki Kimura, Jonathan Natkins, Andrew Pavlo, Alexander Rasin, Stanley Zdonik, Evan PC Jones et al. H-store: A high-performance, distributed main memory transaction processing system. *Proceedings of the VLDB Endowment*, 1(2): 1496–1499, 2008.

38. Jeremy Kepner. *Parallel MATLAB for Multicore and Multinode Computers*, volume 21. SIAM, Philadelphia, PA, 2009.

39. Jeremy Kepner, Christian Anderson, William Arcand, David Bestor, Bill Bergeron, Chansup Byun, Matthew Hubbell et al. D4m 2.0 schema: A general purpose high performance schema for the accumulo database. In *High Performance Extreme Computing Conference*, pp. 1–6. IEEE, 2013.

40. Jeremy Kepner, William Arcand, William Bergeron, Nadya Bliss, Robert Bond, Chansup Byun, Gary Condon et al. Dynamic distributed dimensional data model (d4m) database and computation system. In *2012 IEEE International Conference on Acoustics, Speech and Signal Processing*, pp. 5349–5352. IEEE, 2012.

41. Jeremy Kepner, William Arcand, David Bestor, Bill Bergeron, Chansup Byun, Vijay Gade-pally, Matthew Hubbell et al. Achieving 100,000,000 database inserts per second using accumulo and d4m. In *IEEE High Performance Extreme Computing*, 2015.

42. Jeremy Kepner, William Arcand, David Bestor, Bill Bergeron, Chansup Byun, Vijay Gadepally, Matthew Hubbell et al. Lustre, hadoop, accumulo. In *IEEE High Performance Extreme Computing*, 2015.

43. Rakesh Kumar, Neha Gupta, Shilpi Charu, and Sunil K Jangir. Manage big data through newsql. In *National Conference on Innovation in Wireless Communication and Networking Technology–2014, Association with THE INSTITUTION OF ENGINEERS (INDIA)*, 2014.

44. Avinash Lakshman and Prashant Malik. Cassandra: A decentralized structured storage system. *ACM SIGOPS Operating Systems Review*, 44(2):35–40, 2010.

45. Doug Laney. 3D data management: Controlling data volume, velocity and variety. *META Group Research Note*, 6, 2001.

46. Yucheng Low, Joseph E Gonzalez, Aapo Kyrola, Danny Bickson, Carlos E Guestrin, and Joseph Hellerstein. Graphlab: A new framework for parallel machine learning. arXiv preprint arXiv:1408.2041, 2014.

47. Peter Mell and Tim Grance. The nist definition of cloud computing. *National Institute of Standards and Technology*, 53(6):50, 2009.

48. Richard C Murphy, Kyle B Wheeler, Brian W Barrett, and James A Ang. Introducing the graph 500. Cray User's Group (CUG), 2010.

49. Juan Piernas, Jarek Nieplocha, and Evan J Felix. Evaluation of active storage strategies for the lustre parallel file system. In *Proceedings of the ACM/IEEE Conference on Supercomputing*, p. 28. ACM, New York, 2007.

50. Dan Pritchett. Base: An acid alternative. *Queue*, 6(3):48–55, 2008.

51. Albert Reuther, Jeremy Kepner, William Arcand, David Bestor, Bill Bergeron, Chansup Byun, Matthew Hubbell et al. Llsupercloud: Sharing HPC systems for diverse rapid prototyping. In *High Performance Extreme Computing Conference*, pp. 1–6. IEEE, 2013.

52. Mohammed Saeed, Mauricio Villarroel, Andrew T Reisner, Gari Clifford, Li-Wei Lehman, George Moody, Thomas Heldt et al. Multiparameter intelligent monitoring in intensive care ii (mimic-ii): A public-access intensive care unit database. *Critical Care Medicine*, 39:952–960, May 2011.

53. Siddharth Samsi, Vijay Gadepally, and Ashok Krishnamurthy. Matlab for signal processing on multiprocessors and multicores. *IEEE Signal Processing Magazine*, 27(2):40–49, 2010.

54. Scott M Sawyer, David O'Gwynn, An Tran, and Tao Yu. Understanding query performance in accumulo. In *High Performance Extreme Computing Conference*, pp. 1–6. IEEE, 2013.

55. Rukhsana Shahnaz, Anila Usman, and Imran R Chughtai. Review of storage techniques for sparse matrices. In *9th International Multitopic Conference*, pp. 1–7. IEEE, 2005.

56. Nikita Shamgunov. The memsql in-memory database system. In *IMDM@ VLDB*, 2014.

57. Michael Stonebraker. Errors in database systems, eventual consistency, and the cap theorem. *Communications of the ACM*, BLOG@ACM, 2010. http://cacm.acm.org/blogs/ blog-cacm/83396-errors-in-database-systems-eventual-consistency-and-the-cap-theorem/ fulltext.

58. Michael Stonebraker. Newsql: An alternative to nosql and old sql for new oltp apps. *Communications of the ACM*, 2012. http://cacm.acm.org/blogs/blog-cacm/109710-new-sql-an-alternative-to-nosql-and-old-sql-for-new-oltp-apps/fulltext.

59. Michael Stonebraker, Jacek Becla, David J DeWitt, Kian-Tat Lim, David Maier, Oliver Ratzesberger, and Stanley B Zdonik. Requirements for science data bases and scidb. In *CIDR*, volume 7, pp. 173–184, 2009.

60. Michael Stonebraker, Paul Brown, Donghui Zhang, and Jacek Becla. Scidb: A database management system for applications with complex analytics. *Computing in Science & Engineering*, 15(3):54–62, 2013.

61. Michael Stonebraker and Ugur Cetintemel. One size fits all: An idea whose time has come and gone. In *Proceedings of the 21st International Conference on Data Engineering*, pp. 2–11. IEEE, 2005.

62. Michael Stonebraker, Gerald Held, Eugene Wong, and Peter Kreps. The design and implementation of ingres. *ACM Transactions on Database Systems (TODS)*, 1(3): 189–222, 1976.

63. Michael Stonebraker and Lawrence A Rowe. *The Design of Postgres*, volume 15. ACM, New York, 1986.

64. Michael Stonebraker and Ariel Weisberg. The voltdb main memory dbms. *IEEE Data Engineering Bulletin*, 36(2):21–27, 2013.

65. Ankit Toshniwal, Siddarth Taneja, Amit Shukla, Karthik Ramasamy, Jignesh M Patel, Sanjeev Kulkarni, Jason Jackson et al. Storm@twitter. In *Proceedings of the ACM SIGMOD International Conference on Management of Data*, pp. 147–156. ACM, New York, 2014.

66. Hiroshi Wada, Alan Fekete, Liang Zhao, Kevin Lee, and Anna Liu. Data consistency properties and the trade-offs in commercial cloud storage: The consumers' perspective. In *CIDR*, volume 11, pp. 134–143, 2011.

67. Feiyi Wang, Sarp Oral, Galen Shipman, Oleg Drokin, Tom Wang, and Isaac Huang. Understanding lustre filesystem internals. Technical Report for Oak Ridge National Laboratory, Oak Ridge, TN, Report No. ORNL/TM-2009/117, 2009.

68. Jim Webber. A programmatic introduction to neo4j. In *Proceedings of the 3rd Annual Conference on Systems, Programming, and Applications: Software for Humanity*, pp. 217–218. ACM, New York, 2012.

69. Matei Zaharia, Mosharaf Chowdhury, Michael J Franklin, Scott Shenker, and Ion Stoica. Spark: Cluster computing with working sets. In *Proceedings of the 2nd USENIX Conference on Hot Topics in Cloud Computing*, volume 10, p. 10, 2010.

Chapter 3

Performance Evaluation of Protocols for Big Data Transfers

Se-young Yu

Nevil Brownlee

Aniket Mahanti

CONTENTS

Lately, IP traffic volumes have been increasing rapidly [14]. Recent scientific works in various areas generate vast volumes of data each day. The Large Hadron Collider (LHC) generated 13 PB in 2010 (36 TB each day) [8] to study particle physics. The Sloan Sky Digital Survey (SSDS) aims to map objects in the universe to study. It generated 5 TB of data every year (14 GB each day) [25] for 5 years. That data is made public and available in the SDSS website.* The Square Kilometre Array (SKA) telescope project will study astronomical physics using its multiple antennas. SKA expects to gather 1 EB of data per day for the next few decades. These scientific endeavors are collaborative studies among international researchers and require the transfer of data between their research sites. CSIRO ASKAP, who run 36 SKA antennas, transfer 2.5 Gb/s continuously at (105 TB per day) to their Pawsey Supercomputing Centre.†

The data gathered from these studies can be processed locally to reduce its size, but the processed data still needs to be transferred. The size of the processed data differs as these projects use different techniques that can be applied to each specific research, but its order of magnitude is still large enough to be classified as *big data*. For example, SSDS provides public access to their data set even though it is larger than 100 TB.

A content delivery network with large data centers located around the world is another entity that requires big data transfer for data migration, updates, and backups. As cloud computing becomes common, the capacity of the data centers and both the intra-network and inter-network of those data centers increases.

Applications of such research are diverse, but they have one common requirements—they all need fast *big data* transfer systems.

The speed of data transfer depends on many factors, but they can be classified into the following:

1. *Network speed*: Higher network speed allows a link to carry more data on the wire per unit of time. Increasing the link speed provides more capacity to network activities using the links. Also, the packet forwarding devices within the path should have enough processing speed and transmission speed to ensure that packets are not lost but forwarded with minimum delay.

---

*http://www.sdss.org/.
†http://www.atnf.csiro.au/projects/askap/ASKAP_SW_0017_v1.0_distribution.pdf.

2. *Host processing speed*: Each host of a transfer must have the capability to read and write data at high speed. The hosts are responsible for locating data to send and pass to a transfer protocol. At the receiver, the host gets the data from the transfer protocol and saves it to a storage location.

3. *Transfer protocol speed*: When a sender and a receiver exchange packets, their transfer protocol defines a mechanism to transfer packets. It defines how to initiate the transfer, how to prepare data, how fast the sender can send packets, how to detect congestion of the network link, how to avoid further congestion, how to recover from packet losses, how to reassemble the data, and how to terminate the transfer.

Improving the network speed and host processing speed can be achieved by having carefully tuned faster processing hardware and network infrastructure. Those provide more capability to users by allowing more data on the wire, and by preparing and storing data efficiently. However, there are technical limits to how fast the network speed can advance at any given time. The capacity of the link needs to be shared between users of the link. The number of users increases as the link is interconnected with other links, and the available capacity per user decreases if more users demand efficient data transfers. When there is more demand from users than the link can handle, it becomes congested and the protocols start losing their packets.

A transfer protocol handles the movement of packets at both ends of the path between two hosts. The current transfer protocols implement functions that read data from a file system, write to a socket using either transmission control protocol (TCP) or user datagram protocol (UDP), and write to a file system. While handling the data from source and to destination, they apply a number of techniques to accelerate the transfer. These techniques are meant to provide better data transfer in different network circumstances.

In a short-haul network there are fewer routers, which reduces the chance of having packet losses. The shorter cable distance reduces round trip time (RTT), this allows the sender to keep less data in its transport buffer to ensure reliability. Also when a packet is lost, the sender is notified within a shorter time.

In a long-haul network, there is more chance of losing a packet not just because there are more routers in the path, but potentially because there are more network activities to share the available capacity. Longer RTT causes a sender to have more data held in its buffer, so in case of packet losses it can resend the packet from that buffer.

When congestion is noticed because of the packet loss it causes, a common protocol action is to reduce its packet sending speed, to avoid congestive collapse. When this happens on a shared link, it is likely that multiple hosts will reduce their sending speed, freeing available capacity and leading to a momentary link underutilization. To increase utilization of the link again, protocols start recovering their sending speed until the next congestion event.

Traditional TCP, using this approach, works well in links with small bandwidth-delay product (BDP), but it is not efficient enough to utilize links with large capacity and high RTT (long fat pipe) well because it takes longer to recover from a packet loss [34]. Recent TCP implementations such as CUBIC TCP [30] and scalable TCP [34] provide faster congestion window growth, which recovers the congestion window size faster after packet losses. On the other hand, UDP-based protocols implement their own congestion control to avoid having slow sending rate caused by slow congestion window growth.

More aggressive congestion-avoidance algorithms allow faster data sending rates, which generates more congestion on the link; fairness becomes an issue with these protocols. It is possible for an aggressive congestion-avoidance algorithm to monopolize the available link capacity by increasing its congestion window faster than other protocols. To avoid this unfairness, it should yield the available capacity to other protocols when there is congestion by slowing down its own throughput.

The traditional fairness definition focus on how different protocols or network flows share the available capacity, but network activities have different objectives. For example, a high-speed data transfer needs to utilize the available capacity as much as possible, but at the same time, voice over IP (VoIP) requires relatively low utilization but less congestion. It does not seem fair for a high-speed data transfer flow to reduce its sending speed to allow more capacity to the VoIP flows. However, causing less congestion causes fewer problems for other protocols on the same link.

In this chapter, we focus on measuring the induced congestion of high-speed data transfer protocol on the link. We measure RTT changes and packet losses caused by high-speed data transfers, and study the implications of the different behaviors observed in the protocols.

For every network path, properties such as available capacity, RTT, and packet loss probability change dynamically depending on the amount of network activity. To adapt to such dynamic changes of available capacity and congestion, some transfer protocols use the congestion avoidance provided by TCP [1,31,35], while others implement their own congestion control on top of UDP [19,26,32,43].

TCP can be optimized for long fat pipes by using a better congestion-avoidance algorithm, but that would not change TCP's slow start. Two problems with TCP slow start are raised by Ha et al. [29,55]. CUBIC TCP proposed a new approach to avoid slow start, while protocols using UDP do not have this problem because they do not use slow start.

High-speed data transfer protocols also provide additional features, such as multiple I/O threads, multiple transfer flows, data striping, and data pipelining. The multiple I/O threads feature reads files and prepares packets using multiple threads for better CPU utilization. Multiple flows open up parallel data connections between a sender and a receiver, to utilize the link more by sending out data using multiple streams simultaneously. Data striping reads a block of data across multiple I/O threads. Data pipelining is used to optimize server processing to transfer many small size files by reducing the number of sending commands to a single command [7].

A transfer host can be optimized using many techniques: using faster disk system, using Jumbo frames, and allocating large enough socket buffers from the OS kernel. Faster disk systems allow a transfer protocol to read data and write data faster, Jumbo frames reduce the packet header processing time, and larger socket buffers allow increased congestion window size, so that it is large enough to utilize all the available capacity in a long fat pipe.

These features are provided in a number of protocols used for high-speed transfers. There are many studies on implementing better congestion-avoidance algorithms and on achieving better link utilization for multiple transfer flows with Jumbo frames. However, there is a lack of study on characterizing the improvements for these protocols using such features.

In addition, there are studies comparing the efficiency of protocols using TCP and UDP, but there is little study of their behavior in transferring big data under different network conditions. Network conditions, such as those where available capacity and congestion levels are different for different links, and change from time to time within the same link.

Understanding the effect of these features and techniques are important because it helps to inform users as to how to choose the features that best suits their network environment. This chapter aims to help users of big data high-speed transfer protocols by identifying common limitations of high-speed data transfers and providing solutions to improve their efficiency with less interference and other network activities in the same path.

The rest of this chapter is organized as follows. Section 3.1 discusses several open source big data transfer protocols analyzed in this chapter. Section 3.2 discusses related work. We describe our local gigabit experimental network setups and performance evaluation results of the big data protocols in Section 3.3. Our experimental multigigabit in-lab network setup and performance evaluation results of big data transfer protocols are discussed in Section 3.4. Our national network setup and performance evaluation of big data transfer protocols are discussed in Section 3.5. Factors affecting performance and fairness of the protocols in the national testbed are discussed in Section 3.6. Section 3.7 concludes this chapter.

3.1 Background

This section discusses the details of big data transfer protocols analyzed in this chapter.

3.1.1 GridFTP

GridFTP [1] is an extension to file transfer protocl (FTP) [49] that includes enhancements suited for a grid environment. GridFTP can use parallel data transfer, data striping, and/or redundant data to have better access to the stored data. GridFTP is capable of transferring data with concurrent TCP or a single UDP (through UDT implemented within the GridFTP framework) sessions, multiple senders and receivers, and automatic TCP buffer tuning. These techniques can improve performance of high-speed data transfers between hosts over a long fat pipe. GridFTP has been used for many data-driven applications such as remote access to climate simulation data [11].

3.1.2 FDT

FFDT [35] is a high-speed data transfer protocol that uses TCP at the transport layer. To achieve high transfer rates, FDT uses concurrent threads and TCP connections during the transfer. The aggregate throughput of multiple TCP streams has been shown to be higher compared to UDP-based protocols [15]. FDT is integrated in the MonALISA framework [35] for monitoring, control, and optimization of complex systems, and it has been used to transfer data between CERN in Switzerland and Manhattan Landing Internet Exchange in New York and Starlight Exchange in Chicago and SARA Exchange in Amsterdam, the Netherlands [13].

3.1.3 High-performance SSH

SSH [4] provides an interactive shell for connecting to a remote host and SSH implementation usually bundles SFTP and SCP within its implementation to provide ways to transfer files.

High-performance SSH (HPN-SSH) [52] is an application protocol built on top of TCP, modifying OpenSSH* and SFTP [23] buffer sizes to allow a larger congestion window size. It can be used with different TCP implementations as its transport protocol and it provides encryption features to secure a connection between two hosts.

3.1.4 UDT

UDT [26] is a UDP-based connection-oriented data transfer protocol with its own congestion control mechanism. From the start of a transfer, UDT tries to increase throughput as much as

---

* www.openssh.org.

possible by keeping its inter-packet interval to as small as possible with maximum window size of 16 mss. Instead of using AIMD, UDT uses AIMD with decreasing increases (DAIMD) [56], which updates its congestion window as

$$cw_{new} = w + \begin{cases} \frac{1}{mss} & \text{if } B \leq C \\ \max(10^{(\text{ceil}(\log_{10}((B-C)*mss*8))} \times \frac{0.0000015}{mss}, 1/mss) & \text{if } B > C \end{cases} \qquad (3.1)$$

where:
    B is estimated link capacity
    C is current throughput

Once a NAK is received, UDT uses DAIMD to maximize sending rate by increasing window size slowly when the sending rate is close to the available bandwidth measured at the start. UDT changes its packet sending period at fixed intervals (10 ms is the default), rather than every RTT. According to UDT: Breaking the Data Transfer Bottleneck,* UDT is used for transferring different types of data such as large outer space images to astronomers and molecular chemistry research data.

3.1.5 Reliable blast UDP

Reliable blast UDP (RBUDP) [32] is a UDP-based protocol, it aims to utilize the available capacity as much as possible by sending out data at a user-specified rate. It has minimal reliability control, where the receiver sends a list of lost packets when the sender sends a DONE message via a TCP control channel. RBUDP does not consider any fairness, and it updates its sending rate for the next retransmission:

$$R_{new} = R * (0.95 - p) \qquad (3.2)$$

where p is the loss rate calculated from the receiver. Because RBUDP does not consider fairness, it can monopolize the network and causes congestion collapse. We advise users to use RBUDP only when there is traffic control available in the link. Also, it is not suitable for sending large files because it has to keep a tally for every packet, therefore using it with large files will cause the tally to grow and consume much memory.

3.1.6 Tsunami

Tsunami [43] is another UDP-based high-speed data transfer protocol with its own reliability and congestion control algorithms, using rate control and an adjustable error threshold. Tsunami splits a file into numbered blocks, and divides each block into IP packets before sending them. Its rate control sets the inter-block delay. A Tsunami client receives the blocks and puts each block into a ring buffer. It manages a list of lost blocks and sends the list after 50 blocks and predefined update period. The server adjusts the inter-block delay in proportion to the error rate. Tsunami has been used for electronic very long baseline interferometry (e-VLBI) data transfer from Onsala and Metsahovi, both in Sweden, to the Bonn correlator[†] in Germany.

---

  * http://udt.sourceforge.net/poweredby.html.
  [†] http://mars.hg.tuwien.ac.at/ evga/proceedings/S18_Haas.pdf.

3.2 Related Work

High-speed big data transfer protocols have been evaluated in theory and in simulation previously, however, they have not been analyzed in an experimental setting. An analytical and simulation-based study on these protocols may not reflect realistic performance of each protocol [21]. Using a testbed, we can observe the protocols themselves rather than their models.

Bateman et al. [5] studied behavior of high-speed TCP variants in a simulation study using ns-2 and using a testbed. They found that BIC TCP, CUBIC TCP, Hamilton TCP, high-speed TCP, and scalable TCP show good fairness against TCP new Reno in simulation but less fairness in a testbed network.

Li et al. [36] compared HSTCP, scalable TCP, H-TCP, BIC TCP and FAST TCP in terms of their fairness, efficiency and impact on web traffic. They showed that BIC TCP, FAST TCP and scalable TCP had some unfair behavior in their experiments, and claimed that HSTCP and BIC TCP have slow convergence time when recovering from a packet loss.

Marian et al. [40] also measured performance of FAST TCP, high-speed TCP, H-TCP, BIC, CUBIC, Hybla, TCP-Illinois, Westwood, compound TCP, scalable TCP, and YeAH-TCP using both 1 Gb/s and 10 Gb/s networks. They found that all TCP variance's performance degrade if RTT increases, even if the socket buffer size is larger than BDP.

Ha et al. [27] compared fairness and link utilization of several high-speed TCP variants such as BIC, CUBIC, FAST, HSTCP, H-TCP, and STCP using the dummynet testbed with varying levels of background traffic. They found that all the high-speed protocols showed less fairness to TCP as the delay increased. With background traffic, link utilization was improved because the background traffic filled up the remaining bandwidth. In a later study [28], they evaluated each of the high-speed TCP variants with background traffic and showed their behavior in terms of link utilization and fairness. H-TCP and STCP were found to be the most aggressive in terms of utilizing bandwidth. Although they studied performance of various TCP variants in the presence of background traffic, their study is limited by using only TCP variants.

Cottrell et al. [15] compared throughput, stability, fairness, CPU utilization of several high-speed data transfer protocols such as HSTCP, scalable TCP, and FAST TCP using a 1 Gb/s academic and research network at the Stanford Linear Accelerator center. They found that scalable TCP, BIC, and H-TCP were most efficient while TCP Reno, HSTCP-LP, and HSTCP did not utilize throughput efficiently. UDT performed similarly to the advanced TCP implementations. Although the test network was limited to 1 Gb/s, their results revealed performance and fairness issues with UDT.

Suresh et al. [58] evaluated throughput, fairness, and CPU usage of GridFTP, GridCopy, and UDT with a 2 Gb/s network. They found that UDT performed well compared to the rest of the tested protocols. GridFTP also performed better, except when transferring smaller files. While this study compared the performance of GridFTP with UDT, their testbed was restricted to using a 2 Gb/s network.

Weston et al. [63] transferred astronomy data from New Zealand to Germany and Finland using two UDP-based protocols, namely UDT and Tsunami, via the New Zealand high-speed academic network. They found that UDT showed better performance and better fairness than Tsunami. They also identified fairness issues related to Tsunami. This study, however, did not evaluate any TCP-based protocols.

Yue et al. [66] compared four UDP-based protocols, RBUDP, Tsunami, PA-UDP, and UDT in terms of throughput, RTT, loss rate, fairness, and CPU utilization using a 1 Gb/s link. They found that PA-UDP achieved optimal performance. UDT was most convenient to use because of its parameterless setup.

Lu et al. [38,39] proposed parallel peak link utilization transport (para-PLUT) that uses multiple UDP connections to utilize the network capacity better. They dynamically increase in the number of UDP flows used in para-PULT by estimating bottleneck link capacity, and use more flows if the current throughput is lower than that. They compared PLUT, UDT, and TCP new Reno using a 1 Gb/s network. They found that the multiple flows PLUT outperformed a single flow PLUT, UDT, and TCP new Reno in the 1 Gb/s link.

Park et al. [47] proposed improvement in UDT using parallel UDT flows calculated from an optimum parallel channel number (OPCN) which is dependent on RTT. It showed that the parallel UDT flows improve the single UDT by *up to 644% in 1 Gb/s networks* with varying emulated RTT. An et al. [3] also proposed improvement in UDT with router queue size estimation, but their evaluation showed that their approach only improved the UDT by 16% in 1 Gb/s network. Neither of these studies provided their implementation, so we could not apply their changes to our testbed.

Nam et al. [45] compared scp, rsync [61], bbcp [31], GridFTP, and Globus online's performance using a 10 Gb/s link. They found that bbcp and GridFTP performed best on their 10 Gb/s short-haul link and rsync and scp were least efficient. In addition, their study showed that there is significant improvement when multiple flows are used and larger socket buffer sizes are used.

Our work complements previous work by performing an experimental analysis of four well-known protocols for transferring big data over big pipe networks, using TCP and UDP-based protocols. We also analyze individual protocol behavior under varying levels of congestion-inducing TCP and UDP background traffic. We also assess the fairness of each protocol in terms of induced congestion.

For high-speed data transfer, it is important to consider the effect of having high-speed transfer on the other flows rather than being able to share the capacity equally among all flows. Therefore, we use our experimental testbeds to assess fairness in real high-speed networks. With the local testbed, we generate other bulk transfer flows to observe how each protocol shares the capacity when there is another high-speed data transfer using the same path. With the national testbed, we measure RTT increase to observe induced congestion on the path. The RTT increase is an indication of congestion level on the path, where many packets may be queued in the routers, leading to packet losses. By measuring this, we can observe how much impact these protocols have and how much influence they have on the other network activity.

3.3   Local Gigabit Testbed

In Section 3.1 we described data transfer protocols, with emphasis on their efficiency when used in high-speed networks. From these protocols we chose a selection, based on the following properties:

1. *Performance in large BDP network*: Traditional TCP cannot use the full capacity of a large BDP network and a high-speed network needs to overcome the limitations of large RTT and link capacity.

2. *Free software that runs on GNU/Linux*: The freedom to run, study, redistribute, and modify the software is essential for our study and for the commonwealth of the software community.

3. *Capable of reliable data transmission*: The integrity of scientific data is a must, unreliable transmission harms the integrity of the transmitted data.

4. *Implementation of the software must be stable*: For high-speed data transfer, the quality of implementation is important to provide usability and stability.

To explore the feasibility of each protocol in physical networks, we set up a 1 Gb/s cross-cable testbed as our preliminary setup. We designed the testbed to represent the most simple network, free from unnecessary anomalies. In the testbed, we measured the goodput of our initial choice of protocols with different emulated RTTs. That provided information on the basic usability and efficiency of each protocol in small to large BDP networks.

From what we observed in the cross-cable network, we developed a more realistic testbed, representing another local area network scenario. In this testbed, we added a router and background traffic generator to observe queuing effects of packets in the router, and introduced background traffic to the network. These changes enabled us to measure goodput in a greater variety of network circumstances, including congested networks.

We found that the performance of TCP-based protocols is similar to that of UDP-based protocols in our testbed, without any emulated RTT. With longer emulated RTTs, Tsunami and RBUDP do not degrade their performance, while TCP-based protocols and UDT degrade significantly. With background traffic, Tsunami and HPN-SSH degrade their performance. Tsunami is the least affected by exogenous packet loss while other protocols are heavily affected. Having multiple flows allow HPN-SSH, UDT, and FDT to increase their performance but multiple flows do not increase Tsunami's throughput.

3.3.1   Cross-cable testbed

The testbed consisted of two machines with identical setup. Each of the machines acted as sender and receiver of data, as well as UDP background traffic generator. Figure 3.1 shows the cross-cable gigabit testbed network.

Table 3.1 shows details of the hardware used in the testbed. Both machines used their default TCP socket buffer size, which is 112 kB per socket (and maximum of 128 kB). The transmit queue length was set to 1000 packets, and the TCP congestion-avoidance algorithm used was CUBIC.

With this testbed, we studied usability of the protocols listed in Section 3.1 as well as other protocols, including bbcp [31], PAUDP [19], and saratoga [64]. We found technical difficulties

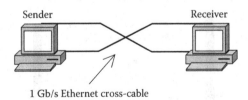

1 Gb/s Ethernet cross-cable

Figure 3.1: Cross-cable gigabit testbed.

Table 3.1   A summary of the three techniques discussed in the chapter

Role	Model	CPU
Sender and receiver	HP d530	Intel Pentium 4 3.2 GHz
Memory	**OS**	**NIC**
1 GB	GNU/ Linux 2.6.31.1	Broadcom BCM5782 Gigabit Ethernet

with the usability of GridFTP, bbcp, paudp, and saratoga. We used HPN-SSH, UDT, Tsunami, FDT, and RBUDP as our choice of protocols after this usability test.

We used dummynet [9] to emulate different RTTs in the network to observe the behavior of each protocol. We generated delays for each packet when they are received at the receiver. We expected to observe their behavior when they are used in hosts connected with different physical distances between them.

We used Iperf [60] to generate background UDP traffic. The reason we chose UDP over TCP is because we could not generate a fixed rate of traffic with TCP. Iperf with UDP was able to generate a steady background traffic of 0.3, 0.5, and 0.7 Gb/s while we were testing the high-speed data transfer protocols.

We measured goodput and CPU usage of each protocol without emulated delay. We also measured goodput of each protocol using emulated delay, and with UDP background traffic applied. We measured the goodput once each with using different factors for each protocol.

Figure 3.2 shows goodput of HPN-SSH, UDT, FDT, Tsunami, and RBUDP, and their CPU consumption on sender and receiver hosts. FDT had the highest throughput among the protocols. Performance of HPN-SSH was less efficient, with higher processing resource consumed. This is because HPN-SSH uses default SSH encryption, that is, the aes128-ctr cipher algorithm with HMAC-MD5 hash algorithm to secure its data channel. We could not ignore the overhead of encryption because we could not disable it in HPN-SSH.

The differences of CPU usage between sender and receiver are relatively small except for UDT and RBUDP. The possible reason for our UDT sender to have significant CPU usage compared to its receiver may come from inefficient implementation of the algorithm, especially when it does in-memory copies between socket buffer and file [66].

Figure 3.3 shows goodput of protocols measured with different delays emulated with dummynet. As the RTT between sender and receiver increases, the goodput degrades for HPN-SSH, UDT, FDT, and RBUDP. As RTT increases, the performance of TCP is degraded because it is limited by TCP socket buffer size from reaching higher throughput. Also, TCP slow start takes longer to reach high throughput and that increases the transfer time. The performance of HPN-SSH and FDT are degraded because they are using TCP as their transport layer protocol. Furthermore, HPN-SSH is CPU-intensive and our emulating delay with dummynet reduces the CPU cycles available for HPN-SSH to process the data transfer; therefore it degrades further

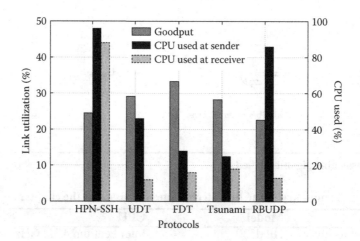

Figure 3.2: Goodput measured for each protocol with CPU consumption.

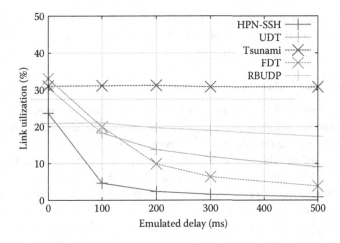

Figure 3.3: Goodput measured for each protocol with delay emulated with dummynet.

compared to FDT. Performance of UDT degrades less compared to the protocols that use TCP, however it still degrades because the flow window size of UDT is limited by UDP's socket buffer size. RBUDP is not affected significantly because it has a minimal congestion-avoidance algorithm, but it also slightly degrades because it also uses TCP as a signaling transport to ensure reliability, which degrades with increasing RTT. Tsunami also uses TCP as a signaling traffic so it is degrades slightly as RTT increases.

Figure 3.4 shows the goodput of each protocol measured with different amounts of UDP background traffic applied to the network. Tsunami is most affected by the UDP background traffic, especially when it is more than 0.3 Gb/s. It reduces its packet sending rate because some of its UDP packets are lost consecutively as it competes with background traffic, so less available network capacity is left for Tsunami. The goodput of HPN-SSH also decreases again, possibly because of CPU cycles used by Iperf.

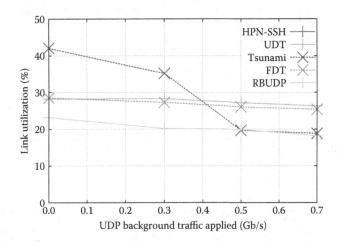

Figure 3.4: Goodput measured for each protocol with UDP background generated by Iperf.

We have measured performance of HPN-SSH and FDT with different TCP congestion-avoidance algorithms (BIC TCP, scalable TCP, and H-TCP), but we found no evidence of performance differences in this testbed.

3.3.2 Gigabit local area network testbed

To the cross-cable testbed described in Section 3.3.1, we added a router and traffic generator to offload the network emulation process and background traffic generation. The sender and receiver only handle the file transfer process, which is dominant in the general use-case.

The router is added to address more realistic delay emulation, and the background traffic generator to generate enough background traffic to cause congestion on the network. We expect to measure the behavior of each protocol when the link is congested with background traffic.

The testbed consisted of four Linux machines; sender, receiver, router, and background traffic generator. Figure 3.5 shows the testbed network. We formed a simple star network in which sender, receiver, and background traffic generator are physically connected to the router.

We used the same machine as in Table 3.1 for a sender and a receiver, and used machines for router and background traffic generator as in Table 3.2. We used Ubuntu 10.04 for the router because dummynet was not able to run on the linux 3.0+ kernel versions in more recent Ubuntu releases. Delay and random packet loss were emulated with dummynet on the router.

We chose to test HPN-SSH, FDT, UDT, and Tsunami based on the performance measured in Section 3.3.1. RBUDP was discarded from the experiment because it does not have a flexible congestion control algorithm, instead it requires a manual process to find its optimal capacity,

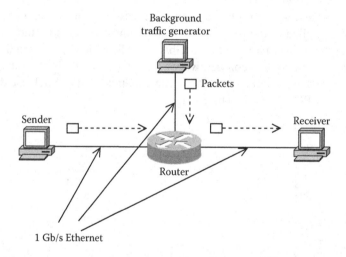

Figure 3.5: Gigabit testbed network.

Table 3.2   Hardware used in the Gigabit testbed

Role	Model	CPU	Memory	NIC	OS
Router and background traffic generator	Supermicro X7SLA	Intel Atom 330 1.60 GHz	2 GB	Intel 82575EB Gigabit Ethernet	GNU/Linux 2.6.32

once the available capacity is set. RBUDP's congestion control algorithm is not flexible enough to adapt when available capacity varies.

We measured the change in goodput in terms of change in RTT, exogenous packet loss, and generated amount of background traffic. We used arbitrary RTT values between 0 and 300 ms, which represent the RTT of a local network and the RTT between New Zealand and US Midwest We used 0, 10, 50, 100, 150, 200, 250, and 300 ms emulated delay. Depending on the delay emulated, we changed the maximum TCP socket buffer size to be equal to BDP, to accommodate enough congestion window size for TCP.

The emulated exogenous packet loss is from 0% to 3%, possible arbitrary values for a poorly maintained network with a physical cable fault. We decided to move an arbitrary file of size 10 GB for this test. The size of the file is chosen to let each protocol run for at least 100 s with throughput that fully utilizes the 1 Gb/s capacity, but not more than 10 min with a lower level (10%) of utilization. This made the repetition of each run possible within a reasonable time frame.

The TCP congestion-avoidance algorithm used was CUBIC, and TCP SACK was disabled as well as the Nagle algorithm. We also ran multiple flows for each protocol with each emulated RTT. We executed 1, 2, 5, and 10 flows of the protocol at the same time and measured aggregated goodput with different emulated RTT. Because the HPN-SSH, UDT, and Tsunami applications do not support using multiple flow transfers in an instances, we had to run multiple instances of HPN-SSH, UDT, and Tsunami, while FDT was used with built-in multiple flow transfers.

We used Iperf to generate TCP and UDP background traffic. Because we cannot rate-control TCP, we used increasing number of background TCP flows from one to five. For UDP, we generated from 0.1 to 0.5 Gb/s background traffic.

Figure 3.6 shows the goodput of each protocol in our testbed with a dummynet emulator in the middle. We emulated delays for packets coming from either direction in the router, to generate 0–300 ms delays to the traffic. HPN-SSH and FDT performance degrades as RTT increases because of their TCP transport layer. Goodput of UDT also degrades because of its congestion control, dependent on RTT. Goodput of Tsunami does not degrade because it has its own rate control.

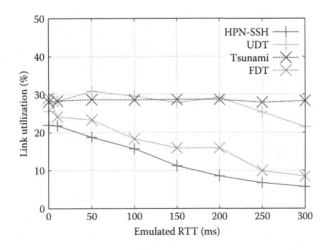

Figure 3.6: Goodput of each protocol vs. emulated RTT.

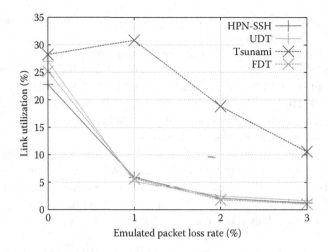

Figure 3.7: Goodput of each protocol vs. emulated exogenous packet loss.

Figure 3.7 shows the goodput of each protocol in our testbed with different exogenous packet loss probability emulated with dummynet in the router. The behavior of HPN-SSH, UDT, and FDT show similar behavior when they react to random packet loss. HPN-SSH and FDT use TCP as their transport layer so their behavior when reacting to network changes is similar. The congestion control algorithm used in UDT reacts in a similar way to CUBIC TCP when packets are lost in its flow. Tsunami is not affected by packet loss rate less than 1% of probability, but its goodput degrades with packet loss probability over 2%.

Figure 3.8 shows aggregated goodput of high-performance SSH over different emulated RTTs. Compared to a single flow, two and five HPN-SSH flows show better performance with increasing RTT. This is because a TCP socket buffer is allocated to each HPN-SSH flow, and the aggregated buffer size is larger as the number of flows increases. Because HPN-SSH is a CPU-intensive application, the performance of 10 flows degrades considerably compared to

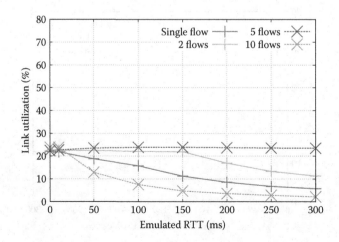

Figure 3.8: Aggregated goodput of HPN-SSH flows vs. emulated RTT.

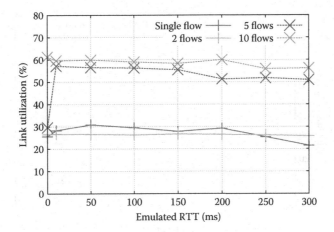

Figure 3.9: Aggregated goodput of UDT flows vs. emulated RTT.

that of 5 flows. As the number of flows increases, flows interfere with each other and cause packet losses.

Figure 3.9 shows aggregated goodput of UDT for different emulated RTTs. Multiple instances of UDT utilize the capacity most among the protocols. Five instances of UDT show a maximum throughput of 590 Mb/s. Although there is a slight increase in goodput for two flows compared to a single flow, goodput for two flows degrades more as RTT increases over 200 ms. The aggregated goodput for five flows and ten flows are much higher than goodput for a single flow, but the difference between 5 flows and 10 flows is small. Multiple flows of UDT also have more aggregated UDP socket buffer space, enabling them to have more goodput compared to the single flow. However, from looking at the trace file with Wireshark, we've seen enormous numbers of packets sent from UDT, giving rise to much transfer overhead. This is an implementation issue and possible limitation of goodput with more than five flows.

Figure 3.10 shows aggregated goodput of Tsunami over different emulated RTTs. There is almost no performance difference between single flow, two flows, and five flows with different

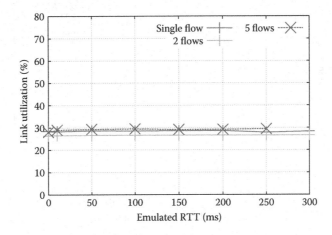

Figure 3.10: Aggregated goodput of Tsunami flows vs. emulated RTT.

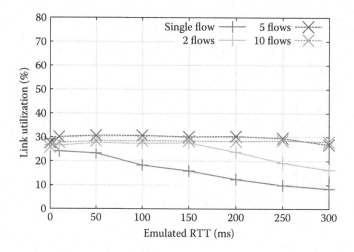

Figure 3.11: Aggregated goodput of FDT flows vs. emulated RTT.

RTTs. Tsunami uses rate control, allowing it to perform transfers without degradation with increasing RTT, and multiple flows of Tsunami are not limited by socket buffer size, but by disk read and write speed.

Figure 3.11 shows aggregated goodput of FDT over different emulated RTTs. As the number of flows increase, degradation of goodput reduced with high RTT. Similar to HPN-SSH, multiple FDT flows allow more aggregated buffer size, so that FDT can transfer faster with longer RTT. FDT did not have any issue with computing resource, so even 10 FDT flows did not have any problem running on the two hosts.

Figure 3.12 shows the goodput of each protocol with TCP traffic generated in the background with Iperf. TCP background traffic did not have any noticeable effect on Tsunami or FDT, while HPN-SSH and UDT had degraded goodput.

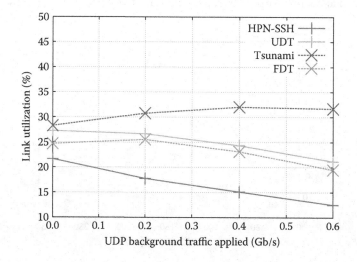

Figure 3.12: Goodput of each protocol with TCP background traffic.

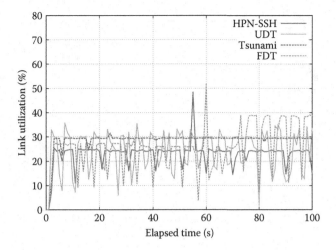

Figure 3.13: Goodput of each protocol with UDP background traffic.

Figure 3.13 shows the goodput of each protocol with background UDP traffic generated with Iperf. Goodput of HPN-SSH, UDT, and FDT decreases as the amount of UDP background traffic increased. Tsunami's goodput did not decrease.

We measured the throughput of each protocol with multiple instances of application and emulated RTTs. We used a Perl script to record average number of bytes transferred for each second from virtual/proc filesystem in GNU/Linux. This minimizes the overhead of the process and disk writes compared to collecting network trace files. Figures 3.14 through 3.25 are generated from the record.

Figure 3.14 shows throughput for a single instance of each protocol over elapsed time, without delay emulation. The throughput of all flows fluctuates, and frequent dips appear throughout the entire time except for Tsunami. The maximum socket buffer size set in the GNU/Linux kernel limited the maximum throughput of HPN-SSH and FDT. Although the

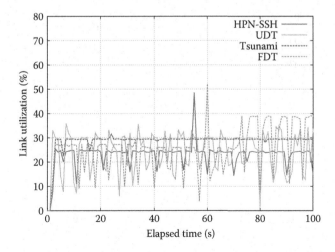

Figure 3.14: Throughput for a single instance of each protocol, without dummynet.

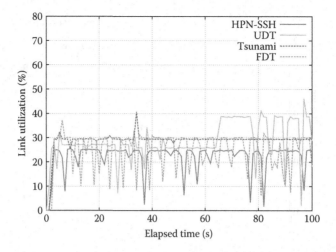

Figure 3.15: Throughput for a single instance of each protocol, 10 ms RTT.

buffer size of UDP is not related to the RTT, UDP buffer size per socket was still a limiting factor for UDT.

From Figure 3.15 we can see the dips for HPN-SSH are deeper because the RTT has increased, so the buffer size required to reach similar throughput increased. This makes TCP wait longer until it frees buffer space to accommodate the next packet. After 70 s elapsed, throughput of FDT increases to 0.4 Gb/s. That is caused by the FDT implementation issue, which tries to resend the same data for a while when it is close to the end of the transmission.

Comparing Figures 3.16 and 3.17, there are almost no changes in HPN-SSH and FDT's behavior across the two figures. However, we can see that multiple instances of UDT utilize more link capacity in our network compared to the link without dummynet. By looking at some UDT trace files from the later experiments with Wireshark, we identified a UDT implementa-

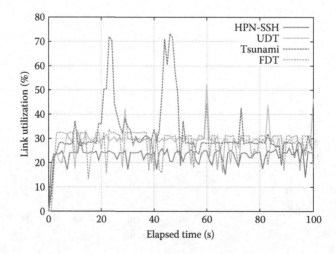

Figure 3.16: Throughput for five instances of each protocol, without dummynet.

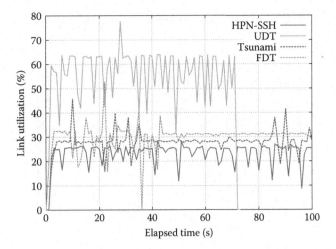

Figure 3.17: Throughput for five instances of each protocol, 10 ms RTT.

tion issue, which reports multiple NAK messages when it is used in a low latency link; that causes its congestion control algorithm to prohibit the increasing function to further grow its congestion window. Throughout Figures 3.17, 3.19, and 3.22 through 3.25, multiple instances of UDT utilized the link well. Although this is a positive behavior of UDT, the data must be manually split and merged at each end to be used with multiple UDT instances.

Figure 3.18 shows throughput of HPN-SSH, UDT, and FDT. Tsunami was not able to perform the transfer with 10 instances due to its instability in implementation. Compared to Figure 3.14, using multiple instances reduces the depth of dips on HPN-SSH's throughput. As the number of flows increases, recovery time from a packet loss reduces, resulting in reduced dips. From Figures 3.21, 3.23, and 3.25, we can see the effect of small TCP socket buffer size on HPN-SSH and FDT, but not on Tsunami and UDT which use UDP. It may be useful to use UDP-based protocols when the data transfer node only supports small buffer sizes.

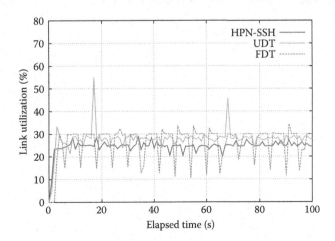

Figure 3.18: Throughput for 10 instances of each protocol, without dummynet.

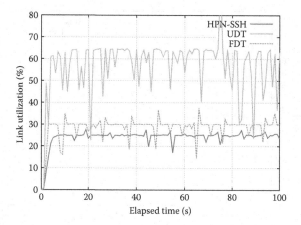

Figure 3.19: Throughput for 10 instances of each protocol, 10 ms RTT.

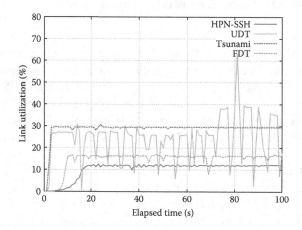

Figure 3.20: Throughput for a single instance of each protocol, 150 ms RTT.

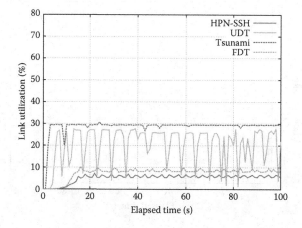

Figure 3.21: Throughput for a single instance of each protocol, 300 ms RTT.

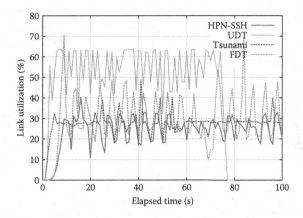

Figure 3.22: Throughput for five instances of each protocol, 150 ms RTT.

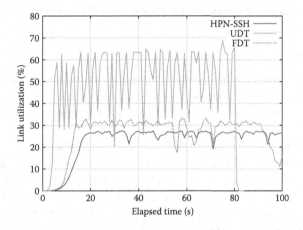

Figure 3.23: Throughput for five instances of each protocol, 300 ms RTT.

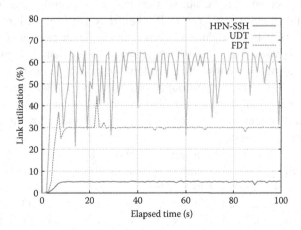

Figure 3.24: Throughput for 10 instances of each protocol, 150 ms RTT.

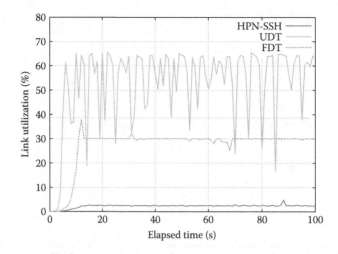

Figure 3.25: Throughput for 10 instances of each protocol, 300 ms RTT.

3.3.3 Discussion

From the experiment, we found that various overheads running the protocols in old processors inflicted undesirable influence on our measurement. As we progressed through experiments, we used more better, more modern hardware and operating systems to minimize the effect of these overheads.

During the experiments, we explored different properties of high-speed data transfer protocols. TCP-based protocols use the same congestion-avoidance algorithm, but had different performance over different RTT for different reasons. HPN-SSH requires more resource, because it does not provide transfer without encryption, hence it shows lower performance compared to FDT which uses the same transport protocol. This is primarily due to the lack of resource in our testbed, which consisted of out dated machines with low-performance processors. UDP-based protocols do not show similar behavior to each other, because their congestion control algorithm is not bound by TCP. Tsunami and RBUDP are less affected by increasing RTT of the link, but UDT is affected more because it uses congestion control that is similar to TCP.

Using multiple flows is effective for TCP when they are used in longer RTT links. This is because multiple TCP flows reduce recovery time for congestion window growth. The aggregated congestion window's size growth is much faster than a single flow, because all of the flows are increasing their congestion window size simultaneously.

Multiple UDT flows increased the throughput drastically; this was beyond our expectation. The cause of this is an implementation issue that we could not track but we suspect each single UDT flow is independent and less affected by each other flows, allowing each flow to have a substantial amount of throughput.

In this network, we were not able to generate enough TCP background traffic to cause each protocol to be affected by reduced available capacity. Although HPN-SSH and UDT degrades, Tsunami and FDT do not, and it is likely that the traffic generated by Iperf was not stable enough to produce a large amount of traffic during the experiment. On the other hand, UDP background traffic causes all protocols except Tsunami to reduce their throughput. This is due to the reduced available capacity and congestion control at work for the three protocols. Tsunami did not have enough flexible congestion control to reduce its data sending rate, so almost no effect was shown.

3.4 In-Lab 10 Gb/s Testbed

Recent capacity of academic networks has been growing beyond gigabit networks to 10 Gb or even 100 Gb. High-speed data transfer protocols are designed to be used in the multi-gigabit networks, however there has been little study of such protocols on actual high-speed networks.

In a 10 Gb/s network, increased capacity allows more network traffic to travel through. This provides shorter completion time for high-speed data transfers, but only if the protocol is efficient enough to utilize the multigigabit capacity.

Using our testbed, we automated the process of running the experiments and measuring the performance of each protocol. With the setup, we ran each protocol with varying RTT, number of flows, and background traffic volume. Also, we measured the ramp-up time of each protocol, and its capacity share against a TCP Reno flow and a CUBIC TCP flow.

3.4.1 Methodology

We set up a simple network with two linux machines, sender and receiver, and two 10 Gb/s network switches. The machines and switches were set up as in Figure 3.26.

Details of the machines used in this network are shown in Table 3.3. Because we had no resource available for a dedicated packet generator, we decided to run nuttcp 6.1.2* in the sender as a background traffic generator. A Dell R320 is capable of generating more than 10 Gb/s of traffic, so we presumed there would be a minimal effect from generating data transfer traffic and the background traffic in the same machine.

A long fat pipe network is usually shared by many hosts, and the traffic they generate can take up much of the available capacity. A high-speed data transfer protocol's behavior can be affected by how much available capacity is left on a link.

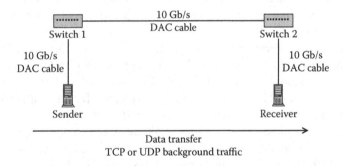

Figure 3.26: In-lab 10 Gb/s testbed network.

Table 3.3   Hardware used in the 10 Gb/s testbed

Role	Model	CPU	Memory	NIC	OS
Sender and receiver	Dell R320	Intel Xeon E5-2407 2.20 GHz	16 GB	Broadcom BCM57810 10 Gb	Ubuntu 12.04
Switches	Pronto P-3290	MPC8541	2 GB	Firebolt-3 ASIC	XorPlus 1.3

---

*http://www.nuttcp.net/.

To induce congestion, a number of concurrent TCP flows were run alongside each protocol to observe their behavior when there is a significant amount of TCP, handling traffic ongestion using AIMD. We did not attempt to generate a fixed number of TCP packets per second in this experiment. This is due to the nature of TCP, which uses its congestion control algorithm to decide how much data it should send at any time. We wanted to work with a realistic amount of background traffic for TCP. Therefore, we decided to increase the number of background TCP flows rather than increasing the background TCP data rate.

We used nuttcp to generate TCP and UDP background traffic for two emulated RTT links. The first link represents data transfer between New Zealand and Western Australia, which corresponds to an RTT of about 75 ms. The second link emulates data transfer between New Zealand and the US Midwest, which corresponds to an RTT of about 150 ms. These locations were selected based on the likelihood of New Zealand scientists sharing astronomy data with the Murchison Radio-Astronomy Observatory in Western Australia and universities in the US Midwest.

With UDP background traffic, we generated 10,000–40,000 1250-byte datagrams to induce congestion on the link with an emulated 75 ms RTT. The packet size and number of packets per second were selected according to the level of congestion likely to be induced. With 10,000 packets per second, we hardly saw packets dropped by congestion while with 40,000 packets per second we observed about 33% of UDP being dropped due to congestion. For an RTT of 150 ms, 9,000–21,000 packets of 1250-byte datagrams per second were sent, which generated a similar level of congestion.

After our background traffic experiment, we decided to perform a test with varying RTT and amounts of background traffic. We measured the goodput of each protocol while running nuttcp to generate 1–10 background TCP flows or 1–10 Gb/s background UDP traffic in a link with RTTs of 0–200 ms. In each set of tests, we used either TCP or UDP, with each of 1–10 different levels of background traffic in one of the 0–200 ms emulated RTTs so that each goodput recorded represented a performance of each protocol in a fixed environment of background traffic and RTT.

We used netem* on the receiver side to emulate delay, because dummynet was not compatible with our more recent linux kernel. High-speed data transfer protocols behave differently with varying RTTs. We emulated the following RTT values: 0, 10, 50, 75, 100, 150, and 200 ms. These values represent RTT for local hosts (0 ms), cross-Tasman path (50 ms), and paths between New Zealand and US Midwest (150 ms). Arbitrary RTT values for hosts in between New Zealand and US East and farther locations (200 ms) were considered as well.

Since we were able to use an up-to-date version of Ubuntu, we resolved the issue of installing GridFTP from source and the Ubuntu repository. GridFTP supports multiple connections so we decided to use GridFTP instead of HPN-SSH because of its capability and better computing resource utilization.

We chose GridFTP, UDT, Tsunami, and FDT in our experiment to understand behavior of each of these protocols over different RTT, background traffic, and using Jumbo frames of 9000-byte maximum transmission units (MTUs) in most of our experiments. In this section we state whether we used 1500-byte MTU throughout each section.

We expected to see less goodput for each protocol because the disk speed of the sender and receiver is slower than the network capacity. To maximize the disk performance, we used the ZFS [6] RAIDZ file system with three physical disks for both systems. That provided higher disk read/write speeds compared to cross-cable testbed, but they were still much less than our 10 Gb/s network capacity. Therefore we decided to run each protocol using

---

*http://www.linuxfoundation.org/collaborate/workgroups/networking/netem.

memory-to-memory copies. To do this, we used /dev/zero and /dev/null for the source and destination files in our experiment. To compare ZFS with a different file system, we set up ext4 using a single hard disk in each machine. We ran each protocol with files in each file system, transferring 30 GB zero-filed files generated with dd.* This file size represents the typical volume of data produced per day from a satellite during a stereoscopic census study [59].

We have seen the effect of small TCP and UDP buffer size in our preliminary analysis. We decided to modify TCP and UDP socket buffer size because it is an important factor for achieving high throughput over a long fat pipe. The maximum TCP window size depends on the TCP socket buffer size. We have tuned the operating system socket buffer size as explained in ESnet web page,[†] as well as other system parameters, except for the TCP congestion-avoidance algorithm. We used CUBIC because it is widely used (e.g., in Android systems [65]) and it is the default congestion-avoidance algorithm in GNU/Linux. In our cross-cable testbed, we found no significant performance improvement from using any of CUBIC TCP, H-TCP, or scalable TCP.

3.4.2   Results

3.4.2.1   Throughput for varying RTT and multiple flows

Figure 3.27 shows the goodput of each protocol with different emulated RTTs. Overall, performance of FDT and GridFTP was better than UDT or Tsunami in this environment. Although the transfer rate decreases more rapidly as RTT increases, FDT and GridFTP showed better performance over UDT and Tsunami, at least for RTTs of 200 ms or less.

For increasing RTT, TCP-based protocols performed much better than UDP-based protocols. The performance of FDT for less than 100 ms RTT was the best among all four protocols. The highest throughput of FDT was 2.34 Gb/s with a 1 ms RTT. However, the throughput of FDT decreased faster compared to GridFTP with TCP for RTTs greater than 100 ms. For RTTs greater than 100 ms, GridFTP with TCP performed best.

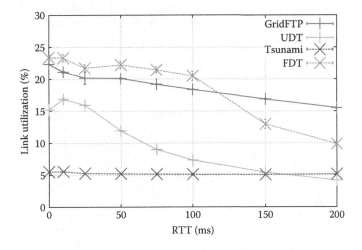

Figure 3.27: Goodput of each protocol in 10 Gb/s testbed.

---

*https://www.gnu.org/software/coreutils/manual/coreutils.html#dd-invocation.
[†] http://fasterdata.es.net/host-tuning/linux/.

UDP-based protocols did not perform well for most of the RTTs we tested. GridFTP with UDT performed poorly for most of the RTTs. Although its author claims that UDT is more efficient with increasing RTT, the rate of decreasing throughput for increasing RTT from 0 to 100 ms was the fastest among all protocols tested. At RTT 200 ms, UDT was the slowest among all the protocols.

Throughput of Tsunami did not decrease much as RTT increased, however its throughput was the lowest among all four protocols for RTTs less than 200 ms. Tsunami showed throughput of 0.55 Gb/s across all the RTTs we tested.

GridFTP and UDT showed slightly higher throughput with lower RTTs. This may imply that GridFTP or FDT have better performance than UDT and Tsunami between hosts that are geographically close. However, as the distance between hosts increases, the throughput of GridFTP and FDT decreases, as does UDT. Tsunami's throughput is not affected by increasing distance, but its throughput is the lowest of all the protocols tested for the hosts with RTT less than 150 ms.

Figure 3.28 shows the change in goodput of GridFTP with varying RTT and number of flow used without background traffic. As number of flows increased, the goodput increased slightly for shorter RTT, because of the increased aggregated congestion window growth rates, but limited by disk read/write speed. As RTT increased, the goodput of GridFTP decreases slightly; having fewer flows made the goodput decrease, but increased again with more than seven flows.

Figure 3.29 shows changes in goodput of FDT for varying RTT and number of flows used without background traffic. Goodput of FDT also did not change much as RTT increased for the same reason as for GridFTP. Increasing the number of flows to more than seven decreases goodput because of the disk read/write process overhead for assembly and disassembly of data to/from multiple data transfer flows.

3.4.2.2 Ramp-up time for each protocol

Figures 3.30 through 3.35 shows the startup transient time for each protocol. The throughput of each protocol is measured at each second during the first 60 s when transferring a 30 GB file using different emulated RTTs.

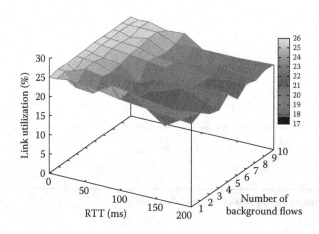

Figure 3.28: Goodput of GridFTP in 10 Gb/s testbed using multiple flows and emulated RTTs.

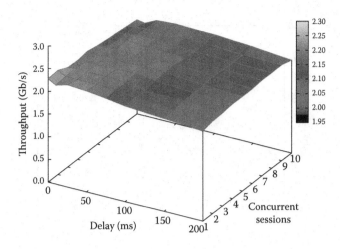

Figure 3.29: Goodput of FDT in 10 Gb/s testbed using multiple flows and emulated RTTs.

From Figures 3.30 through 3.32, FDT shows many dips, caused by slow disk read/write time. The dips are also shown for GridFTP in Figure 3.30, but the maximum and minimum throughput is different because file handling of GridFTP and FDT are different.

As RTT increases, socket buffer size limits the throughput of GridFTP and FDT while RTT-aware congestion control in UDT limits its throughput. From Figures 3.33 through 3.35, the maximum throughput of GridFTP, UDT, and FDT decreases as RTT increases. Tsunami did not change its maximum throughput throughout the experiment. As the maximum throughput decreases, dips in the GridFTP and FDT in short RTT networks disappear because the throughput of the protocols in longer RTT networks never reach the maximum read/write speed of the file system.

GridFTP seems to use larger socket buffer size on longer RTT links compared to FDT. Figure 3.33 shows that, FDT has better steady-state throughput from having larger socket

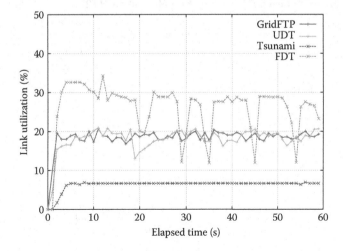

Figure 3.30: Throughput of each protocol for first 60 s.

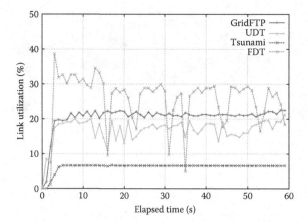

Figure 3.31: Throughput of each protocol for first 60 s, 10 ms RTT.

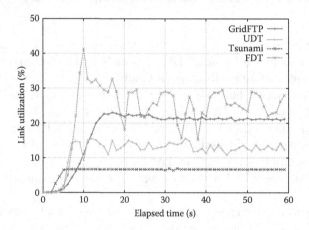

Figure 3.32: Throughput of each protocol for first 60 s, 50 ms RTT.

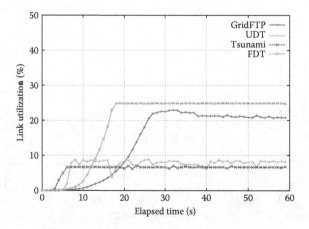

Figure 3.33: Throughput of each protocol for first 60 s, 100 ms RTT.

buffers, but as RTT increases beyond 100 ms the steady-state throughput of FDT decreases significantly compared to that of GridFTP. Although it takes longer to ramp up to its steady-state throughput, GridFTP had higher steady-state throughput than FDT, which means it uses larger socket buffer sizes.

UDT and Tsunami were quicker to increase their throughput from the start of a transfer between distant hosts in high RTT networks. The throughput of GridFTP and FDT increased slowly from their start-up as RTT increased. In Figure 3.34, the times taken for UDP-based protocols were shorter compared to the TCP-based protocols at 150 ms RTT. In particular, Tsunami shows its high throughput from the start of the transfer; it took 1 s to reach 0.5 Gb/s, which was its steady-state throughput. UDT was the second fastest (it took 2 s to reach 0.5 Gb/s). FDT was slower than GridFTP with UDT, while GridFTP with TCP took the longest to reach 0.5 Gb/s. GridFTP and FDT are dependent on TCP's slow start, which takes significant

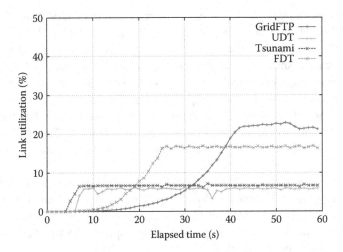

Figure 3.34: Throughput of each protocol for first 60 s, 150 ms RTT.

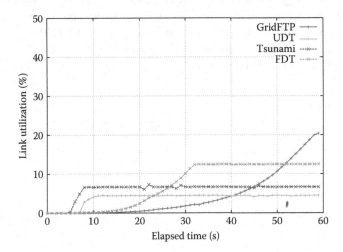

Figure 3.35: Throughput of each protocol for first 60 s, 200 ms RTT.

time to increase its window size from the start in high RTT paths. They take relatively longer than UDP-based protocols, which use their own rate control or congestion control.

Depending on the design and implementation of rate control or congestion control of a UDP-based protocol, it can be more efficient when ramping up throughput from start-up. From what we see in Figure 3.35, Tsunami's rate control and UDT's congestion control are more efficient during the start-up transient time, however they are not efficient at utilizing the long fat pipe. Their maximum throughput is small compared to the capacity available for the transfer. In Figures 3.30 through 3.32, GridFTP, UDT, and FDT have higher maximum throughput with short ramp-up time, however as RTT increases, throughput of the three protocols decreases. With longer RTT, ramp-up times for UDP-based protocols are still quicker than TCP-based protocols.

3.4.2.3   Inter-protocol fairness

We have tested inter-protocol fairness of each protocol with the CUBIC and Reno congestion-avoidance algorithm. Using nuttcp, we generated a single CUBIC TCP and Reno TCP flow, along with a high-speed data transfer flow. We measured the throughput of each flow and analyzed their capacity share. We limited the capacity of the link to 1 Gb/s using netem to remove disk read/write bottlenecks and the TCP buffer bottlenecks for GridFTP, FDT, and UDT. This allowed us to measure capacity share between a TCP flow and high-speed data transfer flow which uses the same congestion-avoidance algorithm but different data handling processes and automatic TCP parameter optimization. The TCP implementation used is TCP Reno and CUBIC, which are the prevalent implementations used in Linux and BSD [42].

We measured the throughput of GridFTP, UDT, and FDT against TCP Reno and CUBIC. We ran Iperf to generate TCP Reno or CUBIC background traffic, waited 30 s, then launched the data transfer protocols. This was based on a likely scenario where the background traffic is already in steady state in a shared link and the high-speed data transfer is started after the background traffic is settled in a bottleneck link.

Figures 3.36 and 3.37 show the capacity share of GridFTP and TCP Reno with various emulated RTTs. Without emulated delay, short RTTs allowed TCP Reno to utilize 90% of the capacity from Iperf generated flows. After GridFTP started, throughput of GridFTP's flow ramps up quickly to take 42% of the capacity and throughput of the background traffic flows reduced to 48%. Considering that they are using the same TCP congestion-avoidance algorithm, this shows that TCP Reno uses a fair share of the capacity in a short RTT link. However, as RTT increases, problems arise with TCP Reno. With only 50 ms RTT, utilization

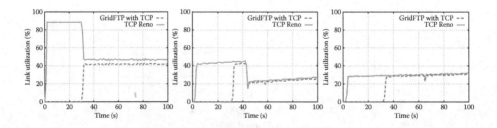

Figure 3.36: Capacity share of GridFTP and TCP Reno, without netem, 50 and 75 ms RTT emulated.

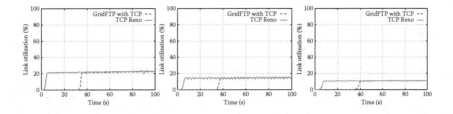

Figure 3.37: Capacity share of GridFTP and TCP Reno, 100, 150, and 200 ms RTT emulated.

of a single TCP Reno flow reduces to 43%. When a high-speed data transfer protocol flow is started, both flows are reduced to 23% utilization after 10 s. With the longer RTT, both background traffic flows and high-speed data transfer flow are not efficient enough to utilize all the available capacity. As RTT increases, throughput of the flows decreases even if there is more capacity available, and because of this, throughput of the background traffic flow does not decrease after high-speed data transfer flow is started.

On the other hand, CUBIC TCP does not experience the same underutilization problem in the same experiment. Figures 3.38 and 3.39 show capacity shares of GridFTP and CUBIC TCP with various emulated RTTs. From near-zero RTT to longer RTTs, a single CUBIC TCP generated from Iperf is capable of utilizing 90% of the capacity. Without netem, and as emulated RTTs becomes shorter, the throughput of background flow and high-speed data transfer flows converges quicker. Although with 50 ms RTT it takes longer convergence time because of packets dropped during the convergence, CUBIC TCP's overall convergence time is shorter when the RTT is shorter. After the throughputs converged, there is very little difference in throughput between the two flows.

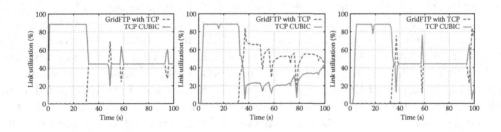

Figure 3.38: Capacity share of GridFTP and CUBIC TCP, without netem, 50 and 75 ms RTT emulated.

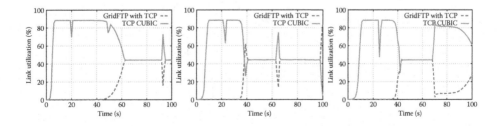

Figure 3.39: Capacity share of GridFTP and CUBIC TCP, 100, 150, and 200 ms RTT emulated.

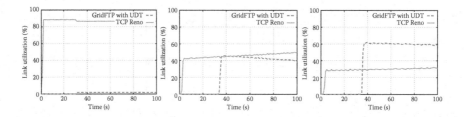

Figure 3.40: Capacity share of UDT and TCP Reno, without netem, 50 and 75 ms RTT emulated.

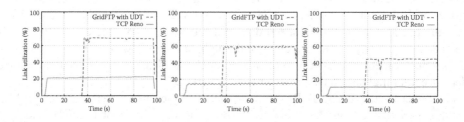

Figure 3.41: Capacity share of UDT and TCP Reno, 100, 150, and 200 ms RTT emulated.

Figures 3.40 and 3.41 show the capacity shares of UDT and TCP Reno with various emulated RTTs. Without RTT emulation, throughput of UDT is suppressed by TCP Reno's throughput. With 90% of link utilization, background traffic did not leave much room for UDT. When UDT started after 30 s, throughput of background traffic reduced slightly to 89%, and UDT took 2% of the available capacity. With longer emulated RTTs, throughput of background traffic reduced, due to reduced performance of TCP Reno, and UDT was able to take the remaining capacity. With 50 ms emulated RTT, the background traffic utilized 46% of the available capacity. At the time UDT started, UDT utilized 46% of the capacity and the background traffic utilized. While background traffic slowly increased its throughput up to 52%, the throughput of UDT decreased to 48%. With 100 ms RTT, the background traffic utilizes 22% and UDT utilizes 70%. With 150 and 200 ms RTT, background traffic's utilization is reduced to 14% and 11%, but the throughput of UDT is also reduced to 60% and 45%. This shows that with longer RTT, neither TCP Reno nor UDT utilize the 1 Gb/s capacity.

Figures 3.42 and 3.43 show the capacity shares of UDT and CUBIC TCP with various emulated RTTs. Because TCP CUBIC increases congestion window more aggressively compared

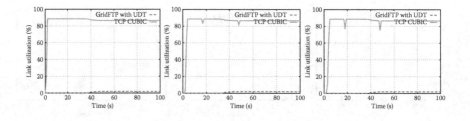

Figure 3.42: Capacity share of UDT and CUBIC TCP, without netem, 50 and 75 ms RTT emulated.

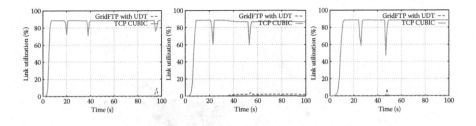

Figure 3.43: Capacity share of UDT and CUBIC TCP, 100, 150, and 200 ms RTT emulated.

to TCP Reno, UDT is not able to take capacity from the TCP CUBIC flow. Throughout the experiment against CUBIC TCP, UDT was not able to take capacity from CUBIC TCP, regardless of RTT.

From what we found, UDT is the least aggressive protocol and CUBIC TCP is the most aggressive. With shorter RTTs (less than 100 ms), UDT can be a complement to the other TCP that are not able to fully utilize available capacity. UDT can take up the capacity left, even with RTT greater than 75 ms, where TCP Reno cannot fully utilize the capacity. However, UDT is not able to utilize the capacity when it is used with RTT longer than 100 ms, or to compete with more aggressive protocols such as CUBIC which utilize the capacity well in any RTT. To compete with CUBIC TCP, it is better to use the same congestion algorithm (i.e., CUBIC) to ensure that the data transfer takes up a fair amount of the available capacity. Fairness comparison between TCP Reno and CUBIC TCP is shown in [30].

3.4.2.4   *Performance of each protocol with varying background traffic and RTT*

After we had completed the previous tests, we decided to expand the test environment to have an experiment environment that has both background traffic and RTT emulation at the same time. We measured the throughput of each protocol with different amounts of TCP and UDP background traffic generated with nuttcp and different emulated RTTs. We faced technical problems with Tsunami, which crashed too often throughout the experiment. We decided to discard Tsunami for this experiment.

Figure 3.44 shows goodput of GridFTP measured with different TCP background traffic generated in different emulated RTTs. Having TCP background traffic reduces the goodput for any emulated RTT, especially for more than two background TCP sessions and RTT longer than 50 ms; in these conditions the goodput of GridFTP reduced to nearly zero. This can be caused by several factors. In a high RTT link, the recovery time of the background TCP flows and GridFTP flows is longer. When a TCP flow experiences packet loss, it reduces its throughput while allowing other flows that did not lose a packet to increase their throughput. The probability of having packet losses across all TCP flows is low at the start of the transfer, when the throughput of the flows are generally low. When the data transfer flow gets a packet loss it reduces its throughput, and further packet loss from congestion induced by other traffic will cause the data transfer flow to reduce its throughput further. Another factor will be the lack of processing power of the machine. The processor is not efficient enough to handle all TCP flows in 10 Gb/s; it tends to generate more background traffic from nuttcp because that is a more lightweight process compared to GridFTP, which takes more processor cycles. These

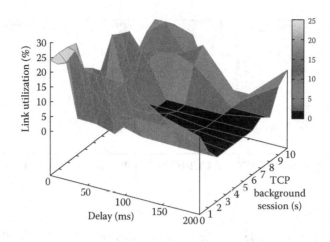

Figure 3.44: Goodput of GridFTP with TCP background traffic, varying RTT.

two factors both affect the data transfer flows, the performance of GridFTP reduced for longer RTT, and increasing TCP background sessions.

Goodput measured with less TCP background traffic does suffer reduced performance, but not as much as goodput for many TCP background flows. That is because GridFTP flows are affected by fewer packet losses from less TCP background traffic.

More goodput is measured against background traffic on shorter emulated RTT links. This is because the recovery time is shorter for the data transfer flow as well as the background flows. It alows the data transfer flow to increase its throughput faster after having a packet loss, so that throughput is already recovered before the next packet loss occurs. As TCP background sessions increase to more than six sessions, nuttcp requires more process time to manage the increasing number of TCP flows and it becomes heavy weight. We noticed that nuttcp with this number of TCP flows becomes unstable. It allows the data transfer flows in GridFTP to achieve higher throughput by generating less background traffic.

Figure 3.45 shows goodput of UDT measured with different TCP background traffic generated in different delay emulated. As it has lower overall goodput without any background traffic, the amount of goodput reduced from having TCP background sessions is relatively small. The behavior of goodput against many TCP background sessions and longer emulated RTTs are similar to the goodput measured in GridFTP. Because UDT uses TCP friendly congestion control, it behaves similarly when background traffic suffers similar congestion to its flows. It is also affected by the lack of processing power and slow recovery time with longer RTT links.

Figure 3.46 shows goodput of FDT measured with different TCP background traffic generated for different emulated RTTs. FDT also shows behavior similar to GridFTP and UDT when it is used against many TCP background sessions with longer RTT. With shorter emulated RTTs, the performance of FDT was less affected by the TCP background flows. This is because FDT uses bigger TCP socket buffers and better handling of the file systems. As the number of TCP flows increases, there is more congestion, but the recovery time is short enough for FDT to reach higher throughput than GridFTP, allowing FDT to sustain higher throughput most of time throughout the data transfer. With many of the TCP background flows, FDT still survived with higher goodput in shorter RTT links, but with longer RTT links the goodput decreased

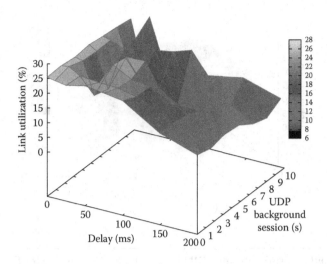

Figure 3.45: Goodput of UDT with TCP background traffic, varying RTT.

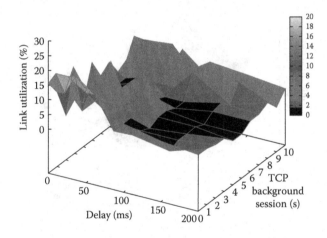

Figure 3.46: Goodput of FDT with TCP background traffic, varying RTT.

to almost 0 Gb/s until the increasing number of background TCP flows causes nuttcp to be unstable.

Figure 3.47 shows goodput of GridFTP measured with different amount of UDP background traffic generated in different emulated RTTs. GridFTP has higher goodput against UDP background traffic in shorter RTT, but degrades when RTT increases. As RTT increases, the recovery time for TCP increases but UDP has no recovery time at all because there is no congestion control in UDP. As UDP background traffic increases in longer RTT networks, it has more effect on the goodput of GridFTP, as recovery time increases and more packet loss events are caused by increasing background traffic.

Figure 3.48 shows goodput of UDT measured with different amounts of UDP background traffic generated in different emulated RTTs. Goodput of UDT is less affected by UDP

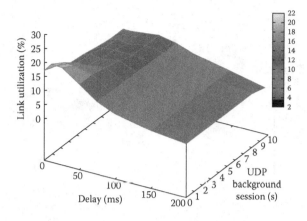

Figure 3.47: Goodput of GridFTP with UDP background traffic, varying RTT.

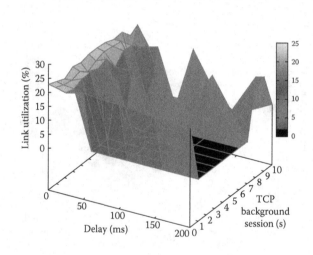

Figure 3.48: Goodput of UDT with UDP background traffic, varying RTT.

background traffic and more affected by increased RTT. An increase in UDP background traffic causes more packet loss for UDT, but UDT is more tolerant of packet loss than TCP, which allows UDT to have less degradation over increased background UDP traffic.

Figure 3.49 shows goodput of FDT measured with different amount of UDP background traffic generated in different emulated RTTs. As stated before, FDT is more effective against background traffic because of its larger TCP socket buffer size in short RTT links. FDT seems to work well with more UDP background traffic and utilize capacity better than GridFTP.

3.4.2.5 Disk system speed effect

To measure the performance difference between different file systems and memory transfers, we ran each protocol with ext4 in a single disk, ZFS RAID with three hard disks, and from

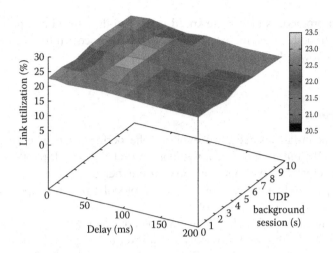

Figure 3.49: Goodput of FDT with UDP background traffic, varying RTT.

/dev/zero to /dev/null. We also measured sequential file read/write performance of ext4 and ZFS RAID using bonnie++.*

Table 3.4 shows the performance of each protocol in ext4, ZFS RAID, and memory. With the ext4 file system, our three protocol's goodputs are limited by slow disk write performance. GridFTP shows best performance among the three protocols in ext4 file system. With the ZFS RAID, GridFTP and FDT are limited by disk write performance. UDT's implementation issue slows the UDT's goodput, while other two protocols are efficient enough to reach to the ZFS RAID's maximal write speed. Without the disk speed limitation, GridFTP and FDT utilizes more than 82% of the link capacity.

3.4.2.6   Multiple process threads effect

We ran GridFTP, UDT, and FDT with different number of process threads. Increasing the number of process threads allows simultaneous I/O processes using multiple threads. Each thread is able to read data from disk and prepare packets to be sent to the network. We used one to six process threads for each protocol to measure the performance difference of increasing the number of process threads.

From the experiment, we found no difference in performance for all protocols, in both disk-to-disk and memory transfers using different number of process threads. This is because the

Table 3.4   Performance of each protocol using ext4 and ZFS RAID

File System	Bonnie++ Read (Gb/s)	Bonnie++ Write (Gb/s)	GridFTP (Gb/s)	UDT (Gb/s)	FDT (Gb/s)
ext4	1.59	0.98	0.89	0.86	0.86
ZFS RAID	3.24	2.30	2.02	1.62	1.86
Memory	N/A	N/A	8.38	1.36	8.28

---

*http://www.coker.com.au/bonnie++.

limitation of each protocol is disk write speed in disk-to-disk speed and packet overhead for memory transfer. In both cases, increasing the number of process threads cannot improve the disk write speed or lower the packet overhead.

3.4.3 Discussion

Using our different in-lab testbeds, we explored the performance of each protocol in real 10 Gb/s networks. We ran each protocol with varying RTT and multiple flows, varying background traffic, different file systems, and varying number of process threads. We measured goodput, throughput, and ramp-up time of each protocol from the experiments and found behavior differences between the protocols.

From the experiments, we found TCP-based protocol's performance to be better than that of UDP-based protocols. UDP-based protocols are quicker to ramp-up their throughput, and the congestion-avoidance algorithm used in UDT is less aggressive than TCP Reno as CUBIC TCP.

There was not enough processing power for the sender to generate background traffic and run the data transfer protocols in our 10 Gb/s network, so we had to introduce a background traffic generator. Tsunami was not stable in the presence of high-volume background traffic. We found that UDT was not aggressive enough to maintain high-speed traffic against TCP background traffic.

GridFTP's goodput is highest among the protocols when there is no background traffic. Because it uses TCP as transport layer protocol, its ramp-up time takes longer than UDP-based protocols because of the TCP slow start. Increasing TCP background traffic causes throughput to degrade until the background traffic reaches 6 Gb/s, after 6 Gb/s there is no change in throughput as they share that capacity with the same throughput.

FDT's goodput is slightly lower than GridFTP without background traffic when the RTT is shorter than 100 ms. With increasing RTT, FDT's performance degrades significantly. FDT's ramp-up time with increasing RTT and against increasing TCP background traffic are similar to GridFTP because they both use the same TCP.

UDT's goodput is slightly lower than GridFTP without background TCP in short RTT links. With increasing RTT, the performance degrades significantly. UDT cannot utilize the 10 Gb/s capacity without the disk bottleneck because of its implementation issue. UDT's congestion control is TCP friendly. It reduces sending window when there are consecutive packet losses. Its congestion control is less aggressive than TCP Reno and CUBIC. Increasing TCP background traffic degrades UDT's throughput, allowing background traffic to have increasing throughput as RTT increases. At the same time, UDT's background TCP drops packets more as background traffic increases. We noticed that UDT has an implementation issue which generates many NAK packets at the start of a transfer, causing low link utilization even if there is no congestion on the link.

Tsunami's goodput is lowest among all the protocols except for the link with 200 ms RTT. There are almost no changes in goodput with increasing RTT because its rate control algorithm does not change goodput with increasing RTT. The usability of Tsunami is poor. It caused our host machine to freeze and not to respond occasionally, for no apparent reason. We also found that Tsunami drops out from the experiment when it is used against background traffic in our 10 Gb/s testbed.

We did not test GridFTP's pipelining feature because our focus was to transfer a large dataset rather than many small files. In addition, Rajendran et al. [51] showed that the pipelining feature did not improve the performance of GridFTP in their test.

From our analysis, GridFTP is the best protocol to be used in a 10 Gb/s link without background traffic. With background traffic, FDT is almost as efficient as GridFTP. UDT and Tsunami do not have good enough usability to be used in a 10 Gb/s network for any RTT we considered.

3.5 National Testbed

3.5.1 Methodology

We built a testbed between Auckland and Wellington in New Zealand. Collaborating with REANNZ,[*] we used REANNZ's 10 Gb/s national network between the University of Auckland and Victoria University of Wellington. Figure 3.50 shows the network path between the two machines. The average RTT of the link was 5.49 ms with 2 hops. There was almost no difference between different times in a day and there was only a negligible difference between 0%~20% load on the network.

The capacity of the network had seldom been fully used,[†,‡] leaving more than 7 Gb/s of available capacity most of the time when we performed the experiment.

Figure 3.51 shows the detail of the machines used in the experiment. Each machine is tuned as in the Linux Tunning Guide,[§] except that CUBIC TCP is used as the TCP congestion-avoidance algorithm. From our previous experiments, we found that CUBIC was as efficient as other high-speed TCP congestion-avoidance algorithm implementations. We used ZFS RAID0 with two HDDs in the Auckland machine, while software RAID0 with five HDDs in the Wellington machine. Because the machine in Wellington has the faster disk system, we send data from Auckland to Wellington to maximize disk-to-disk transfer performance.

Figure 3.50: National testbed.

[*] http://www.reannz.co.nz.

[†] http://weathermap.reannz.co.nz/node.php?src=anx01&int=128.

[‡] http://weathermap.reannz.co.nz/node.php?src=anx02&int=41.

[§] http://fasterdata.es.net/host-tuning/linux/.

Location	Model	CPU	Memory	NIC	OS
University of Auckland	Supermicro X8DTT	2*Intel XeonX 5650 2.67GHz	24 GB	Intel 82599ES 10-Gigabit	CentOS6
Victoria University of Wellington	Supermicro X8DTT	2*Intel XeonX 5650 2.67GHz	96 GB	InfiniBand Mellanox MT26428	CentOS6

Figure 3.51: Hardware used in the national testbed.

Unfortunately, we are not allowed to change the kernel TCP socket buffer size, so it was set to the default which is 16,777,216 bytes.

A file size of 30 GB is used for any transfer. The file size represents the typical volume of data produced per day from a satellite during a stereoscopic census study. Smaller file size are too short to be influenced much by the short burst of the transfer, while larger file sizes took too much time to be recorded.

We chose GridFTP, UDT, and FDT as the high-speed data transfer protocols to test based on their performance in our previous experiments. Unfortunately, we could not use the UDT bundle integrated in GridFTP because we could not install the bundle in CentOS, so we had to use a stand-alone UDT implementation. As a result, we could not use UDT with memory-to-memory transfer.

To test each protocol with basic performance over the national link, we measured goodput of each protocol with 1500- and 9000-byte (Jumbo) frames.

Because we do not have control over background traffic, we decided to observe RTT changes caused by the high-speed data transfer protocols as a measure of congestion. The REANNZ network is a shared link supporting research activities, therefore we must avoid generating background traffic in addition to our high-speed data traffic because it can potentially generate congestion that may cause other users activities to malfunction.

Increase in RTTs are measured by observing RTT without our data transfer and during the data transfer. We used the ping utility* to send Internet control message protocol(ICMP) [48] packets from the sender to receiver each second and measure the RTT increases during the data transfer period. The RTT change observed represents impact of the data transfer to the other traffic in the shared network.

3.5.2 Results

3.5.2.1 Single flow

Figure 3.52 shows the single flow performance of each protocol for disk and memory transfers with and without Jumbo frames. We observe that disk I/O throttles the performance of the transfer protocols. To remove that limitation, we measured memory-to-memory transfer performance for GridFTP and FDT. With the memory-to-memory transfers, we tested with and without Jumbo frames to measure the effect of large frame size.

---

*http://www.gnu.org/software/inetutils/manual/inetutils.html#ping-invocation.

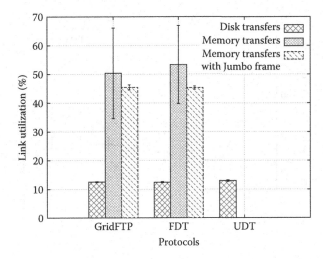

Figure 3.52: Goodput of each protocols with single flow.

This demonstrates the importance of faster disk systems when high-speed data transfer systems are use a in short-haul 10 Gb/s network. There is almost no different in behavior between GridFTP and FDT, because they use the same CUBIC TCP for their transport layer protocol. Using Jumbo frames decreased the performance of GridFTP and FDT slightly, but decreased the fluctuation of the goodput. This is because while enabling Jumbo frames, the maximum TCP socket buffer size is reduced to 6.4 MB, which restrained the maximum TCP congestion window size. This limits the throughput to the maximum of 5.1 Gb/s and because of that limit of maximum throughput, there are less packets lost compared to the non-Jumbo flows which does not have the limit.

3.5.2.2   *Multiple flows*

With a single flow, each protocol reaches the disk write speed limit. Because of that, there was no improvement in goodput when we used the multiple flows for GridFTP and FDT. UDT did not support multiple flows so we discarded it from the experiment.

Figure 3.53 shows goodput of each protocol using multiple transfer flows. Compared to the single flow, using two flows degraded the performance of both protocols. This is because having more flows causes more lost packets, and both flows never stay in TCP self-clocking mode, while a single flow stays in TCP self-clocking mode with throughput more than 8 Gb/s between 1 and 12 s depending on the available capacity of the link.

For more than two flows, goodput increases as number of flows increases for GridFTP and FDT. This is because more flows increases overall congestion window growth rate for the aggregated flow. With the increase-congestion window growth rate, data rates increase rapidly so the overall goodput increases.

With multiple transfer flows, goodput of FDT is less than for GridFTP. This is because FDT takes more time to initiate its transfer due to its longer transfer preparation process.

Figure 3.54 shows goodput of each protocol using multiple flows with Jumbo frames. Having multiple flows allow aggregated TCP socket buffer size to grow beyond the limit a single TCP flow has, but the number of lost packets increases, dropping the throughput of the

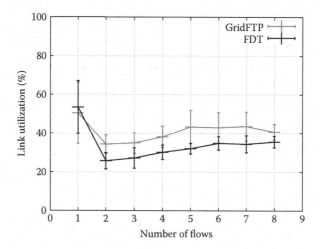

Figure 3.53: Protocol goodput for multiple flows.

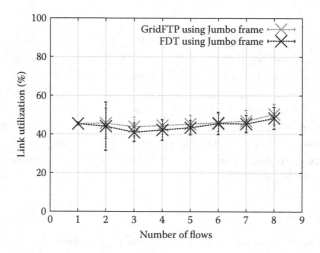

Figure 3.54: Protocol goodput for multiple flows using Jumbo frames.

flows. As a result, the overall goodput of the protocol does not increase, but goodput varies for each run of the experiment.

Figure 3.55 shows increased RTT when using multiple flows. RTT increases less when two transport flows are used instead of a single flow. This is because a single flow has higher throughput and causes few steep rises to RTT. When using a single flow, large burst of packets fill a router's queue and excessive packets that arrive after the buffer is full gets dropped. When this happens, RTT increases as queuing delay of each packets increases on the routers in between the path. Because the single flow stayed longer at 8 Gb/s, which is optimal throughput, it causes more queuing delay for the packets sent. Having more flows increases the frequency of reaching the maximum available bandwidth for the aggregated data rate of the flows, but they do not stay more than 3 s at the maximum until they reduce the data rate due to packet

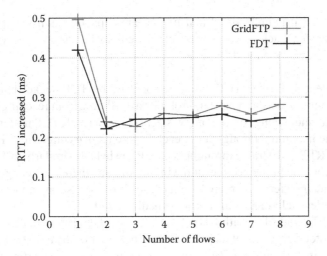

Figure 3.55: RTT increased for multiple flows.

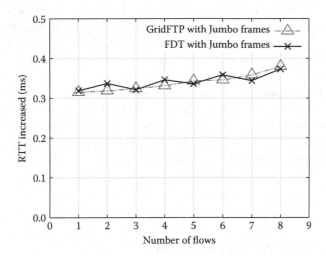

Figure 3.56: RTT increased for multiple flows using Jumbo frames.

losses. Because the congestion window size drops frequently, RTT increases less but increases the rise in frequency as they reach the maximum more often.

Figure 3.56 shows increases in RTT using multiple flows and Jumbo frames. Limited TCP socket buffer size causes TCP flow to underutilize the link. This allows to have almost no congestion on the path and so be free from packet loss. Having multiple flows starts to increase the congestion and RTT increases as the number of flows increases.

3.5.3 Discussion

In the previous section, we have studied the behavior of GridFTP, FDT, and UDT in terms of goodput, data rate, and RTT changes using increasing number of flows and Jumbo frames.

From Section 3.4, it is understood that file system capability is the major bottleneck for high-speed file transfer. Each protocol was able to send as fast as the file system process the file writes. This bottleneck should be removed prior to any other improvement in short-haul high-speed data transfer.

The next bottleneck for the TCP-based protocol is the TCP socket buffer size. Having limited socket buffer size acts as a barrier, prohibits a TCP flow from reaching higher data rates by limiting the maximum congestion window size. Having a small TCP buffer can deliberately be used to limit the maximum throughput of each flow to reduce congestion. TCP socket buffer size can be set to (RTT × desirable throughput). TCP can behave similarly to TCP self-clocking mode when it reaches the limited congestion window size and there is no congestion on the link. That forces the data transfer to be more fair; it causes less congestion and less packet losses to the network activities including itself in a short-haul network.

We used GridFTP and FDT with CUBIC TCP so that the performance difference was the outcome of the protocol implementation. The actual behavior of the transfer flows do not differ because both are using the same CUBIC flow. FDT has more delay in starting the transfer when executing it from the command line.

By using Jumbo frames, TCP congestion window size growth increases. Although it may seem to increase the data rate more rapidly, this does not give more goodput at the end of the transfer because it causes more congestion. Increase in congestion window size growth rate allows a TCP flow to reach the maximum available capacity in shorter time, leads to packet losses, and pushes TCP into congestion-avoidance mode. These changes make the TCP flow with Jumbo frame more aggressive on the shared link, and less fair in terms of causing more congestion.

Having multiple transport flows also allows the aggregated congestion windows size to grow faster than a single flow. As the number of flows increases, the aggregated congestion window size growth rate increases. This makes the packet loss event more frequent, leading to more congestion. With limited TCP socket size per flow, having multiple flows allows the aggregated flow to increase congestion window size beyond the limit, but leads to more packet losses.

3.6 Factors Affecting Performance

In this section, we focus on the implications of what we have found from the various experiments and further discuss the technical difficulties we faced during the experiment. During our experiments, we have found a few issues that need to be addressed to understand results of this chapter. We emphasize that this chapter is an empirical study of transfer protocol implementations in both a real-world environment and in a fully controlled environment. We are aware that there may be differences in the protocol design and its implementation; we made decisions for creating our system to represent most common real-world applications.

We do not focus on optimizing the performance of each system based on its throughput, but instead study on the behavior of each system in different circumstances including different RTTs and levels of background traffic, with limited socket buffer sizes. We designed our experiments to represent the likely situations that users of the systems in the real world may face, to ensure that they are aware of the behavior of each system, so as to make their decision on which system best suits their needs. The stability in paths with background traffic, and the need to not disrupt too much other traffic on those paths, are important to demonstrate that the protocols can be used in real-world situations.

3.6.1 Fairness issues

We are interested in utilizing available capacity with minimal intervention to existing flows. Because the primary focus of the high-speed data transfer protocols is to utilize the capacity in long fat pipes, sharing the same capacity with the other flows with low capacity required flows such as VoIP, may not seem fair.

RBUDP [32] and other UDP-based protocols do not consider fairness because they only aim to fully utilize link capacity. Combined with dedicated network or bandwidth reservation techniques for a bulk transfer, this approach may be beneficial. RBUDP sends the datagrams at the fixed rate until it has sent out all the datagrams, which may be able to fully utilize the available capacity during a transfer. However, many research sites use shared network paths, and using RBUDP with a fixed rate may cause other traffic to be starved of capacity in the same path.

Our approach to measuring fairness of these protocols is to measure induced congestion caused by the high-speed data transfer protocols. Both TCP-based protocols and UDP-based protocols have congestion control mechanisms to avoid a congestive collapse and allow other protocols to have a portion of capacity available. We measured the impact of these protocols on the network when it is used with large-size socket buffer, Jumbo frames, and multiple transfer flows.

In our local testbed, we emulated different levels of competing background traffic to measure capacity share between the high-speed data transfer flow and background traffic. TCP-based protocols are able to secure the available bandwidth well while allowing 60% of available capacity to the background traffic in a 10 Gb/s network.

With a 1 Gb/s bottleneck link, UDT was able to utilize the available capacity well when competing TCP Reno traffic was not using all the available capacity in high RTT link. When TCP Reno was used with UDT in a shorter RTT link, UDT yielded most of the capacity to the TCP Reno. In a 10 Gb/s network, UDT was not able to utilize the available capacity well, instead yields most of the bandwidth to the background traffic.

Multiple transfer flows allow each flow to increase the congestion window size simultaneously, which leads to faster congestion window growth in the aggregated congestion window, causing more packet loss events. In addition, the flows become synchronized over time and therefore cause packet losses at the same time. Because each flow reaches its maximum share of the available capacity roughly at the same time, they cause a higher level of congestion at that time.

3.6.2 Jumbo frames

In GNU/Linux kernel network stack, blocks of data are passed to IP stack by applications. IP stack enqueues pointers to the blocks of the driver queue so that the driver queue holds pointers to the socket kernel buffers (SKB) [18], which contain packet data. A NIC dequeues from the driver queue for transmission [12].

Jumbo frames reduce CPU overhead by reducing the overhead for processing packet headers [24,44,53]. A CPU with limited processing capability on a 1 Gb/s network, and even a faster processor using a 100 Gb/s link benefits from using Jumbo frames because of the reduced overhead and able to achieve more throughput [44].

In addition, Jumbo frames increase the congestion window size growth for TCP because they increase the initial congestion window size and congestion window size increments when TCP is in congestion-avoidance mode. Rutherford [54] showed that longer MTU size performed better than shorter MTU with even larger MTU size super Jumbo frames (9,000–16,000 bytes) on 1–100 Gb/s links.

The CRC32 frame check algorithm used in Ethernet frames is capable of detecting errors in Jumbo frames [41], but having larger packets means more process overhead to generate and check CRC32 in Ethernet frames [44]. By using a TCP offload engine (TOE) to offload checksum calculation, CPU usage can be decreased when the frame sizes are larger [22]. This reduces the CRC32 overhead created by using Jumbo frames for bulk transfers.

A TOE [16] offloads packet processing to the network hardware in NIC, to reduce CPU overhead when it processes TCP segments. When it is used, the IP stack can forward pointers to a large size SKB to the driver queue and eventually to the offload engine in NIC hardware. TOE breaks the data from the SKB into MTU-sized segments, prepares TCP headers for the segments, and sends the segments to the NIC for transmission. At the receiving end, TOE receives a series of segments and form a larger SKB and then passes them to CPU to handle the data to the application.

A TOE can also offload checksum calculation to the network hardware in NIC. Foong et al. [22] showed that TOE reduces CPU load when it is used to offload TCP checksum calculation for large segments. Because TCP checksum is a per-byte operation, offloading the TCP checksumming to NIC reduces the load from the OS.

Jumbo frames increase RTT because they increase the transmit delay for each packet. For flows using small size packets such as ICMP, having a bulk transfer flow in the same machine using Jumbo frames increases RTT because Jumbo frames have larger transmit delay than non-Jumbo frames. Therefore, small size packets that are in the same transmit queue have longer queuing delay. In our national testbed, we have observed increase in RTT for ICMP packets using ping alongside high-speed transfers.

In our 10 Gb/s testbed, we did not measure CPU usage for each protocol because we used modern processors that should be capable of generating traffic to fully utilize 10 Gb/s links. In fact, we were able to generate full 10 Gb/s to test traffic locally without having processors limiting the throughput.

However, using Jumbo frames increased the congestion window size growth rate for TCP in our national testbed. Jumbo frames increase both the initial congestion window size [2] and maximum segment size. Initial congestion window size is increased from $1,460 \times 3 = 4,380$ bytes to $8,960 \times 2 = 17,920$ bytes. This increases congestion window size growth in slow start [22].

Jumbo frames increased RTT for ICMP packets in the same path in our national testbed. Figures 3.55 and 3.56 show the RTT increase when Jumbo frames are used. Jumbo frames increase the transmission delay for each packet because larger packets have longer time to transmit.

When using Jumbo frames, the size of a SKB [18] increases to take larger block of data for increased MTU. For a bulk transfer flow, it will use the MTU-size SKB most of the time while SKB for an ICMP packet will stay small because of small packet size. Driver queue in the NIC has pointers to these SKBs and the ICMP packet has to wait until the larger packets that are in front of it in the queue to be transmitted. Therefore, the ICMP's packet has longer queuing delay because it takes longer to send larger packets in front of the ICMP's SKB. This is why RTT increases more for a single flow with Jumbo frames than for multiple flows without Jumbo frames, even though the single flow with Jumbo frames did not have any congestion.

We used TCP segmentation offload (TSO) [37] in our NIC's TOE to offload packet processing to the NIC. This allows the IP stack in GNU/Linux to forward SKB [18] that are even larger than MTU because it offloads the process of making packets to the NIC. This increases the RTT further for ICMP packets because the SKB for the bulk transfer flows has larger size [57].

The byte queue limit (BQL) algorithm [33] addresses this issue by calculating the number of bytes to be queued in the transmit queue automatically. BQL adjusts the amount of data that can be queued at any time, avoiding the transmit queue to be empty or overflow. In recent research [10], BQL reduced the RTT drastically in simulation, however we were not able to use it in our research because it is only supported in recent GNU/Linux kernel versions > 3.3.

3.6.3 Multiple connections

Many TCP-based high-speed data transfer protocols support multiple transfer flows. Because TCP's congestion control tries to allocate equal capacity to each flow, using multiple flows is a way to take a higher portion of available capacity in a bottleneck link [50].

When multiple TCP flows are used, each of the flows increases their congestion window size simultaneously. This allows the aggregated congestion window size to grow N times faster when N flows are used. With the faster congestion window size growth, the recovery time for TCP's congestion avoidance reduces. If a transfer machine has a limited socket buffer size for TCP, then using multiple flows is a way to have more aggregated socket buffer.

Using multiple TCP flows also provide more aggregated congestion window size. If the maximum congestion window size for a single TCP flow is limited, then having multiple TCP flows with the same limit gives more aggregated congestion window size for the aggregated flow.

Both effects when using multiple TCP flows leads to more aggregated throughput compared to a single TCP flow's throughput. Damjanovic et al. [17] extended Padyhe's single TCP throughput formula [46] to model aggregated throughput for multiple TCP flows.

However, increased congestion window growth rates and aggregated throughput leads to unfairness to the path. The other TCP flows in the same bottleneck suffer more packet losses caused by multiple TCP flows. These multiple flows generate more packet losses and have more impact on the RTT increases.

Because of the unfairness of multiple TCP flows, Welzl et al. [62] proposed MulTFRC which modifies TCP friendly rate control (TFRC) [20] for multiple TCP-friendly flows. It uses packet loss intervals to measure levels of congestion when calculating the average sending rate for each TCP flow, instead of measuring one congestion per RTT as in TFRC. This allows MulTFRC to have smooth throughput changes over time for aggregated TCP-friendly flows.

3.6.4 Technical difficulties

Our work is based on the empirical study of a selection of high-speed data transfer protocols. Because we focus on measuring the performance of each protocol under different circumstances in 10 Gb/s paths, there were a number of issues we were not able to address.

To have more control over the many factors in the experiments while still approximating real-world situations, we decided to create a testbed rather than doing a simulation. On the other hand, we are well aware that we are testing the implementation of protocols, not protocols themselves. Although the design of these protocol sounds fair, the actual implementation of the design may include flaws or lack of optimization that become apparent in a on 10 Gb/s network.

We could not make use of Tsunami in our 10 Gb/s testbed because it crashed too often and causes system to halt when we generated background traffic.

We found that the performance of UDT is inefficient in 10 Gb/s networks. We tried to use GridFTP bundled with UDT and also the stand-alone version of UDT, but found that it is difficult to achieve improved performance. The recent stable GridFTP version claims to support UDT as its transport protocol, but since we did not have sufficient privilege to install any software on the machine in Wellington on our national testbed, we had to use stand-alone

UDT, which does not support multiple stream transfer on the application level nor reading from the /dev/zero stream on GNU/Linux.

There are always challenges associated with conducting experiments. These challenges entail unpredictable behavior of some applications used in the experiments. The key is to have a fine balance between running well-designed experiments and coming up with appropriate solutions to manage these issues.

In our local testbed, we found that nuttcp become unstable when it is running in the same machine with high-speed data transfer protocols, especially when we tried to create too many TCP sessions in a high RTT link. This may have particularly affected the average throughput of the high-speed transfer protocols when they are used with more than four TCP background traffic flows. We tried repeating experiments to reduce uncertainty in our results, but we were not able to stabilize nuttcp behavior or that of other traffic generators. We improved this experiment by using a dedicated traffic generator. With no background traffic, we could observe a slightly lower throughput of GridFTP with UDT in links with RTT lower than 50 ms. We suspect that this effect is due to its implementation rather than the protocol itself.

Because we do not have any access to the router within the path, we were not able to gather link information from the path. We were not able to measure the amount of background traffic when we used national testbeds, therefore we were not able to assess fairness in terms of the capacity share in the bottleneck. For the same reason, we were not able to measure total amount of packet losses in the network.

Instead we decided to measure the RTT changes within the data transfer flows themselves and used ping to measure RTT of the path to measure impact of the high-speed data transfer to the other flows in the same path. This allows us to observe the possible impact on the path caused by the transfer, as a measure of fairness.

We only had limited access time for the machines used in national testbed and we had to focus on measuring performance and fairness in limited environment for national testbed. We were not able to test with large TCP congestion windows with Jumbo frames.

3.7 Conclusions

We have analyzed different high-speed big data transfer protocols to address issues with high-speed data transfer protocols. From our experiments, we found a number of bottlenecks causing inefficiency in a high-speed data transfer in the network environment, in the protocol itself, and in the hosts.

We found that the primary bottleneck for high-speed data transfer protocol is file system performance. By using faster file systems, we were able to improve the performance of each protocol significantly. When disk reading/writing was removed, further improvement on the goodput was achieved.

Limited TCP socket buffer size restricts TCP's congestion window size growth. The congestion window size in TCP determines how much data can be sustained in-flight, and that affects the data rate of the flow. Having a TCP socket buffer larger than $2 \times BDP$ allows a single TCP flow to utilize the capacity well in long distance 10 Gb/s network.

Having Jumbo frames increases MSS for TCP and because congestion avoidance increases congestion window size by the number of bytes it receives, Jumbo frames increase the congestion window size growth rate. This leads TCP flows to increase their congestion window more aggressively and allows them to reach higher throughput than without Jumbo frames. However, increased aggressiveness causes more impact on the path, because it generates bursts of packets when it increases the congestion window size.

In our preliminary 1 Gb/s network, the UDP-based protocols worked better than the TCP-based protocols. This is because UDT and Tsunami are free from TCP slow start and congestion avoidance and they can implement their own congestion control. As a result, their ramp-up times for high RTT links are significantly faster than TCP slow start. Also, Tsunami was able to sustain high throughput for increasing RTT because its congestion control algorithm is independent of RTT. However in a 10 Gb/s network, both UDT and Tsunami are not suitable for high-speed transfer because of their implementation quality. In our testbeds, Tsunami and UDT showed unreliable performance.

Using multiple transfer flows allows protocols to have more aggregated TCP congestion window size when a single flow is limited by the maximum TCP socket buffer size. Because the maximum TCP socket buffer size is applied to individual flows, having multiple flows allows the aggregated congestion window size to grow beyond that limit. This allows the overall goodput to increase significantly than that for a single flow. In addition, because multiple TCP flows increase their congestion window size simultaneously, the aggregated congestion window size increases more aggressively. This increases the RTT significantly for flows using the same path, causing more impact on the path and possibly to other network activity.

Having more I/O threads for the applications does not provide significant performance improvement because their performance is limited by the file read/write rate of the file system. Enabling data striping also did not improve the performance because data striping is already done by the file system in ZFS RAID.

References

1. W. Allcock, J. Bester, J. Bresnahan, A. Chervenak, L. Liming, and S. Tuecke. GridFTP: Protocol extensions to FTP for the Grid. Global Grid Forum GFD-RP Document, No. 20, 2003.

2. M. Allman, V. Paxson, and E. Blanton. TCP congestion control. RFC 5681, RFC Editor, September 2009. Published: Internet requests for comments. http://www.rfc-editor.org/rfc/rfc5681.txt.

3. D. An, J. Park, G. Wang, and G. Cho. An adaptive UDT congestion control method with reflecting of the network status. In *International* Conference on Information Networking, pp. 492–496, Kota Kinabalu, Malaysia, February 2012.

4. D.J. Barrett, R.E. Silverman, and R.G. Byrnes. *SSH, The Secure Shell. The Definitive Guide*. O'Reilly Media, Sebastopol, CA, 2005.

5. M. Bateman, S. Bhatti, G. Bigwood, D. Rehunathan, C. Allison, T. Henderson, and D. Miras. A comparison of TCP behaviour at high speeds using ns-2 and Linux. In *Proceedings of the 11th Communications and Networking Simulation Symposium*, pp. 30–37, Ottawa, Canada, 2008. ACM, New York.

6. J. Bonwick and B. Moore. Zfs: The last word in file systems. 2007. http://wiki.illumos.org/download/attachments/1146951/zfs_last.pdf.

7. J. Bresnahan, M. Link, R. Kettimuthu, D. Fraser, and I. Foster. Gridftp pipelining. In *Proceedings of the TeraGrid Conference*, Madison, WI, June 2007.

8. G. Brumfiel. Down the petabyte highway. *Nature*, 469(20):282–283, 2011.

9. M. Carbone and L. Rizzo. Dummynet revisited. *SIGCOMM Computer Communication Review*, 40(2):12–20, April 2010.

10. T.B. Cardozo, A.C. da Silva, A.B. Vieira, and A. Ziviani. Bufferbloat systematic analysis. In *International Telecommunications Symposium*, pp. 1–5, Sao Paulo, Brazil, August 2014.

11. A. Chervenak, E. Deelman, C. Kesselman, B. Allcock, I. Foster, V. Nefedova, J. Lee et al. High-performance remote access to climate simulation data: A challenge problem for data grid technologies. *Parallel Computing*, 29(10):1335–1356, 2003.

12. G. Chuanxiong and Z. Shaoren. Analysis and evaluation of the TCP/IP protocol stack of LINUX. In *International Conference on Communication Technology Proceedings, WCC—ICCT 2000*, volume 1, pp. 444–453, Beijing, China, 2000.

13. C. Cirstoiu, R. Voicu, and N. Tapus. Framework for high-performance data transfers optimization in large distributed systems. In *Proceedings of the International Symposium on Parallel and Distributed Computing*, pp. 385–392, Krakow, Poland, July 2008.

14. Cisco Systems Inc. *Cisco Visual Networking Index: Forecast and Methodology, 2013–2018*, Cisco Systems, San Jose, CA, June 2014.

15. R. Cottrell, S. Ansari, P. Khandpur, R. Gupta, R. Hughes-Jones, M. Chen, L. McIntosh, and F. Leers. Characterization and evaluation of TCP and UDP-based transport on real networks. *Annales Des Telecommunications*, 61(1–2):5–20, 2006.

16. A. Currid. TCP offload to the rescue. *Queue*, 2(3):58–65, May 2004.

17. D. Damjanovic, M. Welzl, M. Telek, and W. Heiss. Extending the TCP steady-state throughput equation for parallel TCP flows. Technical Report 2, DPS NSG Technical Report, University of Innsbruck, Institute of Computer Science, Innsbruck, Austria, 2008.

18. M. David. How SKBs work. http://vger.kernel.org/~davem/skb.html.

19. B. Eckart, X. He, and Q. Wu. Performance adaptive UDP for high-speed bulk data transfer over dedicated links. In *Proceedings of the IEEE International Symposium on IPDPS 2008*, pp. 1–10, Miami, FL, April 2008.

20. S. Floyd, M. Handley, J. Padhye, and J. Widmer. TCP friendly rate control (TFRC): Protocol specification. RFC 5348, RFC Editor, September 2008. Published: Internet requests for comments. http://www.rfc-editor.org/rfc/rfc5348.txt.

21. S. Floyd and V. Paxson. Difficulties in simulating the internet. *IEEE/ACM Transactions on Networking*, 9(4):392–403, August 2001.

22. A.P. Foong, T.R. Huff, H.H. Hum, J.P. Patwardhan, and G.J. Regnier. TCP performance re-visited. In *IEEE International Symposium on Performance Analysis of Systems and Software*, pp. 70–79, Austin, TX, March 2003.

23. J. Galbraith and O. Saarenmaa. SSH file transfer protocol. Internet-draft draft-ietf-secsh-filexfer-13, IETF secretariat, July 2006. Published: Working draft. http://www.ietf.org/internet-drafts/draft-ietf-secsh-filexfer-13.txt.

24. N.M. Garcia, M.M. Freire, and P.P. Monteiro. The ethernet frame payload size and its effect on IPv4 and IPv6 traffic. In *International Conference on Information Networking*, pp. 1–5, Busan, South Korea, January 2008.

25. J. Gray, A.S. Szalay, A.R. Thakar, P.Z. Kunszt, C. Stoughton, D. Slutz, and J. van den Berg. Data mining the SDSS SkyServer database. arXiv preprint cs/0202014, 2002.

26. Y. Gu and R.L. Grossman. UDT: UDP-based data transfer for high-speed wide area networks. *Computer Networks*, 51(7):1777–1799, 2007.

27. S. Ha, Y. Kim, L. Le, I. Rhee, and L. Xu. A step toward realistic performance evaluation of high-speed TCP variants. *Elsevier Computer Networks (COMNET) Journal, Special issue on PFLDNet*, February 2006.

28. S. Ha, L. Le, I. Rhee, and L. Xu. Impact of background traffic on performance of high-speed TCP variant protocols. *Computer Networks*, 51(7):1748–1762, May 2007.

29. S. Ha and I. Rhee. Taming the elephants: New {TCP} slow start. *Computer Networks*, 55(9):2092–2110, 2011.

30. S. Ha, I. Rhee, and L. Xu. CUBIC: A new TCP-friendly high-speed TCP variant. *SIGOPS Operating Systems Review*, 42(5):64–74, 2008.

31. A.B. Hanushevsky. Peer-to-peer computing for secure high performance data copying. Technical report, Stanford Linear Accelerator. Center, Menlo Park, CA, 2002.

32. E. He, J. Leigh, O. Yu, and T.A. DeFanti. Reliable blast UDP: Predictable high performance bulk data transfer. In *Proceedings of the IEEE International Conference on Cluster Computing*, p. 317, Washington, DC, 2002. IEEE Computer Society.

33. T. Herbert. bql: Byte queue limits. Patch posted to the Linux kernel network development mailing list, 2011.

34. T. Kelly. Scalable TCP: Improving performance in highspeed wide area networks. *SIGCOMM Computer Communication Review*, 33(2):83–91, 2003.

35. I. Legrand, H. Newman, R. Voicu, C. Cirstoiu, C. Grigoras, C. Dobre, A. Muraru, A. Costan, M. Dediu, and C. Stratan. MonALISA: An agent based, dynamic service system to monitor, control and optimize distributed systems. *40 Years of CPC: A Celebratory Issue Focused on Quality Software for High Performance, Grid and Novel Computing Architectures*, 180(12):2472–2498, December 2009.

36. Y-T. Li, D. Leith, and R.N. Shorten. Experimental evaluation of TCP protocols for high-speed networks. *IEEE/ACM Transactions on Networking*, 15(5):1109–1122, October 2007.

37. S.B. Lindsay. *Network Adapter with TCP Windowing Support*. Google Patents, September 2004. US Patent 6,788,704.

38. X. Lu, Q. Wu, N.S.V. Rao, and Z. Wang. On performance-adaptive flow control for large data transfer in high speed networks. In *IEEE 28th International Performance Computing and Communications Conference*, pp. 49–56, Scottsdale, AZ, December 2009.

39. X. Lu, Q. Wu, N.S.V. Rao, and Z. Wang. On parallel UDP-based transport control over dedicated connections. In *Global Telecommunications Conference*, IEEE, pp. 1–5, Miami, FL, December 2010.

40. T. Marian, D.A. Freedman, K. Birman, and H. Weatherspoon. Empirical characterization of uncongested optical lambda networks and 10GbE commodity endpoints. In *IEEE/IFIP International Conference on Dependable Systems and Networks*, pp. 575–584, Chicago, IL, June 2010.

41. M. Mathis. Arguments about Internet MTU. https://www.psc.edu/~mathis/MTU/arguments.html.

42. I. McDonald and R. Nelson. Congestion control advancements in Linux. In *linux.conf.au (lca)*, volume 89, Dunedin, New Zealand, 2006.

43. M. Meiss. Tsunami: A high-speed rate-controlled protocol for file transfer, 2009. http://citeseerx.ist.psu.edu/viewdoc/download?doi=10.1.1.113.4551&rep=rep1&type=pdf.

44. D. Murray, T. Koziniec, K. Lee, and M. Dixon. Large MTUs and internet performance. In *IEEE 13th International Conference on High Performance Switching and Routing*, pp. 82–87, Belgrade, Serbia, June 2012.

45. H.A. Nam, J. Hill, and S. Parete-Koon. The practical obstacles of data transfer: Why researchers still love scp. In *Proceedings of the 3rd International Workshop on Network-Aware Data Management*, pp. 1–8, Denver, CO, 2013. ACM, New York.

46. J. Padhye, V. Firoiu, D. Towsley, and J. Kurose. Modeling TCP throughput: A simple model and its empirical validation. In *ACM SIGCOMM Computer Communication Review*, volume 28, pp. 303–314. ACM, New York, 1998.

47. J. Park, D. An, and G. Cho. An adaptive channel number tuning mechanism on parallel transfer with UDT. In *International Conference on Information Networking*, pp. 346–350, Bangkok, Thailand, January 2013.

48. J. Postel. Internet control message protocol. STD 5, RFC Editor, September 1981. Published: Internet requests for comments http://www.rfc-editor.org/rfc/rfc792.txt.

49. J. Postel and J. Reynolds. File Transfer Protocol. STD 9, RFC Editor, October 1985. Published: Internet requests for comments http://www.rfc-editor.org/rfc/rfc959.txt.

50. L. Qiu, Y. Zhang, and S. Keshav. Understanding the performance of many TCP flows. *Computer Networks*, 37(3):277–306, 2001.

51. A. Rajendran, P. Mhashilkar, H. Kim, D. Dykstra, G. Garzoglio, and I. Raicu. Optimizing large data transfers over 100gbps wide area networks. In *Proceedings of the IEEE/ACM International Symposium*, Delft, the Netherlands, November 2012.

52. C. Rapier and B. Bennett. High speed bulk data transfer using the SSH protocol. In *Proceedings of the ACM Mardi Gras Conference*, pp. 1–7, Los Angeles, CA, January 2008. ACM, New York.

53. S. Ravot, Y. Xia, D. Nae, X. Su, H. Newman, J. Bunn, and O. Martin. A practical approach to TCP high speed WAN data transfers. In *Proceedings of the 1st Workshop on Provisioning & Transport for Hybrid Networks*, San José, CA, 2004.

54. W. Rutherford, L. Jorgenson, M. Siegert, P. Van Epp, and L. Liu. 16000–64000 B pMTU experiments with simulation: The case for super jumbo frames at Supercomputing '05. *Optical Switching and Networking*, 4(2):121–130, 2007.

55. H. Sangtae and R. Injong. Hybrid slow start for high-bandwidth and long-distance networks. In *Proceedings of the 6th PFLDNet Workshop*, Manchester, UK, 2008.

56. R.N. Shorten, D.J. Leith, J. Foy, and R. Kilduff. Analysis and design of AIMD congestion control algorithms in communication networks. *Automatica*, 41(4):725–730, April 2005.

57. D. Siemon. Queueing in the Linux network stack. *Linux Journal*, 2013(231), July 2013.

58. J. Suresh, A. Srinivasan, and A. Damodaram. Performance analysis of various high speed data transfer protocols for streaming data in long fat networks. In *Proceedings of the International Conference on ITC*, pp. 234–237, Kochi, India, March 2010.

59. P. Teodoro, A. Hutton, B. Frezouls, A. Montmory, J. Portell, R. Messineo, M. Riello, and K. Nienartowicz. Data management at gaia data processing centers. In Luis M. Sarro, Laurent Eyer, William O'Mullane, and Joris De Ridder, editors, *Astrostatistics and Data Mining*, volume 2 of Springer Series in Astrostatistics, pp. 107–115. Springer, New York, 2012.

60. A. Tirumala, F. Qin, J. Dugan, J. Ferguson, and K. Gibbs. Iperf: The TCP/UDP bandwidth measurement tool. htt p://dast. nlanr. net/Projects, 2005.

61. A. Tridgell and P. Mackerras. *The rsync Algorithm*. Australian National University, Canberra, Australia, 1996.

62. M. Welzl, D. Damjanovic, and S. Gjessing. MulTFRC: TFRC with weighted fairness. Internet-draft draft-irtf-iccrg-multfrc-01, IETF Secretariat, July 2010. Published: Working draft http://www.ietf.org/internet-drafts/draft-irtf-iccrg-multfrc-01.txt.

63. S. Weston, T. Natusch, and S. Gulyaev. Radio Astronomy data transfer using KAREN network. In *Proceedings of the General Assembly and Scientific Symposium on IEEE URSI*, pp. 1 –4, Istanbul, Turkey, August 2011.

64. L. Wood, W.M. Eddy, W. Ivancic, J. McKim, and C. Jackson. Saratoga: A Delay-Tolerant networking convergence layer with efficient link utilization. In *International Workshop on Satellite and Space Communications*, pp. 168–172, Salzburg, Austria, September 2007.

65. D.C. Wyld, J. Zizka, and D. Nagamalai. Advances in computer science, engineering and applications: Proceedings of the second international conference on computer science, engineering and applications (ICCSEA 2012), May 25–27, 2012, New Delhi, India, volume 2. In *Advances in Intelligent and Soft Computing*. Springer, New York, 2012.

66. Z. Yue, Y. Ren, and J. Li. Performance evaluation of UDP-based high-speed transport protocols. In *Proceedings of the IEEE International Conference on ICSESS*, pp. 69–73, Beijing, China, July 2011.

Chapter 4

Challenges in Crawling the Deep Web

Yan Wang

Jianguo Lu

CONTENTS

In the era of big data, the vast majority of the data are not from the surface web, the web that is interconnected by hyperlinks and indexed by most general-purpose search engines such as Google. Instead, the trove of valuable data often reside in the deep web, the web that is hidden behind query interfaces. In contrast to the surface web that can be accessed by following hyperlinks embedded in web pages, the deep web documents may not contain hyperlinks, and they can be accessed by queries only. Examples of the deep web data sources include various online social networks such as Twitter and Facebook, and virtually every large website that provides a query interface enabling either manual or programmable searches. It is believed that the deep web is larger than the surface web in orders of magnitudes. Hence, crawling the deep web is an essential step in big data acquisition, and has attracted attentions from both academia and industry, including Google [1] and Microsoft [2].

This chapter surveys the important approaches to the deep web crawling, raises major challenges in the area, and outlines the solutions to these challenges. The deep web is abstracted as a graph, and the crawling problem is modeled using random graph theories. We classify the deep web into several categories, each category has its unique challenges in crawling. In model M_0, documents have zero variation of being captured; in model M_h, documents have heterogeneous capture probabilities; and in model M_r, documents are ranked and only top k are returned. For each model, we will delineate the cost of crawling, and methods to improve the crawling performance. This chapter serves as the reference for researchers and practitioners in the deep web crawling.

4.1 Introduction

The searchable web forms and programmable web APIs permeate the daily lives of ordinary web users as well as professional web programmers. The trove of the data hidden behind these query interfaces constitutes the deep web [1,3–5]. In contrast to the surface web that is connected by hyperlinks, the deep web cannot be crawled by following the hyperlinks embedded in web pages. Instead, documents in the deep web can be retrieved using queries only. For this reason, it is also called the hidden web [6–8].

The deep web is considered full of rich content that is much bigger than the surface web [5]. Nowadays, almost every website comes with a search box. Many of them, such as twitter.com, provide in addition a programmable web API. First of all, it would be nice if those deep web documents were search engine visible. Not surprisingly general search engines, such as Google [1,9,10] and Bing [2] try to index some of these un-crawled territories. In addition, numerous applications want to tap into the rich deposit of data to build distributed search engines [11], data integration applications [12], vertical portals [13], and so on. While the deep web data providers are happy to serve the data to ordinary users and even application programs, they may not want to be overloaded with automated crawlers whose target is to index or even worse to download the entire database to set up their own operation. Thus, an intriguing question that is of interest to both deep web data providers and crawlers is how difficult it is to harvest most of the data records inside a deep web data source by sending appropriate queries.

The difficulties of crawling deep web data sources are usually derived from the different stages of crawling. Generally speaking, the crawling process contains four steps, including (1) locating and estimating the deep web data sources, (2) understanding the HTML forms of each data source, (3) selecting appropriate queries, and (4) retrieving the returns and extracting the relevant content if it is imbedded inside HTML pages. In Step 1, the main challenges are to efficiently find the entries of deep web data sources from trillion surface web pages and accurately estimate their properties, such as their sizes [14,15]. In Step 2, the key challenge is the automated recognition of proper search interfaces for accepting queries [16–19]. In Step 3, the challenge is to select queries that can return most contents of deep web data sources with minimal network cost [1,2,20–22]. In Step 4, the challenge is to train an agent to automatically extract the specific contents from returned documents [23,24]. Currently, with the prevalence of programmable web APIs and the maturity of information extraction technique, the challenges in other steps become easier to be solved compared to the past. Query selection problem in Step 3 has played a key role in improving the quality of crawling deep web data sources.

It is proved that, without an elaborate query selection method, only random queries will cause more than nine times repeated retrieval and many of them are large easy-captured documents, which consume too much network bandwidth. If considering the effect of ranking support plus return limit (only part of matched results can be returned), the performance of crawlers could be extremely exacerbated. Thus, with the different conditions, we categorize all deep web data sources into four models—M_0, M_h, M_r, and M_{hr}—corresponding to different crawling hardness.

M_0 is an ideal model, it assumes that documents in a data source have the same probabilities of be matched by random queries. The subscript represents zero variation of being captured. In this model, even random queries can reach good results. If the assumption of M_0 is removed, it turns into another more complicated model M_h. In this model, large documents will be repeatedly retrieved with random queries such that well-crafted query selection methods are needed to reduce redundancies. Here the subscript h means heterogeneity. From M_0, it becomes model M_r when the ranking condition plus the return limit (only top k documents can be returned) are added into it. Under this model, queries should satisfy some criteria, otherwise, documents ranked high will be repeatedly retrieved while documents ranked low could not be returned at all. M_{hr} is the most complicated one of the four models. It have all features of M_h and M_r, that is, documents have variation of being captured and are ranked with the return limit. Data sources belonging to this model are difficult to exhaustively crawl even with carefully designed query selection methods.

In this chapter, first of all, we separately define the crawling process and cost by using bipartite graph and *overlapping rate*. Then, all deep web data sources are categorized into the four models with the definitions. For each model, we will analyze how difficult it is to crawl from the viewpoint of query selection and introduce the related methods to improve the crawling performance. Finally, some conclusions and future work are presented.

4.2 Problem

4.2.1 Deep web crawling

Crawling of the surface web has been well studied since the advent of the Web, and has become a mature technique used in industry. Deep web crawling, given its similar goal and similar crawling process, seems to be a trivial problem that can be solved by borrowing what we have learnt in the surface web crawling. Why is deep web crawling an issue in the first place?

The deep web can be modeled as a bipartite graph $G = (D, Q, E)$, where nodes are divided into two separate sets D and Q, that is, the set of documents D and the set of queries Q, where $|D| = m$ and $|Q| = n$. Every edge in E ($|E| = v$) links a query and a document. There is an edge between a query and a document if the query occurs in the document.

The deep web crawling problem is to find the queries so that they can cover all the documents. If we regard queries as URLs in surface web pages, the deep web crawling process is

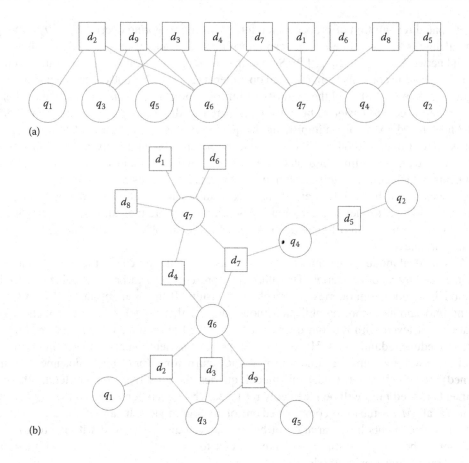

Figure 4.1: Deep web as a bipartite graph. (a) Bipartite graph and (b) same graph as (a) in spring model layout.

very much the same as the surface web crawling: starting from some seed URLs (terms), we obtain the web pages. From those web pages more URLs (terms) are collected and subsequently issued to the web server again. This process repeats until all the pages are traversed. Figure 4.1 gives an example of a deep web that is represented as a bigraph, where $D = \{d_1, d_2, \ldots, d_9\}$ and $Q = \{q_1, q_2, \ldots, q_7\}$. The brute-force crawling process can be illustrated as follows. Suppose that the seed query is q_1. After q_1 is sent, document d_2 is retrieved and more queries q_3 and q_6 are found from the document. We can put these newly found queries in a queue and send these queries in sequel. The process is repeated until all the documents are found.

The algorithm works fine for the surface web but not for the deep web due to the following key difference: in the surface web crawling each URL will return only one page while in deep web crawling each term will return multiple pages. A surface web page has a unique identifier, that is, the URL, while a deep web document has many terms to access it. Because of the multiplicity of the returns, it is inevitable that many of the returns are redundant, and there is no way to exclude these duplicates locally before retrieving them from the server. Such redundancy is costly because it occupies network communications.

Essentially the brute-force method needs to send out all the queries that can be found. Suppose the graph is connected and we can find all the queries, the average cost per document retrieved is

$$\langle d \rangle = \frac{v}{m}.$$

When the queries are single terms, the cost is the average length of the documents. More formally, we can define the volume of a graph through Definition 4.1.

Definition 4.1 Volume   Suppose there are m number of documents and n number of queries. Let d_i denote the degree of the document node i, for $i \in \{1, 2, \ldots, m\}$, and f_j the degree (or document frequency) of the query node j, for $j \in \{1, 2, \ldots, n\}$. The volume v of the documents D is

$$v = \sum_{i=1}^{m} d_i = \sum_{j=1}^{n} f_j.$$

The mean degree of D is

$$\langle d \rangle = v/m.$$

If the brute-force crawling policy is used, all the queries will be issued and the total cost is the volume of the graph. The redundancy is huge in general due to the fact that documents can have many queries, thus they can be retrieved many times. If a query can consist of multiple terms, the number of queries in a document is significantly larger than the number of terms inside a document. The brute-force algorithm will induce roughly $\langle d \rangle$ *overlapping rate*, that is, on average each document is retrieved $\langle d \rangle$ number of times, where $\langle d \rangle$ is the average degree of the documents. This redundancy is the source of the problems in deep web crawling.

Since sending all the queries is not an option, the task here is to select a subset of the queries $Q_s \subseteq Q$, which can cover all documents in D with minimal redundancy.

4.2.2 Performance measurement

The performance of a crawler, for both the surface web and the deep web, is normally measured as the coverage or weighted coverage over the cost to reach that coverage [8,25,26]. In the deep web, the cost can be the cost of sending queries [27], the costs for retrieving resulting pages (each query corresponds to a resulting page containing the URLs of matched documents for downloading), the cost of downloading documents, or some combinations of all based on some weighting scheme [25].

In fact, although the cost of downloading documents is much higher than the others, it should not be used for evaluating a deep web crawler since it is constant, for example, downloading a certain percentage of documents inside a data source consumes constant network transmission for any crawler. In reality, all URLs of retrieved documents are stored in local and this makes sure that no repeated downloading occurs.

For the costs of sending queries and retrieving resulting pages, they grow linearly with the number of matched documents. Usually data sources paginate the results and return only a small number of matches, say p of them, for each query. If documents beyond the first p results are required, the same query needs to be sent again. For instance, in Bing web service [28], if a query q has 200 matches, they will be paginated into 20 sections, each *query session* with keyword q can only obtain one of the sections. In order to obtain all the 200 matches, for the same query, 20 query sessions are required. With the pagination, the cost of a deep web crawling can be measured in terms of the accumulative total number of matched documents by queries sent.

If the accumulative number of total matched documents is counted as the crawling cost, it will be decided by not only the performance of a crawler but also the sizes of data sources. To eliminate the effect of size, the number of total matched documents must be normalized by the unique documents retrieved, that is, the overlapping rate, which is used to measure the performance of a deep web crawler. For instance, suppose that there are two data sources A and B containing 1,000 and 100,000 documents, respectively. For crawling A, the cost caused by a crawler is 5,000 (matched documents) with 100% coverage. Meanwhile, for B, there are 500,000 matched documents retrieved by the same crawler with 100% coverage. Then, the performance of the crawler on A and B is identical since both overlapping rates from A and B are 5.

More formally, given a document-term bipartite graph $G = (D, Q, E)$, a set of queries $Q_s \subseteq Q$ selected by a crawler forms a subgraph denoted by $G_s = (D_s, Q_s, E_s)$, where $D_s \subseteq D$ is the set of the matched documents, $E_s \subseteq E$ is the set of edges that connects the queries and documents. Let $n_s = |Q_s|$, $m_s = |D_s|$, and $v_s = |E_s|$. The performance of the crawler is measured by the overlapping rate defined as follows:

$$\langle d_s \rangle = \frac{v_s}{m_s}. \tag{4.1}$$

Again, the task here is to find a Q_s such that $\langle d_s \rangle$ is minimal while satisfying the constraint $m_s = m$, that is, all the documents are covered by the selected queries with the minimal overlapping rate.

Example 4.1 Subgraph *Figure 4.2 shows a solution $Q_s = \{q_2, q_6, q_7\}$ and the corresponding subgraph.*

$$\langle d_s \rangle = \frac{v_s}{m_s} = \frac{1 + 5 + 5}{9} = \frac{11}{9}.$$

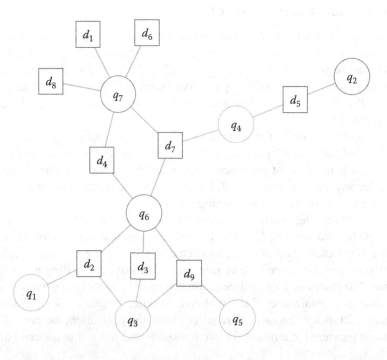

Figure 4.2: A solution of the deep web crawling for the data source shown in Figure 4.1. The darker nodes and edges are in the solution graph. The lighter ones are for illustration purpose and they are not part of the graph.

The other solution $Q'_s = \{q_2, q_3, q_5, q_7\}$ has a lower average degree:

$$\langle d_s \rangle = \frac{v_s}{m_s} == \frac{1+3+1+5}{9} = \frac{10}{9}.$$

Although Q'_s has more queries, it is considered a better one because of its low redundancy.

4.2.3 Models of textual deep web data sources

"The science and practice of deep web crawling is in its infancy" [26]. Crawling a deep web data source is a challenging task, even for textual content with one input field. There is not a panacea that works for all kinds of data sources—different kinds of data sources behave differently and need their own way of crawling. We can classify data sources into several models as follows:

Model M_0: Every document has the same probability of being captured, and all the matched documents are returned.

Model M_h: Heterogeneous data source where documents have varying capturing probabilities.

Model M_r: Ranked data source that has a return limit.

Model M_{hr}: Heterogeneous and ranked data sources.

The relationship between these models can be depicted by Figure 4.3, where $M_x \rightarrow M_y$ indicates that M_y is harder to crawl than M_x. To be more precise, in order to reach the same coverage by using random queries, M_y will require higher cost than M_x. The figure shows that models M_h and M_r are harder to crawl than M_0, while M_h and M_r are two orthogonal dimensions that add the complexity to the problem. Model M_{hr} inherits the complexity from both M_h and M_r. Therefore it is the most challenging case in our classification.

Although the assumptions of M_0 can hardly occur in real applications, it is the starting point for understanding other models as illustrated in Figure 4.3. In addition, the result of M_0 also serves as a lower bound for crawling cost when random queries are used. Model M_h is rather common in data sources, because there are unranked data sources, and more importantly, because many data sources are not very large. When a data source is of moderate size, many

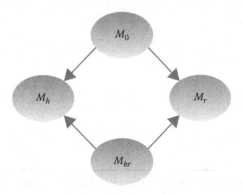

Figure 4.3: Models of data sources.

queries match less than *k* documents. Therefore the data source will return most of the matches even though the matches are ranked.

4.3 Crawling on Model M_0 and M_h

4.3.1 Hardness analysis

There is a diminishing return when crawling continues. Typically the relationship between the harvest and the cost can be depicted in a shape like Figure 4.4, albeit the exact diminishing speed is hardly studied in theory. On the side of empirical study, different diminishing speeds are reported in various experiments [25]. We find that the diminishing speed is in fact solely dependent on the stage of the crawling. More precisely, we give the following theorem.

Theorem 4.1
Let P denote the percentage of the data that has been harvested, and ⟨d⟩ the average degree of the collected documents with respect to the random queries. Then

$$\langle d \rangle \approx \frac{-ln(1-P)}{P}.$$ (4.2)

when $f_j << m$ and the documents are homogeneous, that is, they are of the same probability of being matched. The equation becomes exact when $m \to +\infty$.

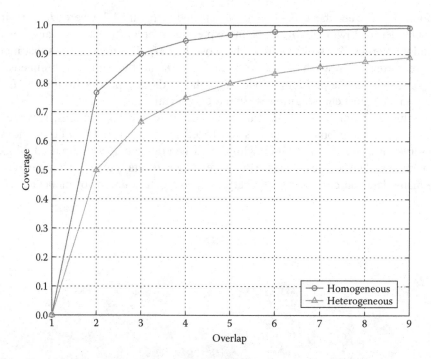

Figure 4.4: Coverage *P* as a function of overlapping, that is, the average degree ⟨*d*⟩ with respect to the query set. Two curves are drawn from Equations 4.6 and 4.7, depicting the homogeneous documents M_0 and heterogeneous documents M_h, respectively.

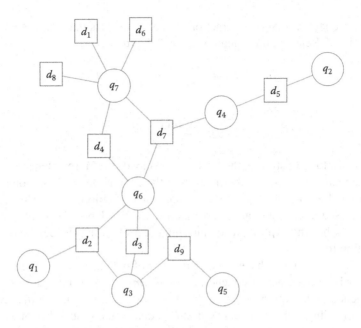

Figure 4.5: Random graph when queries are selected randomly.

Equation 4.2 reveals that the percentage of the documents retrieved is solely dependent on the average degree of the graph, or the overlapping rate. We use Figure 4.5 to explain the relationship between $\langle d \rangle$ and P. The figure is created when three queries are randomly selected. In this graph the darker edges and nodes are the ones selected. They constitute the graph under construction while the lighter edges and nodes are drawn so that we can view what the original graph is. In this graph eight documents are covered, among them d_9 is covered twice. Hence, the mean degree $d = 9/8$. Since the original graph has nine document nodes, the percentage of the documents being covered is $P = 8/9$. Note that this small example is for illustration purpose only—the data is too small to coincide with Equation 4.2.

4.3.1.1 Derivation

Let us first consider the most simple scenario when every document has the same probability of being connected to an edge. At the outset, there are m document nodes and zero edge. When query j is issued, or equivalently, f_j number of edges are added randomly to the graph, the probability of a document i not being covered by any edge is $(1 - 1/m)^{f_j}$. Without loss of generality, let us assume that the return of a query is small relative to the corpus size, that is, $f_j \ll N$. In this case the sampling with replacement can be approximated by the sampling with replacement. After n number of queries are fired, v number of random edges are added to the graph. The probability of document i is still isolated as follows:

$$S = \left(1 - \frac{1}{m}\right)^v \approx e^{-\frac{v}{m}} \tag{4.3}$$

Note that Equation 4.3 becomes exact when $m \to +\infty$. S can be also interpreted as the fraction of the nodes that are isolated in the graph. Let u denote the number of documents that are covered by some queries. u/m is the fraction of the documents that are already captured, that is, $P = u/m = 1 - S$. Rearranging Equation 4.3 we obtain:

$$ln(1 - P) = -\frac{v}{m} = -\frac{v}{u}\frac{u}{m} = -\langle d \rangle P$$

Hence

$$\langle d \rangle = -\frac{ln(1 - P)}{P}$$

Theorem 4.1 is useful to estimate the cost of downloading. More often we need to use the inverse of the equation to estimate the fraction of the data harvested based on the cost. Unfortunately Equation 4.2 does not have a simple analytical solution for P. Although Lambert W function can be used to approximate the solution when P is large, for smaller P there is a big discrepancy even when the expansion contains hundreds of terms. Hence, there is a need to find an approximation for P.

In real applications queries have varying number of matches. Given a sequence of queries $(q_1, q_2, ..., q_k)$, whose document frequencies are $(m_1, m_2, ..., m_k)$. For query q_j, a document i that is not matched by the query is $1 - m_j/N$. For now we are assuming that every document has an equal probability of being matched. Let $P(i)$ denote the probability of a document i that is captured after these k number of queries, u the number of documents retrieved, and P the fraction of the documents retrieved. $u = PN$. Then

$$P = \sum_{i=1}^{N} \frac{P_i}{N} = P_i$$

$$= 1 - \prod_{j=1}^{k} (1 - \frac{m_j}{N})$$

$$\approx 1 - e^{\frac{\sum_{j=1}^{k} m_j}{N}}$$

$$= 1 - e^{\frac{\sum_{j=1}^{k} m_j}{u} \frac{u}{N}}$$

$$= 1 - e^{OR \times P} \tag{4.4}$$

Although it is derived from a simplified document graph where only *dis legomena* (terms that occur only twice) are used, the result remains the same for other queries as long as the query cardinality is much smaller than N. In the derivation above, 2 is the cardinality of the query and it can be replaced by any other number, and the result remains true. Hence, we have the following by applying Taylor expansion:

$$\langle d \rangle = -\frac{ln(1 - P)}{P} = 1 + P/2 + P^2/3 + \cdots \tag{4.5}$$

A more accurate approximation of Equation 4.2 is obtained empirically by running regression on the data generated from Equation 4.2:

$$P \approx 1 - \langle d \rangle^{-2.1} \tag{4.6}$$

4.3.1.2 Implications

An observation we can make based on Equation 4.5 is that when P is very small, $\sum_{i=2}^{\infty} P^i/(i+1)$ is neglectable compared with $P/2$. Hence, P increases almost linearly with $\langle d \rangle$. With the increase of P, the cost increases at a faster speed. If we are harvesting as much data as possible from many data sources, instead of exhaustively siphoning all the data records from one single data source, Equation 4.2 gives a guideline as for when it is the good time to jump to another data source for a fixed crawling resource.

Since $\langle d \rangle$ can be calculated easily from the crawling history, Equation 4.6 is particularly useful to estimate how much data have been downloaded and when the crawling process will stop. Another surprising observation we can make is that large queries induce the same overlapping rate as small queries, since the document frequencies f_j of the queries do not occur in the equation. While it is true that large queries will save the number of queries, it will not save the duplicates retrieved.

From Figure 4.5 it can be seen that $\langle d \rangle$ is rather small to capture a high percentage of the documents. As a rule of thumb, in order to harvest 50% of the documents, the overlapping rate is only about 1.3. If on average each page is accessed three times, you can deduce that those accessed pages constitute 90% of the total population. This result is rather disturbing for data providers—it seems that people can download most of the data with ease.

Fortunately, the assumption for Theorem 4.1, that is, all the documents can be obtained randomly with uniform distribution, does not hold in general. In real deep web data sources, web pages are of different sizes, causing them of different probabilities of being captured by queries. One opinion regarding the size of documents on the web is that it follows a power law with an exponent around two, with the minimal value around 1k. For such data, it is shown that [29]

$$P = 1 - \langle d \rangle^{-1} \tag{4.7}$$

The comparison between Equations 4.7 and 4.6 is illustrated in Figure 4.4. Roughly speaking, to obtain 90% of the data the average degree is around 9 in the case of heterogeneous data, and 3 for homogeneous data. In fact, the return limit has a much more impact on downloading performance as we will explain in Section 4.4.

4.3.2 Related work

Although the cost for crawling unranked data sources is not exorbitantly high by random queries, there are still rooms for improving. In the past decade, there is a line of research for optimizing the quality of crawling data sources in M_h. According to the underlying methodologies of these methods, they can be roughly categorized into three different groups: (1) the methods based on approximation algorithms for minimum set covering problem [20,25,30,31]; (2) the methods based on machine learning [21,22,32,33]; and (3) the methods based on heuristic rules [1,9,27]. In the rest of this section, we introduced the related methods accordingly.

4.3.2.1 Methods based on approximation algorithms

It is well known that the crawling problem can be modeled as a set covering problem. The universe is the set of all the documents, and each query, or the documents that contain the query, is a subset. The constraint is that all the documents need to be covered. Let an $m \times n$ binary matrix A represent the document-query matrix, where

$$a_{ij} = \begin{cases} 1, & \text{if query } j \text{ matches document } i; \\ 0, & \text{otherwise.} \end{cases}$$

With the matrix, the set covering problem tailored to our application can be defined through Definition 4.2:

Definition 4.2 Set covering problem   Given an $m \times n$ binary matrix A. Let **c** be the cost vector and each element stands for the cost for each query. Then the set covering problem is to find a solution x which is an n-column binary vector that minimizes

$$c^T x \tag{4.8}$$

subject to

$$Ax \geq 1 \tag{4.9}$$

where

$$x_j = \begin{cases} 1, & \text{if column } j \text{ is in the solution;} \\ 0, & \text{otherwise.} \end{cases}$$

In [25], the authors first modeled the crawling process as the set-covering problem. The document-query matrix is constructed from all downloaded documents and thus it is incremental iteratively. The cost in [25] consists of sending a query q_j, retrieving the hyperlinks of the matched documents, and downloading them. It is shown as follows:

$$c_j = c_q \cdot 1 + c_r \cdot f_j + c_d \cdot \delta_j$$

where c_q, c_r, and c_d are the average costs for sending a query, retrieving a hyperlink, and downloading a document, respectively, and δ_j is the number of new documents returned by q_j. With the matrix and the cost model, a greedy-based approximation algorithm is proposed to generate promising queries iteratively (Algorithm 4.1).

Algorithm 4.1: The crawling method of [25]

> **Input**: Q'=some seed queries
> **Output**: documents D
> 1 D = all documents returned by queries in Q';
> 2   **while** *stop condition is not satisfied* **do**
> 3 $\quad$ select q_j in Q' maximizing $\hat{\delta}_j/\hat{c}_j$ as the next query;
> 4 $\quad$ download all the new documents D' that match q_j;
> 5 $\quad$ $D = D \cup D'$;
> 6 $\quad$ add all the new terms in D' to Q';
> 7 $\quad$ update $\hat{\delta}_j$ and $\hat{c}_j$ of all terms in Q';
> 8 $\quad$ remove q_j from Q';
> 9   return D;

Algorithm 4.1 maximizes the new documents per unit cost at each iteration in order to improve the performance of crawling. The selection function is $\hat{\delta}_j/\hat{c}_j$, which is used to select promising queries, where $\hat{\delta}_j$ and $\hat{c}_j$ are the estimates of δ_j and c_j, respectively. Note that δ_j is unknown until q_j is sent to the target data source and thus δ_j and c_j need to be estimated. In [25], the Zipf's-law-based estimator from [11] is used to do the estimation. For the stop condition, it could be that all (or a certain percentage) documents are retrieved or others.

The approximation algorithms of [20,30,31,34] are based on the SCP model as well. But the sizes of their document-query matrices are fixed, which is constructed from a big enough uniform sample, not incrementally downloaded documents. Compared to [25], the authors of [20,30,31] believe that (1) the downloading and updating steps in Algorithm 4.1 are time-consuming and impractical and (2) a big enough sample can provide enough information to produce promising queries once and for all. In [20], it is empirically proven that an approximate 3000-document sample can generate good enough queries to crawl deep web data sources containing more than one million documents.

In [20], the crawling process is iterative as well. Given the sample of the target data source, a greedy algorithm is directly applied to it in order to procedure all queries once and for all, that is, all promising queries selected by the greedy algorithm first fully cover the sample with the minimal (or near minimal) overlapping rate and then they are mapped into the target data source for retrieving documents. The empirical study shows that the results of the methods in [20] and [25] are very close to each other, moreover, the former is much efficient than the latter. Note that, in [20], δ_j and $\hat{c}_j$ are changed to the number of new covered rows and the cardinality of the query in the matrix derived from the sample, and the selection function still is $\hat{\delta}_j / \hat{c}_j$.

In [30,31], the authors further improve the work of [20] by introducing the weight for documents. In [20,25], all new documents have the same contribution to δ_j, that is, each document is counted as 1 in the selection function. However, different documents have different importance for crawling. Intuitively, the short documents only containing high-frequency terms should be more important than the long documents containing many low frequency terms since retrieving such short documents will result in more such redundant retrieval especially in the later crawling process. The weight of document is used to differentiate such difference and inserted into the selection function for improving the quality of queries. The experiments in [30] show that the performance can be improved more than 30% on standard testing datasets.

4.3.2.2 Methods based on machine learning

From the aforementioned methods, we can see that δ_j play an important role in generating promising queries. Since δ_j is unknown until the query q_j is sent to the original data source, the accurate estimate $\hat{\delta}_j$ for each query in the sample or downloaded documents becomes the key to optimize the performance of the crawling. Either the Zipf's-law-based estimator implemented in [25] or directly using δ_j in the sample to replace δ_j in the original data source [20,30,31] is solely based on the dfs of queries in the downloaded documents. However, the methods [21, 22,32,33] provided machine-learning-based estimators to estimate δ_j.

The basic idea of [22,32,33] is that, except the document frequency of each term in downloaded documents, some other *features* are also helpful to estimate δ_j, such as linguistic features (e.g., POS, length, and language of each query), other statistical features (e.g., TF and RIDF [35]), and HTML format features (e.g., tag and location in document).

Table 4.1 shows part of the selected features used in [32]. After the features are decided, each feature is given a weight and then the machine-learning technique is used to learn the relationship between the weights and δ_j. Once the proper value for each weight is found, the so-called *harvest model* is constructed. With the harvest model, δ_j for each new candidate query can be estimated. Such estimation could be more accurate than the ones solely dependent on the document frequencies of terms inside the sample. Meanwhile, it can alleviate the dependence on full-text search since Zipf's-law-based estimators need the support of full-text search.

In [32], the 10 features $F_j = (fe_j^1, \ldots, fe_j^{10})$ for each query q_j are selected to construct the harvest model (part of which is shown in Table 4.1). After feature selection, the sampling method in [36] is used to create a sample data source D. After that, a training data set Tr can be obtained automatically by simulating the crawling process on D, that is, like the approach in [20], a set of queries are generated to fully cover D (the selection function here is $\max(\delta_j)$

Table 4.1   Summary of the notations

ID	Description
fe_1	The document frequency of each query
fe_2	The number of redundant documents retrieved by each query
fe_3	The sum of occurrence of all terms in the documents retrieved by each query
fe_4	POS of each query
fe_5	The length of each query
$\cdots$	$\cdots$

not $\max(\delta_j/c_j)$ used in [20]). Meanwhile, at each iteration, each pair (F_j, δ_j) is inserted into Tr. Finally, each selected feature is given a weight λ and the problem of the construction of the harvest model is covered to find the function $f(F_j) = \lambda_1 \cdot fe_j^1 + \cdots + \lambda_{10} \cdot fe_j^{10}$ that minimizes the following formula:

$$\text{Min} \left(\sum_{(F_j, \delta_j) \in Tr} |f(F_j) - \delta_j|^2 \right). \tag{4.10}$$

In [32], the least square method is applied to determine the weights. Algorithm 4.2 shows the method for constructing the harvest model, which can be used to crawl data sources and the only change is that the estimate $\hat{\delta}$ is derived from the function $f(F_j)$ not from the Zipf's-law-based estimator.

Algorithm 4.2: The method to constructing harvest model in [32]

Input: Q'=some seed queries, the sample data source D
Output: the value of each λ_i

1 Q' = all terms in D;
2 Tr=empty;
3 **while** *Not all documents in D are marked* **do**
4 $\quad$ select q_j in Q' maximizing δ_j as the next query;
5 $\quad$ mark all the new documents that match q_j;
6 $\quad$ insert the pair (F_j, δ_j) into Tr;
7 $\quad$ remove q_j from Q';
8 calculate each λ_i of Equation 4.10 based on Tr;
9 return each λ_i;

The work in [33] is analogous with [32], however, they have the three obvious difference: (1) The features are different (HTML format features are considered as new features in [33]). (2) Each element in Tr has the same format (F_j, δ_j) but Tr is incremental because it is constructed with crawling the original data source not the sample data source. (F_j, δ_j) of each issued query are inserted into Tr iteratively. (3) For the harvest model, unlike [32] use $f(F_j)$ to estimate δ_j, $\hat{\delta}_j$ is calculated by the equation $\hat{\delta}_j = \sum_{q_i \in Tr} K(F_j, F_i) \cdot \delta_i$, where q_i is one of issued query inserted into Tr and $K(F_j, F_i)$ is a function used to evaluate the distance between F_i and F_j.

Apart from utilizing the features of the issued queries to estimate δ_j, the authors of [22] use the technique of *reinforce learning* to further improve the performance of crawling. So far, all aforementioned methods in this subsection employ $\hat{\delta}_j/\hat{c}_j$ to generate promising queries, which is a greedy policy to select queries and the local optimal query is selected at each iteration. However, it cannot guarantee a global optimal selection. Thus, the authors argue that it may bring better performance while two consecutive queries are considered together.

In [22], a crawler and a target data source are considered as an agent and the environment, respectively. Then its selection strategy will be dynamically adjusted by learning previous querying results and takes account of two-step long reward. The selection function is shown as follows:

$$Q(s_t, q_j) = r(s_t, q_j) + \max_{q_i \in Q'} [r(s_{t+1}, q_i)] \tag{4.11}$$

where:

s_t represents the current state of the crawler

Q' is the set of candidate queries

$r(s_t, q_j) = \hat{\delta}_j/\hat{c}_j$ is the estimated reward of issuing q_j in the current state s_t

Similarly, $r(s_{t+1}, q_i)$ is the estimated reward of issuing q_i in the state s_{t+1} (the state after q_j has been sent). The query with the maximum value of $Q(s_t, q_j)$ will be selected as the next query. In fact, the estimate $\hat{\delta}_j$ of $r(s_t, q_j)$ is exactly same as the one ($\hat{\delta}_j = \sum_{q_i \in Tr} K(F_j, F_i) \cdot \delta_i$) used in [33]. If without the look-ahead part ($\max_{q_i \in Q'} [r(s_{t+1}, q_i)]$), the method in [22] is degenerated to the one of [33].

4.3.2.3 Methods based on heuristic rules

Except $\max(\delta_j/c_j)$, there are other options for query selection, such as maximizing the returns or the new documents returned at each iteration. Here we introduce the work from [1,9,27] as the methods based on heuristic rules.

In [27], the authors first proposed a greedy-based method to siphon deep web data sources by selecting queries with highest frequencies from a sample. The method is composed of two phases: phase 1 selects a set of queries from the HTML search form and randomly issues them to the target data source. By extracting high-frequency terms from the result pages, their algorithm creates a term list. Then it iteratively updates the frequencies of the terms and adds new high-frequency terms into the list by randomly issuing the terms in the list until the number of submissions reaches the threshold. In phase 2, the method uses a greedy strategy to construct a Boolean query to reach the highest coverage, it iteratively selects the term with the highest frequency from the term list, and adds it to a disjunctive query if it leads to an increase in coverage. For example, if 10 terms $(q_1, \ldots, q_{10})$ are selected, the final issued disjunctive query is $q_1 \vee \cdots \vee q_{10}$.

Google [1] provided another heuristic rules to crawling deep web data sources. The policy of query selection is based on TF-IDF, which is the popular measure for the importance of terms in the information retrieval. It first adds the top 25 terms of every new retrieved web page sorted by their TF-IDF values into the query pool Q'. Then, from the query pool, they remove the following two kinds of terms: (1) eliminate the high-frequency terms, such as the terms that have appeared in many web pages (e.g., $> 80\%$) since these terms could be from menus or advertisements and (2) delete the terms which occur only in one page since many of these terms are meaningless words that are not from the contents of the web pages, such as nonsensical or idiosyncratic words that could not be indexed by the search engine. The remaining terms are issued to the target deep web data source as queries and a new set of web pages are downloaded. This is repeated until the crawling condition is satisfied.

The purpose of [9] is to crawl over all entity documents in deep web data sources, such as the documents containing product names and other attributes. Since the authors have all query logs of Google, the query logs (its format is $< query, url_{clicked}, times_{clicked} >$) toward a target data source are collected and only queries that are clicked for at least two times are considered. Then the relevant entity names are extracted from the satisfied log queries. The extraction is based on the Freebase data [37] that provides 22 million entity names. Finally, all extracted entity names will be sent to the target data source as promising queries to retrieve entity documents.

4.4   Crawling on Model M_r and M_{hr}

Many data sources rank the pages and return only top k documents for queries. When the data source is large, most of the queries will match more than k documents, yet the returns are still limited by k. This will make the query selection almost an impossible task—large queries whose cardinalities are larger than k can only trawl the top layer of data, while small queries (queries whose cardinalities are smaller than k) are hard to obtain. More importantly, using small queries may cause the document-query graph disconnected.

4.4.1   Hardness analysis

4.4.1.1   Large queries

A naive way to solve the query selection problem could be using large queries such as popular words or disjunctions of multiple words. In literature large queries are often preferred. For instance, [27] selects large queries directly, [25] uses set covering method that will lead to the selection of large queries due to their cost model, and [1] uses tf-idf to select queries, where term frequency plays a major role in query selection.

This line of work is based on the assumption that all the matched documents are returned to the user. In reality, most data providers rank the matched documents and only return the top k documents, where k typically ranges between 10 and 1000. For instance Google only returns at most 1000 documents, even if the query matches much more pages. We shall not confuse *return limit* with paginated *query sessions*. In the case of Google search, a query can match one million documents, among them only top 1000 can be returned. This 1000 is the return limit. They can be returned, but these 1000 documents are not returned all at once within one query session. Instead, the first *query session* will return 10 matches.

Note that with this limitation, only top k/f_{min} percent of the documents can be harvested [38], where f_{min} is the cardinality of the smallest query. That is to say, the larger the query, the smaller percentage of the documents you can harvest. More formally, we have the following theorem.

Theorem 4.2
Given a statically ranked data source whose return limit is k. If every query returns more than f_{min} number of documents, then the maximal percentage of the documents can be retrieved is

$$\frac{k}{f_{min}} \qquad (4.12)$$

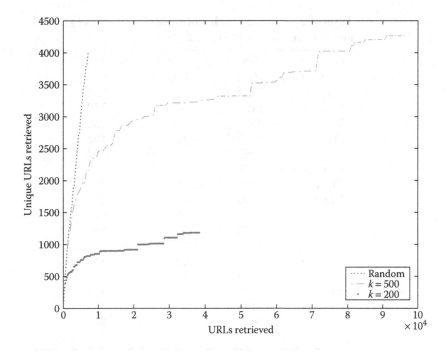

Figure 4.6: In ranked data sources, large queries can only retrieve a portion of the data.

The static ranking can be the page ranking, or rankings based on the time stamp, file size, or alpha-numeric ordering. Although many data sources, especially advanced search engines, use a combination of static ranking (such as page ranking) and relevance ranking, many simple deep websites use static ranking only. Therefore it is worthwhile to study the crawling problem for such ranking policy.

This counter-intuitive phenomenon can be illustrated using Figure 4.6, which shows the growth of the coverage over the number of documents retrieved for the classical 20 news groups corpus (NG20 hereafter)* that contains 19,997 documents. The queries are all the 190 popular words from Webster dictionary with document frequency ranging between 1000 and 5000. More popular words are also tried and they result in even lower coverage. Two return limits, 200 and 500, are experimented. In the case of $k = 200$, in total 38,000 documents were retrieved, among them only 1,183 are distinct ones. In the case of $k = 500$, overall 95,000 documents were retrieved and among them 4,265 are unique.

This experiment shows that even for such a small corpus, the harvest stagnates after a small portion of the documents were retrieved. Notice that in both cases $\langle d \rangle$, the ratio between the total retrieval and the unique documents is very large, indicating that most queries in later stage return only redundant documents. For comparison purpose, we also plotted the coverage for queries randomly selected from Webster, where $k = 200$. Surprisingly, random queries perform much better than popular words due to the large amount of small queries whose document frequencies are within the value of k.

Table 4.2 and Figure 4.7 show that the range of documents retrieved are inversely proportional to the size of the query (f_j). For instance, in the case of $k = 500$, for the first query

---

* Available at http://qwone.com/ jason/20Newsgroups/.

Table 4.2 **Queries, their document frequencies (f_j), and new returns when k=200 and k=500**

Query	Matches (f_j)	New $(k = 200)$	New $(k = 500)$
Understand	1002	200	500
Said	2295	123	353
Until	1159	70	259
Time	4221	46	135
Free	1170	40	196
Year	1681	38	157
Government	1439	20	99
Place	1359	18	91
Right	2950	1	29
Else	1679	16	84
Once	1278	3	56
He	4040	1	9
Available	1326	20	153
Few	2006	4	19
Bit	1289	53	129
System	2118	2	37
Lot	1688	21	44
...			

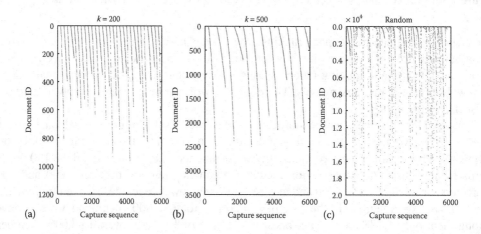

(a)   Capture sequence   (b)   Capture sequence   (c)   Capture sequence

Figure 4.7: Large queries cannot reach the bottom of the deep web. In total 6000 documents are captured, including duplicates. (a) $k = 200$, $df = 1000 \sim 5000$, (b) $k = 500$, $df = 1000 \sim 5000$, and (c) $k = 200$, df is unrestricted.

understand, $f_1 = 1002$, and it covers the documents in the range of 1 and 3300. For the second query *said* which can match 2295 documents (more than twice of the first query), the range of covered documents is between 1 and 1300 (less than half of the first query).

4.4.1.2 Small queries

Since large queries are not effective in crawling ranked data sources, small queries (queries whose cardinalities are smaller than k), seem to be the obvious choice. One may think that

Figure 4.8: Document graph constructed from all the *dis legomena* (terms whose df = 2) in corpus NG20. It shows there is a large component and many small isolated data islands.

by using small queries we can harvest most of the data with ease. However, it is extremely challenging to learn the small queries and their document frequencies. What is worse, the documents may not be covered or connected even if we know a priori all the small queries. By Erdos–Renyi random graph model, a random graph is almost surely connected if the average degree $d > ln(m)$ [39]. In the same vein we have the following theorem.

Theorem 4.3
When $\langle d \rangle = log(m)$, the probability of capturing all the documents is close to 1 when $m \to +\infty$.

When small queries only are used, the average degree may not be larger than $log(m)$, hence the crawling task is doomed to fail. Figure 4.8 shows such a graph constructed from NG20 corpus when *all* the *dis legomena* are used. There are about half of the documents not covered by any query and not drawn in the graph. For covered documents we can see that they are scattered in many small and isolated data islands. It is challenging to learn these queries and the frequencies of these queries.

With the increase of document frequency, the large component becomes bigger and the islands are absorbed by the large component gradually. Only when very large queries are used can all the documents be retrieved. Since the graph is not connected, we can not reach most of the documents even using the brute-force crawling method. If the seed query hits an island, the algorithm stops at the island by retrieving only a few documents. If the seed query hits the large component, we can download all the documents in the large component, leaving all the islands alone.

One misconception is that it is easy to learn most of the (small) queries from a sample set of documents since there are large number of small queries by Zipf's law. Roughly speaking, supposing the vocabulary size is V, there are about $V/2$ number of *hapax legomena* and $V/6$ number of *dis legomena*. Given the sheer size of small queries, one may conclude incorrectly that these queries are good enough to harvest all the documents. However, this is not the case in both theory and practice. According to Heaps' law, the number of terms grows with the size of corpus, although with sublinear speed. Hence, it is impossible to learn the entire vocabulary of a corpus by taking a sample of it. The empirical experiments say that if the sample size is 1%, the fraction of the vocabulary can be learnt is around 10% $(= \sqrt{0.01})$, according to Heaps' law.

4.4.2   Related work

Due to the difficulty in crawling ranked deep web data sources, so far, there are no solid work addressing this problem as far as we know. Now we introduce the two closely related work from [25] and [40] here.

In [25], the authors assumed that all returned documents are ranked randomly and only top k documents can be returned. Under this assumption, they continued to use the crawling method for unranked data sources in [25] to harvest ranked data sources. The only difference is that the ratio of the df of the previous query to k is introduced into the equation for estimating the number of new documents returned by each candidate query. Meanwhile [25] did not provide any experimental results of this method.

Accurately speaking, the crawling method in [40] is not for harvesting all documents inside ranked data sources. Its goal is to retrieve just the top k results for all potential queries with minimal cost, which can enable the applications, such as meta search and content aggregator service. To this end, the following equation is used to select promising queries:

$$r(q_j) = \frac{(1-w) \cdot \hat{\delta}_j + w \cdot RankingReward(q_j)}{c_j}$$

where:
   $RankingReward(q_j)$ is the reward from *derived* ranking for q_j
   w is the weight for balancing the contributions from $RankingReward(q_j)$ and the estimated
      number of new documents $\hat{\delta}_j$

The query with maximum $r(q_j)$ will be selected as the next query. For $RankingReward(q_j)$, the authors use it to evaluate how similar a derived ranking of q_j would be the actual one based on the returned documents from the target data source. If it is more similar with the actual one, the value of $RankingReward(q_j)$ is closer to 0. For each derived ranking for a query, it is constructed by aggregating all orders of the matched documents containing it by the previous queries. If all orders exhibit a strong correlation, $RankingReward(q_j)$ is closer to 0. Note that $\hat{\delta}_j$ is calculated by the estimator in [25] and c_j consists of network communication and bandwidth consume that can measured by f_j.

4.5 Discussions and Conclusions

In deep web crawling a query returns multiple documents that result in duplicates. Reducing this redundancy is a unique problem in deep web crawling, and the source of the challenges in deep web crawling. The major cost is the network traffic which could be measured by the number of queries for small data sources. For large data sources such as online social networks where most of the queries match a large number of documents, the matched documents are often paginated and returned by multiple query sessions. Therefore the query cost is the number of query sessions, which is proportional to the accumulated number of documents retrieved, or the overlapping rate.

In model M_0 the crawling cost (the overlapping rate) is a function of the percentage of the data retrieved, independent on the types of queries (e.g., large or small queries) used. For homogeneous data sources (model M_0) where each document has equal probability of being captured by queries, it is rather easy to harvest most of the data. For heterogeneous data sources (model M_h), the cost to downloading the data is higher, but still affordable.

In ranked data sources (model M_r or M_{hr}) only the top k documents are returned, causing that the documents ranked low may never be returned if large queries are used. Therefore to the contrary of common practice small queries, or the queries whose document frequencies are commensurate to k, should be preferred according to our analytical analysis and empirical experiments. Although large queries are effective in the beginning of the crawling process, they have zero contribution to reach high coverage. For instance, if the document frequency of a query is $2k$, then it can only retrieve the top half of the data source. Therefore it has zero contribution to the retrieval of the remaining half data. To retrieve the other half, smaller queries have to be used.

Ranked data sources require small queries that matches less than k number of documents. Selecting queries of appropriate size is a daunting task because of the well-known over-estimation and large variance problem. For instance, if the sample size is $1/k$ of the total database, every query that occurs once in the sample (*hapax legomenon*) is estimated to have k matches. But in reality its document frequencies can be anywhere between 1 and k. When the total data-base is very large, any feasible sample will be much smaller than $1/k$ of the original size. In that case even the terms only occurring once in the sample may have document frequencies much larger than k. That means even the *hapax legomena* in the sample may return only duplicate documents.

When learning the queries from sample is not feasible, the other alternative is to crawl the documents and collect the new terms in newly retrieved documents. Regardless of the high cost of this approach, it will work only when the graph is connected. Our empirical experiments show that when only small queries are used, the graph is not connected and there are many small isolated islands albeit one large component does exist. This result is supported by the random graph theory.

There are a few implications from our results. For unranked data sources, or small ranked data sources, it is rather easy to download most of the data. For large ranked data sources with a small return limit, it is impossible to download most of the data even with unlimited computing resources. Data providers can design the data source carefully by setting the return limit based on the data source size so that the downloading is infeasible.

References

1. J. Madhavan, D. Ko, L. Kot, V. Ganapathy, A. Rasmussen, and A. Halevy. Google's deep-web crawl. In *Proceedings of the VLDB Endowment*, Vol. 1, No. 2, pp. 1241–1251, 2008.

2. P. Wu, J.R. Wen, H. Liu, and W.Y. Ma. Query selection techniques for efficient crawling of structured web sources. In *Proceedings of ICDE*, pp. 47–56, Atlanta, CA, 2006.

3. B. He, M. Patel, Z. Zhang, and K.C. Chang. Accessing the deep web: A survey. *Communications of the ACM*, 50(5):94–101, May 2007.

4. D. Shestakov, S.S. Bhowmick, and E.P. Lim. Deque: Querying the deep web. *Journal of Data Knowledge Engineering*, 52(3):273–311, 2005.

5. Bright Planet.com. The deepweb: Surfacing hidden value. 2000. http://brightplanet.com.

6. P. Ipeirotis, L. Gravano, and M. Sahami. Probe, count, and classify: Categorizing hidden web databases. In *SIGMOD*, pp. 67–68, Santa Barbara, CA, 2001.

7. S. Raghavan and H.G. Molina. Crawling the hidden web. In *Proceedings of the 27th International Conference on Very Large Data Bases*, pp. 129–138, Rome, Italy, 2001.

8. S.W. Liddle, D.W. Embley, D.T. Scott, and S.H. Yau. Extracting data behind web forms. In *Proceedings of the Joint Workshop on Conceptual Modeling Approaches for E-business: A Web Service Perspective*, pp. 38–49, Tampere, Finland, 2002.

9. Y. He, D. Xin, V. Ganti, S. Rajaraman, and N. Shah. Crawling deep web entity pages. In *Proceedings of WSDM*, pp. 355–364, Rome, Italy, 2013.

10. J. Madhavan, L. Afanasiev, L. Antova, and A. Halevy. Harnessing the deep web: Present and future. In *Proceedings of CIDR*, Asilomar, CA, 2009.

11. P. Ipeirotis and L. Gravano. Distributed search over the hidden web: Hierarchical database sampling and selection. In *VLDB*, Hong Kong, China, 2002.

12. X. Dong and D. Srivastava. Big data integration. In *ICDE*, pp. 1245–1248, Brisbane, Australia, 2013.

13. M. Yang, L. Lim, H. Wang, and M. Wang. Optimizing content freshness of relations extracted from the web using keyword search. In *Proceedings of SIGMOND*, pp. 819–830, New York, NY, 2010.

14. L. Barbosa and J. Freire. An adaptive crawler for locating hidden-web entry points. In *Proceedings of WWW*, pp. 441–450, Banff, Canada, 2007.

15. Y. Wang, J. Liang, and J. Lu. Discover hidden web properties by random walk on bipartite graph. *Information Retrieval*, 17(3):203–228, 2014.

16. M. Alvarez, J. Raposo, A. Pan, F. Cacheda, O. Bellas, and V. Carneiro. Crawling the content hidden behind web forms. In *ICCSA*, pp. 322–333, Kuala Lumpur, Malaysia, 2007.

17. T. Furche, G. Gottlob, G. Grasso, X. Guo, G. Orsi, and C. Schallhart. The ontological key: Automatically understanding and integrating forms to access the deep web. *The VLDB Journal*, 22(5):615–640, 2013.

18. M.C. Moraes, C.A. Heuser, V.P. Moreira, and D. Barbosa. Prequery discovery of domain-specific query forms: A survey. knowledge and data engineering. *IEEE Transactions on Knowledge and Data Engineering*, 25(8):1830–1848, 2013.

19. R. Khare, Y. An, and I. Song. Understanding deep web search interfaces: A survey. *ACM SIGMOD Record*, 39(1):33–40, March 2010.

20. J. Lu, Y. Wang, J. Liang, J. Chen, and J. Liu. An approach to deep web crawling by sampling. In *Proceedings of Web Intelligence*, pp. 718–724, Sydney, Australia, 2008.

21. L. Jiang, Z. Wu, Q. Feng, J. Liu, and Q. Zheng. Efficient deep web crawling using reinforcement learning. In *Proceedings of PAKDD*, pp. 428–439, Hyderabad, India, 2010.

22. Q. Zheng, Z. Wu, X. Cheng, L. Jiang, and J. Liu. Learning to crawl deep web. *Information Systems*, 38(6):801–819, 2013.

23. W. Liu, X. Meng, and W. Meng. Vide: A vision-based approach for deep web data extraction. *IEEE Transactions on Knowledge and Data Engineering*, 22(3):447–460, 2010.

24. N. Kushmerick, D.S. Weld, and R. Doorenbos. Wrapper induction for information extraction. In *Proceedings of the Fifteenth International Joint Conference on Artificial Intelligence (IJCAI)*, pp. 729–737, Nagoya, Japan, August 23–29, 1997.

25. A. Ntoulas, P. Zerfos, and J. Cho. Downloading textual hidden web content through keyword queries. In *Proceedings of the Joint Conference on Digital Libraries*, pp. 100–109, Denver, CO, 2005.

26. C. Olston and M. Najork. Web crawling. *Informatoin Retrieval*, 4(3):175–245, 2010.

27. L. Barbosa and J. Freire. Siphoning hidden-web data through keyword-based interfaces. In *Proceedings of SBBD*, pp. 309–321, Brasilia, Brazil, 2004.

28. Microsoft Inc., Bing Search API and Microsoft Azure Marketplace, December 2015. http://datamarket.azure.com/dataset/bing/search.

29. J. Lu and D. Li. Estimating deep web data source size by capture-recapture method. *Informatoin Retrieval*, 13(1):70–95, 2010.

30. Y. Wang, J. Lu, and J. Chen. Ts-ids algorithm for query selection in the deep web crawling. In *ApWeb*, pp. 189–200, Changsha, China, 2014.

31. Y. Wang, J. Lu, and J. Chen. Crawling deep web using a new set covering algorithm. In *Proceedings of ADMA*, pp. 326–337, Beijing, China, 2009.

32. Y. Dong and Q. Li. A deep web crawling approach based on query harvest model. *Journal of Computational Information System*, 8(3):973–981, 2012.

33. L. Jiang, Z. Wu, Q. Zheng, and J. Liu. Learning deep web crawling with diverse features. In *WI-IAT*, pp. 572–575, Milano, Italy, 2009.

34. Y. Wang, J. Lu, J. Liang, J. Chen, and J. Liu. Selecting queries from sample to crawl deep web data sources. *Web Intelligence and Agent Systems*, 10(1):75–88, 2012.

35. K.W. Church and M. Yamamoto. Using suffix arrays to compute term frequency and document frequency for all substrings in a corpus. *Computational Linguistics*, 27(1):1–30, 2001.

36. Y. Cao, J. Xu, Y. Liu, H. Li, Y. Huang, and H. W. Hon. Adapting ranking svm to document retrieval. In *Proceedings of ACM SIGIR Conference on Research and Development in Information Retrieval*, pp. 186–193, Seattle, WA, 2006.

37. J. Liu, Z.H. Wu, L. Jiang, Q.H. Zheng, and X. Liu. Crawling deep web content through query forms. In *Proceedings of WEBIST*, pp. 634–642, Lisbon, Portugal, March 2009.

38. J. Lu. Ranking bias in deep web size estimation using capture recapture method. *Journal of Data and Knowledge Engineering*, 69(8):866–879, 2010.

39. P. Erdos. On random graphs. *Publicationes Mathematicae*, 6:290–297, 1959.

40. G. Valkanas, A. Ntoulas, and D. Gunopulos. Rank-aware crawling of hidden web sites. *Proceedings of WebDB*, Athens, Greece, 2011.

Chapter 5

Big Data and Information Distillation in Social Sensing

Dong Wang

CONTENTS

Abstract

Social sensing has emerged as a new paradigm of growing interests, where individuals volunteer (or are recruited to) collect and share observations or measurements about the physical world. This opens up unprecedented opportunities and challenges centered on Big Data processing and information distillation in social sensing, where the goal is to distill accurate and credible information from large amounts of unfiltered, unstructured, and unvetted data generated by social sources (e.g., humans or devices on their behalf). Achieving this goal requires multidisciplinary solutions that combine data mining, cyber-physical computing, network science, and statistics. The proliferation of sensors in the possession of the average individual, together with the popularity of online social media (e.g., Facebook and Twitter), heralds an era of Big Data in social sensing that brings together new research challenges reviewed in this chapter.

5.1 Age of Big Data in Social Sensing

We live in the era of *Big Data*. The advent of online social media (e.g., Twitter and Flickr), the ubiquity of mobile Internet access (e.g., 4G and WiFi), and the proliferation of a wide variety of sensors in the possession of common individuals (e.g., smartphones) allow humans to create and disseminate a deluge of information about the physical world. This opens up unprecedented opportunities and challenges in *social sensing*, a key emerging field at the intersection of sensing and Big Data, where the goal is to distill accurate and credible information from large amounts of unfiltered, unstructured, and unvetted data generated by social sources (e.g., humans or devices in their possession). Little is analytically known about data validity in this new sensing paradigm, where sources are noisy, unreliable, erroneous, and largely unknown [46]. This motivates a closer look into recent advances in social sensing with an emphasis on the key problem faced by application designers; namely, how to distill reliable information from the Big Data collected from largely unknown and possibly unreliable sources? Novel solutions that leverage techniques from machine learning, information fusion, estimation theory, and data mining recently offer significant progress on this problem and are reviewed in this chapter.

In situations, where the reliability of sources is known, it is easy to compute the probability of correctness of different observations. For example, one can use, say, Bayesian analysis to fuse data from sources of different (known) degrees of reliability. The distinguishing challenge in social sensing applications is that the reliability of sources is often unknown. For example, much of the chatter on Twitter might come from users who are unknown to the data collection system. Hence, it is hard to assess the reliability of their observations. The same is true of situations, where individuals download a smartphone app that allows them to contribute to a social sensing data collection campaign. If anyone is allowed to participate, the pool of sources is unvetted and the reliability of individual observers is generally unknown to the data collector. It is in this context that the problem of distilling reliable information becomes challenging. The challenge arises from the fact that one neither knows the sources, nor can immediately verify their claims. What can be rigorously said, in this case, about the correctness of collected data? More specifically, how can one jointly ascertain data correctness and source reliability in Big Data social sensing applications? The problem is of importance in many domains and, as such, touches upon several areas of active research.

In sensor networks, an important challenge has always been to derive accurate representations of physical state and physical context from possibly unreliable, non-specific, or weak proxies. Often one trades off quantity and quality. While individual sensors may be less reliable, collectively (using the ingenious analysis techniques published in various sensor network venues) they may yield reliable conclusions. Much of the research in that area focused on physical sensors. This includes dedicated devices embedded in their environment, as well as human-centric sensing devices such as cell phones and wearables. Recent research proposed challenges with the use of humans as sensors. Clearly, humans differ from traditional physical devices in many respects. Importantly to the reliability analysis, they lack a design specification and a reliability standard, making it hard to define a generic noise model for sources. Each human is its own individual with different model parameters that predict how good that individual person's observations are. Hence, many techniques that estimate probability of error for sensors do not apply, since they assume the same error model for all sensors.

Reputation systems is another area of research, where source reliability is the issue. The assumption is that, when sources are observed over time, their reliability is eventually uncovered. Social sensing applications, however, often deal with scenarios, where a new event requires data collection from sources who have not previously participated in other data collection campaigns, or perhaps not been *tested* in the unique circumstances of the current event. For example, a hurricane strikes New Jersey. This is a rare event. We do not know how accurate the individuals who fled the event are at describing damage left behind. No reputation is accumulated for them in such a scenario. Yet, it would be desirable to leverage their collective observations to deploy help in a more efficient and timely manner. How do we determine which observations to believe?

In cyber-physical systems (CPS) research, an important emphasis has always been on ensuring validity and on proving that systems meet specifications [29,36,39]. The topics of reliability, predictability, and performance guarantees receive much attention. While past research on CPS addressed correctness of software systems (even in the presence of unverified code), in today's data-driven world, a key emerging challenge is to ascertain correctness of data (even in the presence of unverified sources). The challenge is promoted by the need to account for the humans in the loop. Humans are the drivers in transportation systems, the occupants in energy management systems, the survivors and first responders in disaster response systems, and the patients in medical systems. It makes sense to utilize their input when trying to assess system state. For example, one can get a more accurate account of current vehicular traffic state and more accurate prediction of its future evolution, if driver input was taken into account in some global, real-time, and automated fashion. This is assuming that the inputs are reliable, which is not always the case. The reliable information distillation problem, if solved, would enable the development of dependable applications in domains of transportation, energy, disaster response, and military intelligence, among others, where correctness is guaranteed despite reliance on the collective observations of untrained, average, and largely unreliable sources.

Techniques reviewed in this chapter are also of relevance to business analytics applications, where one is interested in making sense out of large amounts of unreliable data. These techniques can thus serve applications in social networks, Big Data, and human-in-the-loop systems, and leverage the proliferation of computing artifacts that interact with or monitor the physical world. The goal of this chapter is to review the needed theoretical foundations that exploit advances in social sensing analytics to support emerging data-driven applications.

5.2 A Multidisciplinary Background of Social Sensing

In this section, we review the multidisciplinary background of social sensing. The inspiration for developing mathematical foundations for reliable information distillation systems for social sensing comes from multiple research communities. Many of these communities do not interact. As a result, some opportunities were missed for connecting the dots between their advances. Jointly, these advances offer the needed foundations for leveraging unreliable social sensing sources, while offering collective reliability guarantees on the distilled information.

In the sensor fusion community, well established results exist that describe estimation algorithms using noisy sensors and quantify the corresponding estimation error bounds. It is possible to exploit Bayesian analysis to combine evidence and use estimation theory to rigorously compute confidence intervals as a function of reliability of input sources, even in the presence of noise and uncertainty. These results have typically been applied to the estimation of physical signals and tracking dynamic state such as trajectories of mobile targets. Given a physical model of how a target behaves, and given some observations, theory was developed on how to infer hidden variables that are not directly observed. Since sensor fusion deals with measuring state of the physical world, a key concept that treads through the research is the existence of a unique ground truth (barring, for the moment, the quantum effects and Schrodinger's cat). The existence of a unique ground truth offers a non-ambiguous notion of error that quantifies the deviation of estimated state from ground truth. In turn, the existence of a non-ambiguous notion of error lends itself nicely to the formulation of optimization problems that minimize this error. Such problems were formulated for estimating the state of physical systems, usually given by well-understood models, from noisy indirect observations. However, they have seldom been applied to the estimation of parameters of social sources and reliability of social observations.

Data mining researchers, on the other hand, do not usually exploit physical models of targets. This is because of the nature of the data mining problems. Rather than dealing with well-defined and well-understood objects for which physical dynamic models exist, data mining research tries to understand very large systems, and infer relations that are observed to hold true in the data. Much advances were made in data mining on representing very large datasets as abstract graphs of heterogeneous nodes, and inferring interesting new properties of the underlying systems from the topology of such graphs. This area is referred to as heterogeneous network mining. The analysis techniques described in that space are mostly heuristic, but have the power of producing interesting insights starting with no prior knowledge about the system whose data are collected. Importantly, since the system in question is often very complex and not well-understood, much of the work stops at computing different properties, without defining a notion of error. In many cases, ground truth cannot be defined. For example, when clustering individuals according to their beliefs into liberals and conservatives, it is really hard to define a rigorous and unbiased notion of what ground truth means, and offer a non-ambiguous notion of error. As a consequence of the difficulty in defining error for solutions of data mining problems, few problems are cast as ones of error optimization. Rather, data mining problems are often cast as minimizing internal conflict between observations. Hence, data mining literature does not usually offer bounds on error of data mining algorithms.

Machine learning researchers take a different approach to extracting properties of poorly understood systems. They typically propose a generative model for how the system behaves. Unlike researchers in sensor data fusion who often exploit representations of the exact dynamics of their targets, in machine learning the generative model has hidden parameters that are estimated only empirically. Machine learning literature describes techniques for learning model parameters using algorithms such as expectation maximization (EM).

The above body of results, put together, suggests an approach to reliably distill information in social sensing. Namely, we borrow from data mining the techniques used for knowledge representation. Specifically, we represent sources and observations by graphs that allow us to infer interesting properties of nodes. We then borrow from machine learning the idea of using generative models with hidden parameters to be estimated. Hence, we propose simple models for behavior of individual nodes in those graphs, such as lying, gossiping, or telling the truth. These models allow us to compute likelihood of observations as a function of model parameters of nodes. Finally, once a generative model is present, we are able to use the body of results developed in sensor data fusion to design optimal estimators and assess estimator error and confidence intervals. Specifically, we find model parameters that maximize the likelihood of the specific graph topology borne out from our data. Note specifically that, since the final outcome of this work is to decide which of a large number of social observations are true, we are able to define a rigourous notion of ground truth. After all, the observations we are interested in are those concerning the physical state of the world. Hence, ground truth exists (although is not known). A non-ambiguous notion of error therefore exists as well, and we are able to rigorously cast reliable information distillation in social sensing as an error optimization problem.

5.3 State of the Art

Prior research on social sensing can be mainly classified into three categories (i.e., discount data fusion, trust and reputation systems, and fact-finding techniques) based on whether the prior knowledge of source reliability and claim correctness or credibility is known to the application. We discuss the state-of-the-art techniques in these categories in detail below.

5.3.1 Discount data fusion

When we have prior knowledge about the reliability of the sources but no prior knowledge of the correctness or credibility of the claims (information), one can filter the noise in the claims via fusion. This is a classic case for the target tracking community where disparate sensor systems (the sources) generate tracks that must be combined. The tracks are estimates of the kinematic state of the target, and the reliability of the sources is expressed as a state covariance error for the tracks. The expression for the fused state estimate by accounting for all the sensors as a function of the tracks and error covariances is well known [4,9–11,37]. When the sensor tracks are uncorrelated, the fused track can be interpreted as a weighted average of the senor tracks where the weights are proportional to the inverse of the error covariances for the sensors. The general expression for correlated tracks is slightly more complicated, but it is reasonable to interpret the track fusion process as discounting the tracks based on their reliability followed by a combining process. The difficulty in track fusion is determining which tracks for the various sensors associate for the fusion process. Techniques for track-to-track association do exist [3,27], but they rely on understanding the correlation of tracks from different sensors. Unfortunately, it is not known how to determine this correlation when the tracks are formed for each sensor in a distributed manner.

In the information fusion community, belief theory provides the mechanism to combine evidence from multiple possibly conflicting sources [40,41,59]. The concept of discounting beliefs based upon source reliability before fusion goes back to Shafer [40]. Recently, subjective logic has emerged as a means to reason over conflicting evidence [25]. Subjective opinions are formed from evidence observed from individual sources. When incorporating multiple opinions, the subjective opinions need to be discounted similar to Dempster–Shafer theory before consensus fusion. In essence, this form of discount fusion can be interpreted as a weighted sum of evidence where the weights are proportional to the source reliabilities. The consensus fusion operation in subjective logic assumes the evidence used to form the subjective opinions of the sources are independent. Current research is investigating the proper fusion rule when the sources incorporate correlated evidence.

5.3.2 Trust and reputation systems

When we have prior knowledge of the correctness or credibility of claims (information) but no prior knowledge on the reliability of sources, a lot of work in trust and reputation systems make efforts to assess the reliability of sources (e.g., the quality of providers) [2,7,24,56]. The basic idea of reputation systems is to let entities rate each other (e.g., after a transaction) or review some objects of common interests (e.g., products or dealers), and use the aggregated ratings to derive trust or reputation scores of both sources and objects in the systems [24]. These reputation scores can help other entities in deciding whether or not to trust a given entity or purchase a certain object [2]. Trust and reputation scores can be obtained from both individual and social perspectives [22,23]. Individual trust often comes from experiences of direct interaction with transaction partners while social trust is computed from third-party experiences, which might include both honest and misleading opinions. Different types of reputation systems are being used successfully in commercial online applications [1,16,20,64]. For example, eBay is a type of reputation system based on homogeneous peer-to-peer systems, which allows peers to rate each other after each pair of them conduct a transaction [1,20]. Amazon on-line review system represents another type of reputation systems, where different sources offer reviews on products (or brands, companies) they have experienced [16,64]. Customers are affected by those reviews (or reputation scores) in making purchase decisions. Various techniques and models have also been developed to detect deceitful behaviors of participants [42,61] and identify discriminating attitudes and fraudulent activities [34,58] in the trust and reputation systems to provide reliable service in an open and dynamic environment. Recent work has investigated consistency of reports to estimate and revise trust scores in reputation system [26,38].

5.3.3 Fact-finding techniques

Given no prior knowledge on the reliability of sources and the credibility of their claims (information), there exists substantial work on techniques referred to as *fact-finders* within data mining and machine learning communities that jointly compute the source reliability and claim credibility. The inspiration of fact-finders can be traced to Google's PageRank [6]. PageRank iteratively ranks the credibility of pages on the Web, by considering the credibility of pages that link to them. In fact-finders, they estimate the credibility of claims from the reliability of sources that make them, then estimate the reliability of sources based on the credibility of their claims. *Hubs and authorities* [28] established a basic fact-finder model based on linear assumptions to compute scores for sources and claims they asserted. Yin et al. introduced *TruthFinder* as an unsupervised fact-finder for trust analysis on a providers-facts network [60]. Other fact-finders enhanced these basic frameworks by incorporating analysis on properties [17,30,44] or dependencies within claims or sources [5,14,15,35]. More recent works came up with some new fact-finding algorithms designed to handle the background knowledge [31,32] and multivalued facts [63], provide semantics to the credibility scores [33], and use slot filling systems for multidimensional fact-finding [62]. A comprehensive survey of fact-finders used in the context of trust analysis of information networks can be found in [18].

5.4 Reliable Information Distillation in Social Sensing

This section reviews in great detail a comprehensive analytical framework that optimally (in the sense of maximum likelihood estimation [MLE]) solves the reliable information distillation problem in social sensing and rigorously analyzes the accuracy of the results, offering correctness guarantees on solid theoretical foundations [21,43–45,47–55]. It is the notion of quantified correctness guarantees in social sensing that sets the purpose of the work reviewed in this section apart from other reviews of data mining, social networks, and sensing literature. The developed techniques can be applied to a new range of Big Data and social sensing applications, where assurances of data correctness are needed before such data can be used by the application, in order to meet higher level application goals.

5.4.1 Reliable information distillation problem

To formulate the reliable information distillation problem in social sensing in a manner amenable to rigorous optimization, a social sensing application model was proposed by Wang et al., where a group of M sources, $S_1, ..., S_M$, make individual observations about a set of N claims $C_1, ..., C_N$ in their environment [55]. For example, a group of individuals interested in the appearance of their neighborhood might join a sensing campaign to report all locations of offensive graffiti. Alternatively, a group of drivers might join a campaign to report freeway locations in need of repair. Hence, each claim denotes the existence or lack thereof of an offending condition at a given location.* In this effort, only binary variables are considered, and it is assumed, without loss of generality, that their *normal* state is negative (e.g., no offending graffiti on walls, or no potholes on streets). Hence, sources report only when a positive value is encountered.

Each source generally observes only a subset of all variables (e.g., the conditions at locations they have been to). The goal here is to determine which observations are correct and which are not. As mentioned in the introduction, this work differs from a large volume of previous sensing literature in that no prior knowledge is assumed about source reliability and no prior knowledge is assumed for the correctness of individual observations. Also note that the reviewed work in this chapter assumes imperfect reliability of sources and claims made by sources are mostly true, which is reasonable in a large category of social sensing applications. However, if those assumptions do not hold, the algorithm reviewed later in the chapter could converge to other stationary solutions (e.g., all sources are perfectly reliable or unreliable).

Let the probability that source S_i makes an observation be s_i. Further, let the probability that source S_i is right be t_i and the probability that it is wrong be $1 - t_i$. Note that, this probability depends on the source's reliability, which is not known *a priori*. Formally, t_i is defined as the odds of a claim to be true given that source S_i reports it:

$$t_i = P\left(C_j^t | S_i C_j\right) \tag{5.1}$$

Let a_i represent the (unknown) probability that source S_i reports a claim to be true when it is indeed true, and b_i represent the (unknown) probability that source S_i reports a claim to be true when it is in reality false. Formally, a_i and b_i are defined as follows:

$$a_i = P\left(S_i C_j | C_j^t\right)$$
$$b_i = P\left(S_i C_j | C_j^f\right) \tag{5.2}$$

---

*It is assumed that locations are discretized, and therefore finite. For example, they are given by street addresses or mile markers.

From the definition of t_i, a_i, and b_i, their relationship can be determined by using the Bayesian theorem:

$$a_i = P(S_iC_j|C_j^t) = \frac{P(S_iC_j, C_j^t)}{P(C_j^t)} = \frac{P(C_j^t|S_iC_j)P(S_iC_j)}{P(C_j^t)}$$

$$b_i = P(S_iC_j|C_j^f) = \frac{P(S_iC_j, C_j^f)}{P(C_j^f)} = \frac{P(C_j^f|S_iC_j)P(S_iC_j)}{P(C_j^f)} \tag{5.3}$$

The only input to the algorithm is the social sensing topology represented by a matrix SC, where $S_iC_j = 1$ when source S_i reports that C_j is true, and $S_iC_j = 0$ otherwise. Let us call it the *observation matrix*.

The goal of the algorithm is to compute (1) the best estimate h_j on the correctness of each claim C_j and (2) the best estimate e_i of the reliability of each source S_i. The sets of the estimates are denoted by vectors H and E, respectively. The goal is to find the H^* and E^* vectors that are most consistent with the observation matrix SC. Formally, this is given by

$$< H^*, E^* > = \operatorname*{argmax}_{<H,E>} p(SC|H,E) \tag{5.4}$$

The background bias d will also be computed, which is the overall probability that a randomly chosen claim is true. For example, it may represent the probability that any street, in general, is in disrepair. It does not indicate, however, whether any particular claim about disrepair at a particular location is true or not. Hence, one can define the prior of a claim being true as $P(C_j^t) = d$. Note also that, the probability that a source makes an observation (i.e., s_i) is proportional to the number of claims observed by the source over the total number of claims observed by all sources, which can be easily computed from the observation matrix. Hence, one can define the prior $P(S_iC_j) = s_i$. Plugging these, together with t_i into the definition of a_i and b_i, we get,

$$a_i = \frac{t_i \times s_i}{d}$$

$$b_i = \frac{(1 - t_i) \times s_i}{1 - d} \tag{5.5}$$

So that

$$t_i = \frac{a_i \times d}{a_i \times d + b_i \times (1 - d)} \tag{5.6}$$

5.4.2 EM solution

In this part, we review the solution to the problem formulated in Section 5.4.1 using the EM algorithm. EM is a general algorithm for finding the MLEs of parameters in a statistic model, where the data are *incomplete* or the likelihood function involves latent variables [13]. In the following, we first briefly review the basic ideas and steps of EM.

The EM algorithm is useful when the likelihood expression simplifies by the inclusion of a latent variable. Considering a latent variable Z, the likelihood function is given by:

$$L(\theta; X) = p(X|\theta) = \sum_Z p(X, Z|\theta) \tag{5.7}$$

where $p(X, Z|\theta)$ is $L(\theta|X, Z)$, which is unimodal and easy to solve.

Once the formulation is complete, the EM algorithm finds the MLE by iteratively performing the following steps:

- *Exceptation step (E-step)*: Compute the expected log-likelihood function where the expectation is taken with respect to the computed conditional distribution of the latent variables given the current settings and observed data.

$$Q\left(\theta|\theta^{(t)}\right) = E_{Z|X,\theta^{(t)}}[\log L(\theta;X,Z)] \tag{5.8}$$

- *Maximization step (M-step)*: Find the parameters that maximize the Q function in the E-step to be used as the estimate of θ for the next iteration.

$$\theta^{(t+1)} = \underset{\theta}{\operatorname{argmax}} \, Q\left(\theta|\theta^{(t)}\right) \tag{5.9}$$

The information distillation problem in social sensing fits nicely into the EM model. First, a latent variable Z is introduced for each claim to indicate whether it is true or not. Specifically, a corresponding variable z_j is defined for the jth claim C_j such that $z_j = 1$ when C_j is true and $z_j = 0$ otherwise. The observation matrix SC is denoted as the observed data X, and $\theta = (a_1, a_2, ..., a_M; b_1, b_2, ..., b_M; d)$ is defined as the parameter of the model that needs to be estimated. The goal is to obtain the MLE of θ for the model containing observed data X and latent variables Z. The likelihood function $L(\theta; X, Z)$ is given by

$$\begin{aligned}
L(\theta; X, Z) &= p(X, Z|\theta) \\
&= \prod_{j=1}^{N} \left\{ \prod_{i=1}^{M} a_i^{S_i C_j} (1-a_i)^{(1-S_i C_j)} \times d \times z_j \right. \\
&\quad \left. + \prod_{i=1}^{M} b_i^{S_i C_j} (1-b_i)^{(1-S_i C_j)} \times (1-d) \times (1-z_j) \right\}
\end{aligned} \tag{5.10}$$

where, as we mentioned before, a_i and b_i are the conditional probabilities that source S_i reports the claim C_j to be true given that C_j is true or false (i.e., defined in Equation 5.2). $S_i C_j = 1$ when source S_i reports that C_j is true, and $S_i C_j = 0$ otherwise. d is the background bias that a randomly chosen claim is true. Additionally, it is assumed that sources and claims are independent respectively. The likelihood function above describes the likelihood to have current observation matrix X and hidden variable Z given the estimation parameter θ.

Given the above formulation, substitute the likelihood function defined in Equation 5.10 into the definition of Q function given by Equation 5.8 of EM. The E-step becomes

$$\begin{aligned}
Q\left(\theta|\theta^{(t)}\right) &= E_{Z|X,\theta^{(t)}}[\log L(\theta; X, Z)] \\
&= \sum_{j=1}^{N} \left\{ p(z_j = 1|X_j, \theta^{(t)}) \right. \\
&\quad \times \left[\sum_{i=1}^{M} (S_i C_j \log a_i + (1 - S_i C_j) \log(1 - a_i) + \log d) \right] \\
&\quad + p(z_j = 0|X_j, \theta^{(t)}) \\
&\quad \left. \times \left[\sum_{i=1}^{M} (S_i C_j \log b_i + (1 - S_i C_j) \log(1 - b_i) + \log(1 - d)) \right] \right\}
\end{aligned} \tag{5.11}$$

where X_j represents the jth column of the observed SC matrix (i.e., observations of the jth claim from all sources) and $p(z_j = 1|X_j, \theta^{(t)})$ is the conditional probability of the latent variable z_j to be true given the observation matrix related to the jth claim and current estimate of θ, which is given by

$$
\begin{aligned}
& p(z_j = 1|X_j, \theta^{(t)}) \\
& = \frac{p(z_j = 1; X_j, \theta^{(t)})}{p(X_j, \theta^{(t)})} \\
& = \frac{p(X_j, \theta^{(t)}|z_j = 1)p(z_j = 1)}{p(X_j, \theta^{(t)}|z_j = 1)p(z_j = 1) + p(X_j, \theta^{(t)}|z_j = 0)p(z_j = 0)} \\
& = \frac{A(t, j) \times d^{(t)}}{A(t, j) \times d^{(t)} + B(t, j) \times (1 - d^{(t)})}
\end{aligned} \tag{5.12}
$$

where $A(t, j)$ and $B(t, j)$ are defined as

$$
\begin{aligned}
A(t, j) &= p(X_j, \theta^{(t)}|z_j = 1) \\
&= \prod_{i=1}^{M} a_i^{(t) S_i C_j} \left(1 - a_i^{(t)}\right)^{(1 - S_i C_j)} \\
B(t, j) &= p(X_j, \theta^{(t)}|z_j = 0) \\
&= \prod_{i=1}^{M} b_i^{(t) S_i C_j} \left(1 - b_i^{(t)}\right)^{(1 - S_i C_j)}
\end{aligned} \tag{5.13}
$$

$A(t, j)$ and $B(t, j)$ represent the conditional probability regarding observations about the jth claim and current estimation of the parameter θ given the jth claim is true or false respectively.

Next Equation 5.11 is simplified by noting that the conditional probability of $p(z_j = 1|X_j, \theta^{(t)})$ given by Equation 5.12 is only a function of t and j. Thus, it is represented by $Z(t, j)$. Similarly, $p(z_j = 0|X_j, \theta^{(t)})$ is simply

$$
\begin{aligned}
p(z_j = 0|X_j, \theta^{(t)}) &= 1 - p(z_j = 1|X_j, \theta^{(t)}) \\
&= \frac{B(t, j) \times (1 - d^{(t)})}{A(t, j) \times d^{(t)} + B(t, j) \times (1 - d^{(t)})} \\
&= 1 - Z(t, j)
\end{aligned} \tag{5.14}
$$

Substituting from Equations 5.12 and 5.14 into Equation 5.11, we get

$$
\begin{aligned}
& Q\left(\theta|\theta^{(t)}\right) \\
& = \sum_{j=1}^{N} \left\{ Z(t, j) \times \left[\sum_{i=1}^{M} (S_i C_j \log a_i + (1 - S_i C_j) \log(1 - a_i) + \log d) \right] \right. \\
& \left. + (1 - Z(t, j)) \times \left[\sum_{i=1}^{M} (S_i C_j \log b_i + (1 - S_i C_j) \log(1 - b_i) + \log(1 - d)) \right] \right\}
\end{aligned} \tag{5.15}
$$

The M-step is given by Equation 5.9. θ^* (i.e., $(a_1^*, a_2^*, ..., a_M^*; b_1^*, b_2^*, ..., b_M^*; d^*)$) is chosen to maximize the $Q\left(\theta|\theta^{(t)}\right)$ function in each iteration to be the $\theta^{(t+1)}$ of the next iteration.

To get θ^* that maximizes $Q\left(\theta|\theta^{(t)}\right)$, the derivatives are set to 0: $\frac{\partial Q}{\partial a_i} = 0$, $\frac{\partial Q}{\partial b_i} = 0$, $\frac{\partial Q}{\partial d} = 0$ which yields

$$\sum_{j=1}^{N} \left[Z(t,j) \left(S_i C_j \frac{1}{a_i^*} - (1 - S_i C_j) \frac{1}{1 - a_i^*} \right) \right] = 0$$

$$\sum_{j=1}^{N} \left[(1 - Z(t,j)) \left(S_i C_j \frac{1}{b_i^*} - (1 - S_i C_j) \frac{1}{1 - b_i^*} \right) \right] = 0$$

$$\sum_{j=1}^{N} \left[Z(t,j) M \frac{1}{d^*} - \left((1 - Z(t,j)) M \frac{1}{1 - d^*} \right) \right] = 0 \tag{5.16}$$

Let us define SJ_i as the set of claims the source S_i actually observes in the observation matrix SC, and $\overline{SJ_i}$ as the set of claims source S_i does not observe. Thus, Equation 5.16 can be rewritten as

$$\sum_{j \in SJ_i} Z(t,j) \frac{1}{a_i^*} - \sum_{j \in \overline{SJ_i}} Z(t,j) \frac{1}{1 - a_i^*} = 0$$

$$\sum_{j \in SJ_i} (1 - Z(t,j)) \frac{1}{b_i^*} - \sum_{j \in \overline{SJ_i}} (1 - Z(t,j)) \frac{1}{1 - b_i^*} = 0$$

$$\sum_{j=1}^{N} \left[Z(t,j) \frac{1}{d^*} - \left((1 - Z(t,j)) \frac{1}{1 - d^*} \right) \right] = 0 \tag{5.17}$$

Solving the above equations, the expressions of the optimal a_i^*, b_i^*, and d^* are as follows:

$$a_i^{(t+1)} = a_i^* = \frac{\sum_{j \in SJ_i} Z(t,j)}{\sum_{j=1}^{N} Z(t,j)}$$

$$b_i^{(t+1)} = b_i^* = \frac{K_i - \sum_{j \in SJ_i} Z(t,j)}{N - \sum_{j=1}^{N} Z(t,j)}$$

$$d_i^{(t+1)} = d_i^* = \frac{\sum_{j=1}^{N} Z(t,j)}{N} \tag{5.18}$$

where K_i is the number of claims espoused by source S_i and N is the total number of claims in the observation matrix.

Given the above, The E-step and M-step of EM optimization reduce to simply calculating Equations 5.12 and 5.18 iteratively until they converge. The convergence analysis has been done for EM scheme [57]. In practice, the algorithm runs until the difference of estimation parameter between consecutive iterations becomes insignificant. Since the claim is binary, the decision vector H^* can be computed from the converged value of $Z(t,j)$. Specially, h_j is true if $Z(t,j) \geq 0.5$ and false otherwise. At the same time, the estimation vector E^* of source reliability can also be computed from the converged values of $a_i^{(t)}$, $b_i^{(t)}$, and $d^{(t)}$ based on their relationship given by Equation 5.5.

5.4.3 Confidence bounds quantification

In this subsection, we review a confidence bounds quantification scheme to quantify accuracy of the MLE framework presented in the previous section. In particular, the goal is to demonstrate, in an analytically founded manner, how to compute the confidence interval of each source's reliability. Formally, this is given by

$$\left(\hat{t}_i^{\mathrm{MLE}} - c_p^{\mathrm{lower}}, \hat{t}_i^{\mathrm{MLE}} + c_p^{\mathrm{upper}} \right) \qquad c\% \qquad i = 1, 2, ..., M \tag{5.19}$$

where $\hat{t}_i^{\mathrm{MLE}}$ is the MLE on the reliability of source S_i, $c\%$ is the confidence level of the estimation interval, c_p^{lower} and c_p^{upper} represent the lower and upper bound on the estimation deviation from the MLE $\hat{t}_i^{\mathrm{MLE}}$, respectively. The goal is to find c_p^{lower} and c_p^{upper} for a given $c\%$ and an observation matrix SC. It turns out that the Cramer–Rao lower bounds (CRLBs) of the MLE on the source reliability need to be computed in order to obtain the c_p^{lower} and c_p^{upper}. Therefore, the goal of the reviewed work in this chapter is to (1) derive the actual and asymptotic error bounds that characterize the accuracy of the maximum likelihood estimator and compute its confidence interval, (2) estimate the accuracy of claim classification without knowing the ground truth values of the variables, and (3) derive the dependency of the accuracy of MLE on parameters of the problem space.

In this section, we show how to derive the confidence interval for source reliability through the computation of the CRLB for the estimation parameters (i.e., θ) and by leveraging the asymptotic normality of the MLE. We start with the review of the actual CRLB derivation and identify its scalability limitation. We then review the derivation of the asymptotic CRLB that works for the sensing topology with a large number of sources. Finally, we review the confidence interval on source reliability based on the derived CRLBs.

Let's start with the derivation of the actual CRLB that characterizes the estimation performance of the MLE of source reliability in social sensing. Similarly as in Section 5.4.1, the reliability of sources is assumed to be imperfect and the majority of claims are assumed to be true. In estimation theory, the CRLB expresses a lower bound on the estimation variance of a minimum-variance unbiased estimator. In its simplest form, the bound states the variance of any unbiased estimator is at least as high as the inverse of the Fisher information [19]. The estimator that reaches this lower bound is said to be *efficient*. For notational convenience, the observation matrix SC is denoted as the observed data X and use $X_{ij} = S_i C_j$ for the following derivation.

The likelihood function (containing hidden variable Z) of the MLE we get from EM is expressed in Equation 5.10, where $Z = (z_j | j = 1, 2, ..., N)$ represent the hidden variables. The EM scheme is used to handle the hidden variable and aims to find

$$\hat{\theta} = \underset{\theta}{\mathrm{argmax}} \, p(X|\theta) \tag{5.20}$$

where

$$p(X|\theta) = \prod_{j=1}^{N} \left\{ \prod_{i=1}^{M} a_i^{X_{ij}} (1 - a_i)^{(1 - X_{ij})} \times d \right.$$
$$\left. + \prod_{i=1}^{M} b_i^{X_{ij}} (1 - b_i)^{(1 - X_{ij})} \times (1 - d) \right\} \tag{5.21}$$

By definition of CRLB, it is given by

$$CRLB = J^{-1} \tag{5.22}$$

where

$$J = E[\nabla_\theta \ln p(X|\theta) \, \nabla_\theta^H \ln p(X|\theta)] \tag{5.23}$$

where J is the Fisher information of the estimation parameter, $\nabla_\theta = (\frac{\partial}{\partial a_1}, ..., \frac{\partial}{\partial a_M}, \frac{\partial}{\partial b_1}, ..., \frac{\partial}{\partial b_M})^H$ and H denotes the conjugate transpose operation. In information theory, the Fisher information is a way of measuring the amount of information that an observable random variable X carries about an estimated parameter θ upon which the probability of X depends. The expectation in Equation 5.23 is taken over all values for X with respect to the probability function $p(X|\theta)$ for any given value of θ. Let $\mathcal{X}$ represent the set of all possible values of $X_{ij} \in \{0,1\}$ for $i = 1,2,...,M; j = 1,2,...,N$. Note $|\mathcal{X}| = 2^{MN}$. Likewise, let $\mathcal{X}_j$ represent the set of all possible values of $X_{ij} \in \{0,1\}$ for $i = 1,2,...,M$ and a given value of j. Note $|\mathcal{X}_j| = 2^M$. Taking the expectation, Equation 5.23 can be rewritten as follows:

$$J = \sum_{X \in \mathcal{X}} \nabla_\theta \ln p(X|\theta) \nabla_\theta^H \ln p(X|\theta) p(X|\theta) \tag{5.24}$$

Then, the Fisher information matrix can be represented as follows:

$$J = \begin{bmatrix} A & C \\ C^T & B \end{bmatrix}$$

where submatrices A, B, and C contain the elements related with the estimation parameter a_i, b_i and their cross terms, respectively. The representative elements A_{kl}, B_{kl}, and C_{kl} of A, B, and C can be derived as follows:

$$\begin{aligned} A_{kl} &= E\left[\frac{\partial}{\partial a_k} \ln p(X|\theta) \frac{\partial}{\partial a_l} \ln p(X|\theta) \right] \\ &= E\left[\left(\sum_j \frac{(2X_{kj}-1)Z_j}{a_k^{X_{kj}}(1-a_k)^{(1-X_{kj})}} \sum_q \frac{(2X_{lq}-1)Z_q}{a_l^{X_{lq}}(1-a_l)^{(1-X_{lq})}} \right) \right] \\ &= \sum_j \sum_q E\left[\frac{(2X_{kj}-1)Z_j(2X_{lq}-1)Z_q}{a_k^{X_{kj}}(1-a_k)^{(1-X_{kj})} a_l^{X_{lq}}(1-a_l)^{(1-X_{lq})}} \right] \end{aligned} \tag{5.25}$$

where

$$Z_j = p(z_j = 1|X) = \frac{A_j \times d}{A_j \times d + B_j \times (1-d)}$$

where

$$A_j = \prod_{i=1}^M a_i^{X_{ij}}(1-a_i)^{(1-X_{ij})} \quad B_j = \prod_{i=1}^M b_i^{X_{ij}}(1-b_i)^{(1-X_{ij})} \tag{5.26}$$

Z_j is the conditional probability of the claim C_j to be true given the observation matrix. After further simplification, A_{kl} can be expressed as the summation of only the expectation terms where $j = q$

$$\begin{aligned} A_{kl} &= \sum_j E\left[\frac{(2X_{kj}-1)(2X_{lj}-1)Z_j^2}{a_k^{X_{kj}}(1-a_k)^{(1-X_{kj})} a_l^{X_{lj}}(1-a_l)^{(1-X_{lj})}} \right] \\ &= \sum_{j=1}^N \sum_{X \in \mathcal{X}_j} \frac{(2X_{kj}-1)(2X_{lj}-1) \prod_{\substack{i=1 \\ i \neq k}}^M A_{ij} \prod_{\substack{i=1 \\ i \neq l}}^M A_{ij} d^2}{\prod_{i=1}^M A_{ij} d + \prod_{i=1}^M B_{ij}(1-d)} \end{aligned} \tag{5.27}$$

where

$$A_{ij} = a_i^{X_{ij}}(1-a_i)^{(1-X_{ij})} \quad B_{ij} = b_i^{X_{ij}}(1-b_i)^{(1-X_{ij})} \tag{5.28}$$

Since the inner sum in Equation 5.27 is invariant to the claim index j, $A_{k,l} = N\bar{A}_{k,l}$ where $\bar{A}_{kl}$ is as follows:

$$\bar{A}_{kl} = \sum_{x \in \mathcal{X}_j} \frac{(2X_{kj} - 1)(2X_{lj} - 1) \prod_{\substack{i=1 \\ i \neq k}}^{M} A_{ij} \prod_{\substack{i=1 \\ i \neq l}}^{M} A_{ij} d^2}{\prod_{i=1}^{M} A_{ij} d + \prod_{i=1}^{M} B_{ij}(1-d)} \tag{5.29}$$

It should also be noted that the summation in Equation 5.29 is the same for all j.

By similar calculations, the inverse of the Fisher information matrix is obtained as follows:

$$J^{-1} = \frac{1}{N} \begin{bmatrix} \bar{A} & \bar{C} \\ \bar{C}^T & \bar{B} \end{bmatrix}^{-1}$$

where the klth element of $\bar{B}$, $\bar{C}$ is defined as:

$$\bar{B}_{kl} =$$
$$\sum_{x \in \mathcal{X}_j} \frac{(2X_{kj} - 1)(2X_{lj} - 1) \prod_{\substack{i=1 \\ i \neq k}}^{M} B_{ij} \prod_{\substack{i=1 \\ i \neq l}}^{M} B_{ij}(1-d)^2}{\prod_{i=1}^{M} A_{ij} d + \prod_{i=1}^{M} B_{ij}(1-d)} \tag{5.30}$$

$$\bar{C}_{kl} =$$
$$\sum_{x \in \mathcal{X}_j} \frac{(2X_{kj} - 1)(2X_{lj} - 1) \prod_{\substack{i=1 \\ i \neq k}}^{M} A_{ij} \prod_{\substack{i=1 \\ i \neq l}}^{M} B_{ij} d(1-d)}{\prod_{i=1}^{M} A_{ij} d + \prod_{i=1}^{M} B_{ij}(1-d)} \tag{5.31}$$

Note that the sum of $\bar{A}_{kl}$, $\bar{B}_{kl}$, and $\bar{C}_{kl}$ are over the 2^M different permutations of X_{ij} for $i = 1, 2, ..., M$ at a given j. This is much smaller than the 2^{MN} permutations of $\mathcal{X}$.

This gives the actual CRLB. Note that more claims simply lead to better estimates for θ as the variance decreases as $\frac{1}{N}$. The decrease in variance for the estimates as a function of M is more complicated, which can only be computed numerically. Please note that the actual CRLB computation needs the true values of the estimation parameter. However in real applications, the true values are not known in advance, we substitute the unknown true values for MLEs as an approximation to estimate variances for determining the confidence bounds.

Observe that the complexity of the actual CRLB computation is exponential with respect to the number of sources (i.e., M) in the system. Therefore, it is inefficient (or infeasible) to compute the actual CRLB when the number of sources becomes large. Here, we review the derivation of the asymptotic CRLB for efficient computation in the sensing topology with a large number of sources. The asymptotic CRLB is derived based on the assumption that the correctness of the hidden variable (i.e., z_j) can be correctly estimated from EM. This is a reasonable assumption when the number of sources is sufficient [55]. Under this assumption, the log-likelihood function of the MLE obtained from EM can be expressed as follows:

$$l_{\text{EM}}(x; \theta) = \sum_{j=1}^{N} \left\{ z_j \times \left[\sum_{i=1}^{M} (X_{ij} \log a_i + (1 - X_{ij}) \log(1 - a_i) + \log d) \right] \right.$$
$$\left. + (1 - z_j) \times \left[\sum_{i=1}^{M} (X_{ij} \log b_i + (1 - X_{ij}) \log(1 - b_i) + \log(1 - d)) \right] \right\} \tag{5.32}$$

The Fisher information matrix at the MLE was computed from the log-likelihood function given by Equation 5.32. The converged estimates of a_i and b_i from the EM of Section 5.4.2 were used as the MLE.

Plugging $l_{EM}(x;\theta)$ given by Equation 5.32 into the Fisher information is defined in Equation 5.23, the representative element of Fisher information matrix from N claims was shown as follows:

$$(J(\hat{\theta}_{MLE}))_{i,j} \qquad (5.33)$$

$$= \begin{cases} 0 & i \neq j \\ -E_X\left[\frac{\partial^2 l_{EM}(x;a_i)}{\partial a_i^2}\big|_{a_i=a_i^0}\right] & i = j \in [1,M] \\ -E_X\left[\frac{\partial^2 l_{EM}(x;b_i)}{\partial b_i^2}\big|_{b_i=b_i^0}\right] & i = j \in (M,2M] \end{cases}$$

where a_i^0 and b_i^0 are the true values of a_i and b_i. In the following computation, we estimate them by substituting the known MLEs for the unknown parameter values.

Substituting the log-likelihood function in Equation 5.32 and MLE of θ into Equation 5.33, the asymptotic CRLB (i.e., the inverse of the Fisher information matrix) can be written as follows:

$$(J^{-1}(\hat{\theta}_{MLE}))_{i,j} = \begin{cases} 0 & i \neq j \\ \frac{\hat{a}_i^{MLE} \times (1-\hat{a}_i^{MLE})}{N \times d} & i = j \in [1,M] \\ \frac{\hat{b}_i^{MLE} \times (1-\hat{b}_i^{MLE})}{N \times (1-d)} & i = j \in (M,2M] \end{cases} \qquad (5.34)$$

Note that the asymptotic CRLB is independent of M under the assumption that M is sufficient, and it can be quickly computed from the MLE of the EM scheme.

Finally, we show that the confidence interval of source reliability can be obtained by using the CRLB derived earlier and leveraging the asymptotic normality of the MLE.

The maximum likelihood estimator posses a number of attractive asymptotic properties. One of them is called *asymptotic normality*, which basically states the MLE estimator is asymptotically distributed with Gaussian behavior as the data sample size goes up, in particular [8]

$$(\hat{\theta}_{MLE} - \theta_0) \xrightarrow{d} N(0, J^{-1}(\hat{\theta}_{MLE})) \qquad (5.35)$$

where:

J is the Fisher information matrix computed from all samples

θ_0 and $\hat{\theta}_{MLE}$ are the true value and the MLE of the parameter θ, respectively

The Fisher information at the MLE is used to estimate its true (but unknown) value [19]. Hence, the asymptotic normality property means that in a regular case of estimation and in the distribution limiting sense, the maximum likelihood estimator $\hat{\theta}_{MLE}$ is unbiased and its covariance reaches the CRLB (i.e., an efficient estimator).

From the asymptotic normality of the maximum likelihood estimator [55], the error of the corresponding estimation on θ follows a normal distribution with zero mean and the covariance matrix given by the CRLB. Let us denote the variance of estimation error on parameter a_i as $var(\hat{a}_i^{MLE})$. Recall the relation between source reliability (i.e., t_i) and estimation parameter a_i and b_i is t_i is given by Equation 5.6. For a sensing topology with small values of M and N, the estimation of t_i has a complex distribution and its estimation variance can be approximated [12].

The denominator of t_i is equivalent to s_i based on Equation 5.6.* Therefore, $(\hat{t}_i^{\text{MLE}} - t_i^0)$ also follows a normal distribution with zero mean and variance given by:

$$\text{var}\left(\hat{t}_i^{\text{MLE}}\right) = \left(\frac{d}{s_i}\right)^2 \text{var}\left(\hat{a}_i^{\text{MLE}}\right) \tag{5.36}$$

Hence, the confidence interval that can be obtained to quantify the estimation accuracy of the MLE on source reliability. The confidence interval of the reliability estimation of source S_i (i.e., $\hat{t}_i^{\text{MLE}}$) at confidence level p is given by the following:

$$\left(\hat{t}_i^{\text{MLE}} - c_p \sqrt{\text{var}\left(\hat{t}_i^{\text{MLE}}\right)}, \hat{t}_i^{\text{MLE}} + c_p \sqrt{\text{var}\left(\hat{t}_i^{\text{MLE}}\right)}\right) \tag{5.37}$$

where c_p is the standard score (z-score) of the confidence level p. For example, for the 95% confidence level, $c_p = 1.96$. Therefore, the derived confidence interval of the source reliability MLE can be computed by using the CRLB derived earlier.

5.5 Summary and Discussions

In this chapter, we reviewed a set of recently developed theories and methodologies to distill reliable information from potentially unreliable data in social sensing. Social sensing has emerged as a new paradigm of sensing and data collection due to the proliferation of mobile devices owned by common individuals, fast data sharing and large-scale information dissemination opportunities. A key challenge in social sensing applications lies in *Information Overflow and Data Reliability*. Solutions to address this key challenge is non-trivial given the reliability of participants (sources) is usually unknown *a priori* and there is no independent way to verify the correctness of their measurements. Techniques have been developed to leverage the key insights from data mining, machine learning, data fusion, and estimation theory to address this challenge. More can be done to improve the estimates including understanding the social network channels that connect the sources to each other and to the decision makers. For instance, sources influence each other through direct interaction and through social norms. Furthermore, sources will reveal different thoughts depending on who they perceive will be the audience. For instance, when a person requests a response from a particular source, that source will probably provide a guarded response if he/she does not have a good rapport with the information requester.

References

1. Karl Aberer and Zoran Despotovic. Managing trust in a peer-2-peer information system. In *Proceedings of the 10th International Conference on Information and Knowledge Management*, pp. 310–317, ACM, New York, 2001.

2. Donovan Artz and Yolanda Gil. A survey of trust in computer science and the semantic web. *Web Semantics: Science, Services and Agents on the World Wide Web*, 5(2):58–71, 2007.

---

*The value of s_i is known to be K_i/N no matter the output of the EM (see Equation 5.18). Therefore, s_i can be treated as deterministic, that is, no variance.

3. Yaakov Bar-Shalom and Huimin Chen. Multisensor track-to-track association for tracks with dependent errors. In *Proceedings of the 43rd IEEE Conference on Decision and Control*, Paradise Island, Bahamas, December 2004.

4. Yaakov Bar-Shalom and Xiao-Rong Li. *Multisensor, Multitarget Tracking: Principles and Techniques*. YBS Publishing, Storrs, CT, 1995.

5. Laure Berti-Equille, Anish D Sarma, Xin Dong, Amélie Marian, and Divesh Srivastava. Sailing the information ocean with awareness of currents: Discovery and application of source dependence. In *CIDR'09*, Asilomar, CA, 2009.

6. Sergey Brin and Lawrence Page. The anatomy of a large-scale hypertextual web search engine. In *7th International Conference on World Wide Web*, pp. 107–117, Brisbane, Australia, 1998.

7. Luis Cabral and Ali Hortacsu. The dynamics of seller reputation: Evidence from eBay. *The Journal of Industrial Economics*, 58(1):54–78, 2010.

8. George Casella and Roger L Berger. *Statistical Inference*. Duxbury Press, Pacific Grove, CA, 2002.

9. Huimin Chen, Thia Kirubarajan, and Yaakov Bar-Shalom. Performance limits of track-to-track fusion versus centralized estimation: Theory and application. *IEEE Transactions on Aerospace and Electronic Systems*, 39(2):386–400, April 2003.

10. Chee-Yee Chong, Shozo Mori, William H Barker, and Kuo-Chu Chang. Architectures and algorithms for track association and fusion. *IEEE Aerospace and Electronic Systems Magazine*, 15(1):5–13, January 2000.

11. Chee-Yee Chong, Shozo Mori, and Kuo-Chu Chang. Distributed multitarget multisensor tracking. In Yaakov Bar-Shalom, editor, *Multitarget Multisensor Tracking: Advanced Applications*, pp. 247–295. Artech House, Norwood, MA, 1990.

12. Harald Cramer. *Mathematical Methods of Statistics*. Princeton University Press, Princeton, NJ, 1946.

13. Arthur P Dempster, Nan M Laird, and Donald B Rubin. Maximum likelihood from incomplete data via the EM algorithm. *Journal of the Royal Statistical Society, Series B*, 39(1):1–38, 1977.

14. Xin Dong, Laure Berti-Equille, Yifan Hu, and Divesh Srivastava. Global detection of complex copying relationships between sources. *PVLDB*, 3(1):1358–1369, 2010.

15. Xin Dong, Laure Berti-Equille, and Divesh Srivastava. Truth discovery and copying detection in a dynamic world. *VLDB*, 2(1):562–573, 2009.

16. Randy Farmer and Bryce Glass. *Building Web Reputation Systems*. O'Reilly Media, Sebastopol, CA, 2010.

17. Alban Galland, Serge Abiteboul, Amélie Marian, and Pierre Senellart. Corroborating information from disagreeing views. In *WSDM*, pp. 131–140, New York, 2010.

18. Manish Gupta and Jiawei Han. Heterogeneous network-based trust analysis: A survey. *ACM SIGKDD Explorations Newsletter*, 13(1):54–71, 2011.

19. Robert V Hogg and Allen T Craig. *Introduction to Mathematical Statistics*. Prentice Hall, Englewood Cliffs, NJ, 1995.

20. Daniel Houser and John Wooders. Reputation in auctions: Theory, and evidence from eBay. *Journal of Economics & Management Strategy*, 15(2):353–369, 2006.

21. Chao Huang and Dong Wang. Link weight based truth discovery in social sensing. In *Proceedings of the 14th International Conference on Information Processing in Sensor Networks*, pp. 326–327. ACM, New York, 2015.

22. Trung D Huynh. Trust and reputation in open multi-agent systems. PhD thesis, University of Southampton, Southampton, England, 2006.

23. Trung D Huynh, Nicholas R Jennings, and Nigel R Shadbolt. An integrated trust and reputation model for open multi-agent systems. *Autonomous Agents and Multi-Agent Systems*, 13(2):119–154, September 2006.

24. Audun Jøsang, Roslan Ismail, and Colin Boyd. A survey of trust and reputation systems for online service provision. *Decision Support Systems*, 43(2):618–644, March 2007.

25. Audun Jøsang, Stephen Marsh, and Simon Pope. Exploring different types of trust propogation. In *Proceedings of the 4th International Conference on Trust Management*, Pisa, Italy, May 2006.

26. Lance Kaplan, Murat Scensoy, and Geeth de Mel. Trust estimation and fusion of uncertain information by exploiting consistency. In *17th International Conference on Information Fusion (FUSION)*, pp. 1–8, Salamanca, Spain. IEEE, 2014.

27. Lance M Kaplan, Yaakov Bar-Shalom, and William D Blair. Assignment costs for multiple sensor track-to-track association. *IEEE Transactions on Aerospace and Electronic Systems*, 44(2):655–677, April 2008.

28. Jon M Kleinberg. Authoritative sources in a hyperlinked environment. *Journal of the ACM*, 46(5):604–632, 1999.

29. Edward A Lee. Cyber physical systems: Design challenges. In *11th IEEE International Symposium on Object Oriented Real-Time Distributed Computing*, pp. 363–369. IEEE, Orlando, FL, 2008.

30. Jeff Pasternack and Dan Roth. Knowing what to believe (when you already know something). In *International Conference on Computational Linguistics*, Beijing, China, 2010.

31. Jeff Pasternack and Dan Roth. Generalized fact-finding. In *Proceedings of the 20th International Conference Companion on World Wide Web*, pp. 99–100. ACM, New York, 2011.

32. Jeff Pasternack and Dan Roth. Making better informed trust decisions with generalized fact-finding. In *Proceedings of the 22nd International Joint Conference on Artificial Intelligence—Volume Three*, pp. 2324–2329. AAAI Press, Barcelona, Spain, 2011.

33. Jeff Pasternack and Dan Roth. Latent credibility analysis. In *Proceedings of the 22nd International Conference on World Wide Web*, pp. 1009–1020. International World Wide Web Conferences Steering Committee, Rio de Janeiro, Brazil, 2013.

34. Isaac Pinyol and Jordi Sabater-Mir. Computational trust and reputation models for open multi-agent systems: A review. *Artificial Intelligence Review*, 40(1):1–25, 2013.

35. Guo-Jun Qi, Charu C Aggarwal, Jiawei Han, and Thomas Huang. Mining collective intelligence in diverse groups. In *Proceedings of the 22nd International Conference on World Wide Web*, pp. 1041–1052. International World Wide Web Conferences Steering Committee, Rio de Janeiro, Brazil, 2013.

36. Ragunathan R Rajkumar, Insup Lee, Lui Sha, and John Stankovic. Cyber-physical systems: The next computing revolution. In *Proceedings of the 47th Design Automation Conference*, pp. 731–736. ACM, New York, 2010.

37. Bobby S Rao and Hugh F Durrant-Whyte. Fully decentralized algorithm for multisensor Kalman filtering. *IEE Proceedings-D*, 138(5):413–420, 1991.

38. Murat Sensoy, Geeth de Mel, Lance Kaplan, Tien Pham, and Timothy J Norman. Tribe: Trust revision for information based on evidence. In *16th International Conference on Information Fusion (FUSION)*, pp. 914–921. IEEE, Istanbul, Turkey, 2013.

39. Lui Sha, Sathish Gopalakrishnan, Xue Liu, and Qixin Wang. Cyber-physical systems: A new frontier. In J.J.P. Tsai and P.S. Yu (eds.), *Machine Learning in Cyber Trust*, pp. 3–13. Springer, New York, 2009.

40. Glenn Shafer. *A Mathematical Theory of Evidence*. Princeton University Press, Princeton, NJ, 1976.

41. Florentin Smarandache and Jean Dezert, editors. *Advances and Applications of DSmT for Information Fusion: Collected Works*. Infinite Study, American Research Press, Rehoboth, NM, 2004.

42. WT Teacy, Jigar Patel, Nicholas R Jennings, and Michael Luck. TRAVOS: Trust and reputation in the context of inaccurate information sources. *Autonomous Agents and Multi-Agent Systems*, 12(2):183–198, March 2006.

43. Dong Wang. On quantifying the quality of information in social sensing. PhD thesis, University of Illinois at Urbana-Champaign, Champaign, IL, 2013.

44. Dong Wang, Tarek Abdelzaher, Hossein Ahmadi, Jeff Pasternack, Dan Roth, Manish Gupta, Jiawei Han, Omid Fatemieh, and Hieu Le. On bayesian interpretation of fact-finding in information networks. In *14th International Conference on Information Fusion*, pp. 1–8, Chicago, IL, 2011.

45. Dong Wang, Tarek Abdelzaher, and Lance Kaplan. Surrogate mobile sensing. *IEEE Communications Magazine*, 52(8):36–41, 2014.

46. Dong Wang, Tarek Abdelzaher, and Lance Kaplan. *Social Sensing: Building Reliable Systems on Unreliable Data*. Morgan Kaufmann, Waltham, MA, 2015.

47. Dong Wang, Tarek Abdelzaher, Lance Kaplan, and Charu C Aggarwal. Recursive fact-finding: A streaming approach to truth estimation in crowdsourcing applications. In *33rd International Conference on Distributed Computing Systems*, pp. 530–539, Philadelphia, PA, July 2013.

48. Dong Wang, Tarek Abdelzaher, Lance Kaplan, Raghu Ganti, Shaohan Hu, and Hengchang Liu. Exploitation of physical constraints for reliable social sensing. In *IEEE 34th Real-Time Systems Symposium*, pp. 212–223, Vancouver, Canada, 2013.

49. Dong Wang, Mohamed Tanvir Amin, Tarek Abedlzaher, Dan Roth, Clare Voss, Lance Kaplan, Stephen Tratz, Jamal Laoudi, and Douglas Briesch. Provenance-assisted classification in social networks. *IEEE Journal of Selected Topics in Signal Processing*, 8(4):624–637, 2014.

50. Dong Wang, Tanvir Amin, Shen Li, Tarek Abdelzaher, Lance Kaplan, Siyu Gu, Chenji Pan et al. Humans as sensors: An estimation theoretic perspective. In *13th ACM/IEEE International Conference on Information Processing in Sensor Networks*, pp. 35–46, Berlin, Germany, April 2014.

51. Dong Wang and Chao Huang. Confidence-aware truth estimation in social sensing applications. In *12th Annual IEEE Communications Society Conference on Sensor, Mesh and Ad Hoc Communications and Networks*, pp. 336–344, Seattle, WA, June 2015.

52. Dong Wang, Lance Kaplan, and Tarek Abdelzaher. Maximum likelihood analysis of conflicting observations in social sensing. *ACM Transactions on Sensor Networks*, 10(2):Article 30, January, 2014.

53. Dong Wang, Lance Kaplan, Tarek Abdelzaher, and Charu C Aggarwal. On scalability and robustness limitations of real and asymptotic confidence bounds in social sensing. In *9th Annual IEEE Communications Society Conference on Sensor, Mesh and Ad Hoc Communications and Networks*, pp. 506–514, Seoul, Korea, June 2012.

54. Dong Wang, Lance Kaplan, Tarek Abdelzaher, and Charu C Aggarwal. On credibility estimation tradeoffs in assured social sensing. *IEEE Journal on Selected Areas in Communications*, 31(6):1026–1037, 2013.

55. Dong Wang, Lance Kaplan, Hieu Le, and Tarek Abdelzaher. On truth discovery in social sensing: A maximum likelihood estimation approach. In *11th ACM/IEEE Conference on Information Processing in Sensor Networks*, pp. 233–244, Beijing, China, April 2012.

56. Yao Wang and Julita Vassileva. A review on trust and reputation for web service selection. In *27th International Conference on Distributed Computing Systems Workshops*, pp. 25–25. IEEE, Toronto, Canada, 2007.

57. Chien-Fu Jeff Wu. On the convergence properties of the EM algorithm. *The Annals of Statistics*, 11(1):95–103, 1983.

58. Li Xiong and Ling Liu. Peertrust: Supporting reputation-based trust for peer-to-peer electronic communities. *IEEE Transactions on Knowledge and Data Engineering*, 16(7):843–857, 2004.

59. Ronald R Yager, Janusz Kacprzyk, and Mario Fedrizzi, editors. *Advances in the Dempster-Shafer Theory of Evidence*. John Wiley & Sons, New York, 1994.

60. Xiaoxin Yin, Jiawei Han, and Philip S Yu. Truth discovery with multiple conflicting information providers on the web. *IEEE Transactions on Knowledge and Data Engineering*, 20:796–808, June 2008.

61. Bin Yu and Munindar P Singh. Detecting deception in reputation management. In *Proceedings of the 2nd International Joint Conference on Autonomous Agents and Multiagent Systems*, pp. 73–80. ACM, Melbourne, Australia, 2003.

62. Dian Yu, Hongzhao Huang, Taylor Cassidy, Heng Ji, Chi Wang, Shi Zhi, Jiawei Han, Clare Voss, and Malik Magdon-Ismail. The wisdom of minority: Unsupervised slot filling validation based on multi-dimensional truth-finding. In *25th International Conference on Computational Linguistics*, pp. 1567–1578, Dublin, Ireland, 2014.

63. Bo Zhao, Benjamin IP Rubinstein, Jim Gemmell, and Jiawei Han. A Bayesian approach to discovering truth from conflicting sources for data integration. *Proceedings of the VLDB Endowment*, 5(6):550–561, February 2012.

64. Weijun Zheng and Leigh Jin. Online reputation systems in web 2.0 era. In Matthew L. Nelson, Michael J. Shaw, and Troy J. Strader (eds.), *Value Creation in E-Business Management*, volume 36 of *Lecture Notes in Business Information Processing*, pp. 296–306. Springer, Berlin, Germany, 2009.

Chapter 6

Big Data and the SP Theory of Intelligence*

J. Gerard Wolff

CONTENTS

*© 2014 IEEE. Reprinted, with permission and with minor revisions, from "Big data and the SP theory of intelligence," J. G. Wolff, *IEEE Access*, 2, 301-315, 2014, DOI: 10.1109/ACCESS.2014.2315297, bit.ly/1FEk0LX.

Abstract

This chapter is about how the *SP theory of intelligence* and its realization in the *SP machine* may, with advantage, be applied to the management and analysis of big data. The SP system—introduced in the chapter and fully described elsewhere in this book—may help to overcome the problem of variety in big data: it has potential as a *universal framework for the representation and processing of diverse kinds of knowledge* (UFK), helping to reduce the diversity of formalisms and formats for knowledge and the different ways in which they are processed. It has strengths in the unsupervised learning or discovery of structure in data, in pattern recognition, in the parsing and production of natural language, in several kinds of reasoning, and more. It lends itself to the analysis of streaming data, helping to overcome the problem of velocity in big data. Central in the workings of the system is lossless compression of information: making big data smaller and reducing problems of storage and management. There is potential for substantial economies in the transmission of data, for big cuts in the use of energy in computing, for faster processing, and for smaller and lighter computers. The system provides a handle on the problem of veracity in big data, with potential to assist in the management of errors and uncertainties in data. It lends itself to the visualization of knowledge structures and inferential processes. A high-parallel, open source version of the SP machine would provide a means for researchers everywhere to explore what can be done with the system and to create new versions of it.

6.1 Introduction

Big data—the large volumes of data that are now produced in many fields—can present problems in storage, transmission, and processing, but their analysis may yield useful information and useful insights.

This chapter is about how the *SP theory of intelligence* and its realization in the *SP machine* (Section 6.2) may, with advantage, be applied to big data. Naturally, in an area like that, problems will not be solved in one step. The ideas described in this chapter provide a foundation and framework for further research (Section 6.12).

Problems associated with big data are reviewed quite fully in *Frontiers in Massive Data Analysis* [12] from the US National Research Council, and there is another useful perspective, from IBM, in *Smart Machines: IBM's Watson and the Era of Cognitive Computing* [9]. These and other sources are referenced at appropriate points throughout the chapter.

In broad terms, the potential benefits of the SP system, as applied to big data, are in the following areas:

- *Overcoming the problem of variety in big data*: Harmonizing diverse kinds of knowledge, diverse formats for knowledge, and their diverse modes of processing via a universal framework for the representation and processing of knowledge.

- *Learning and discovery*: The unsupervised learning or discovery of 'natural' structures in data.

- *Interpretation of data*: The SP system has strengths in areas such as pattern recognition, information retrieval, parsing and production of natural language, translation from one representation to another, several kinds of reasoning, planning, and problem solving.

- *Velocity—analysis of streaming data*: The SP system lends itself to an incremental style, assimilating information as it is received, much as people do.

- *Volume—making big data smaller*: Reducing the size of big data via lossless compression can yield direct benefits in the storage, management, and transmission of data, and indirect benefits in several of the other areas discussed in this chapter.

- *Additional economies in the transmission of data*: There is potential for additional economies in the transmission of data, potentially very substantial, by judicious separation of 'encoding' and 'grammar'.

- *Energy, speed, and bulk*: There is potential for big cuts in the use of energy in computing, for greater speed of processing with a given computational resource, and for corresponding reductions in the size and weight of computers.

- *Veracity—managing errors and uncertainties in data*: The SP system can identify possible errors or uncertainties in data, suggest possible corrections or interpolations, and calculate associated probabilities.

- *Visualization*: Knowledge structures created by the system, and inferential processes in the system, are all transparent and open to inspection. They lend themselves to display with static and moving images.

These topics will be discussed, each in its own section, below. But first, the SP theory and SP machine will be introduced.

6.2   Introduction to the SP Theory and SP Machine

The SP theory, which has been under development for several years, aims to simplify and integrate observations and concepts across artificial intelligence, mainstream computing, mathematics, and human perception and cognition, with information compression as a unifying theme.

The theory is conceived as an abstract brain-like system that, in an 'input' perspective, may receive *New* information via its senses, and compress some or all of it to create *Old* information, as illustrated schematically in Figure 6.1. In the theory, information compression is the mechanism both for the learning and organization of knowledge and for pattern recognition, reasoning, problem solving, and more.

In the SP system, all kinds of knowledge are represented with *patterns*: arrays of atomic symbols in one or two dimensions.

At the heart of the system are processes for compressing information by finding good full and partial matches between patterns and merging or 'unifying' parts that are the same. More specifically, all processing is done via the creation of *multiple alignments*, like the one shown in Figure 6.2.*

The close association between information compression and concepts of prediction and probability [11] means that the SP system is intrinsically probabilistic. Each SP pattern has an associated frequency of occurrence, and for each multiple alignment, the system may calculate associated probabilities [17, Section 3.7] (reproduced in [20, Section 4.4]). Although the SP system is fundamentally probabilistic, it can, if required, be constrained to operate in the clockwork style of a conventional computer, delivering all-or-nothing results [17, Chapter 10]. There is also potential to increase the reliability of decisions by the use of error-reducing redundancy. Although, superficially, that may seem to conflict with the notion of *computing as compression*, the two things are independent, as described in [17, Section 2.3.7].

An important idea in the SP program is the *DONSVIC* principle [20, Section 5.2]: the conjecture, supported by evidence, that information compression, properly applied, is the key to the discovery of 'natural' structures, meaning the kinds of things that people naturally

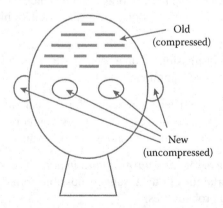

Figure 6.1: Schematic representation of the SP system from an 'input' perspective. (Reproduced from Wolff, J.G., *Information*, 4, 283, 2013, Figure 1, with permission.)

---

*The concept of multiple alignment in the SP system [20, Section 4] and [17, Section 3.4] is borrowed from that concept in bioinformatics, but with important differences.

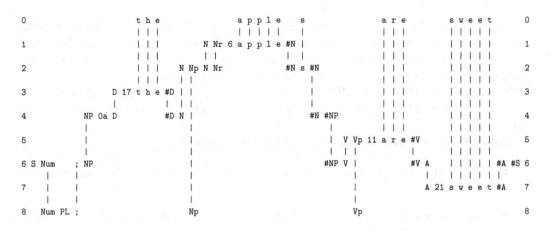

Figure 6.2: A multiple alignment created by the SP computer model that achieves the effect of parsing a sentence ('t h e a p p l e s a r e s w e e t'). (Reproduced from Wolff, J.G., *Information*, 5, 1, 2014, Figure 1, with permission.)

recognize, such as words, objects, and classes of objects. Evidence to date suggests that the SP system does indeed conform to that principle.

The SP theory is realized in a computer model, SP70, which may be regarded as a first version of the SP machine.* It is envisaged that the SP computer model will provide the basis for the development of a high-parallel, open source version of the SP machine, as described in Section 6.12.

The theory has things to say about several aspects of computing and cognition, including unsupervised learning, concepts of computing, aspects of mathematics and logic, the representation of knowledge, natural language processing, pattern recognition, several kinds of reasoning, information storage and retrieval, planning and problem solving, and aspects of neuroscience and of human perception and cognition.

There is a relatively full account of the SP system in [17], an extended overview in [20], an account of its existing and expected benefits and applications in [24], a description of its foundations in [23], and an introduction to the system in [19]. More information may be found via www.cognitionresearch.org/sp.htm.

6.3 Overcoming the Problem of Variety in Big Data

The manipulation and integration of heterogeneous data from different sources into a meaningful common representation is a major challenge. [12, p. 76]

Over the past decade or so, computer scientists and mathematicians have become quite proficient at handling specific types of data by using specialized tools that do one thing very well. ... But that approach doesn't work for complex operational challenges such as managing cities, global supply chains, or power grids, where many interdependencies exist and many different kinds of data have to be taken into consideration. [9, p. 48]

---

*The latest version, SP71, which may be downloaded via a link near the bottom of www.cognitionresearch.org/sp.htm, has only minor changes compared with SP70.

The many different kinds of data include: the world's many languages, spoken or written; static and moving images; music as sound and music in its written form; numbers and mathematical notations; tables; charts; graphs; networks; trees; grammars; computer programs; and more. With many of these kinds of data, there are several different computer-based formats, such as with static images: JPEG, TIFF, WMF, BMP, GIF, EPS, PDF, PNG, PBM, and more. And, normally, each kind of data, and each different format, needs to be processed in its own special way.

Some of this diversity is necessary and useful. For example,

■ The cultural life of a community is often intimately connected with the language of that community.

■ Notwithstanding the dictum that "A picture is worth a thousand words," natural languages, collectively, have special strengths.

■ Ancient texts are of interest for historical, cultural, and other reasons.

■ With techniques and technologies as they have developed to date, it often makes sense to use different formalisms or formats for different purposes.

■ Overzealous standardization may stifle creativity.

Nevertheless, there are several reasons, described in the next subsection, for trying to develop a *universal framework for the representation and processing of diverse kinds of knowledge* (UFK). Such a system may help to reduce unnecessary diversity in formalisms and formats for knowledge and in their modes of processing. But it is likely that many existing systems would continue in use for the kinds of reasons mentioned above, perhaps with translations into UFK form, if or when that proves necessary.

6.3.1 Reasons for developing a universal framework for the representation and processing of knowledge

Of the reasons described here for developing a UFK, some relate fairly directly to issues with big data (Sections 6.3.1.1, 6.3.1.2, and 6.3.1.4), while the rest draw on other aspects of computing, engineering, and biology.

6.3.1.1 Discovery of structure in data

If we are trying to discover patterns of association or other structures in big data (Section 6.4), a diversity of formalisms and formats is a handicap. Let us imagine, for example, how an artificial learning system might discover the association between lightning and thunder. Detecting that association is likely to be difficult if

■ Lightning appears in big data as a static image in one of several formats, like those mentioned above; or in a moving image in one of several formats; or it is described, in spoken or written form, as any one of such things as *firebolt, fulmination, la foudre, der Blitz, lluched, a big flash in the sky*, or indeed *lightning*.

■ Thunder is represented in one of several different audio formats; or it is described, in spoken or written form, as *thunder, gök gürültüsü, le tonnerre, a great rumble*, and so on.

The association between lightning and thunder will be most easily detected via the underlying meanings of the forms that have been mentioned. We may suppose that, at some level, knowledge about lightning has an associated code or identifier, something like 'LTNG', and that knowledge about thunder has a code or identifier such as 'THDR'. Encodings like those would cut through much of the complexity of surface forms and allow underlying associations, such as 'LTNG THDR', to show through.

It seems likely that at least part of the reason that people find it relatively easy to recognize, without being told, that there is an association between lightning and thunder is that, in our brains, there is some uniformity in the way different kinds of knowledge are represented and processed, without awkward inconsistencies (Section 6.3.1.7).

6.3.1.2 Interpretation of data

If we are trying to recognize objects in images, do scene analysis, or otherwise interpret what the images mean, it would make things simpler if we did not have to deal with the diversity of formats for images mentioned earlier. Likewise for other kinds of data.

6.3.1.3 Data fusion

In many fields, there is often a need to combine diverse sources of information to create a coherent whole. For example, in a study of the migration of whales, we may have, for each animal, a stream of information about the temperature of the water at each point along its route, another stream of information about the depths at which the animal is swimming, information about the weather at the surface at each point, information about dates and times, and so on.

If we are to weld those streams of information together, it would not be helpful if the geographical coordinates for different streams of information were to be expressed in different ways, perhaps using the Greenwich meridian for temperatures, the Paris meridian for depths, the Universal Transverse Mercator (UTM) system for weather, and some other scheme for the dates and times.

In short, there is a clear need to adopt a uniform system for representing the data—geographical coordinates in this example—that are needed to fuse separate but related streams of information to create a coherent view.

6.3.1.4 Understanding and translation of natural languages

In our everyday use of natural languages we recognize that meanings are different from the words that express them and that, very often, two or more distinct sequences of words may mean the same thing or have the same referent: *the capital of the United States* means the same as *Washington, D. C.*; *Ursus maritimus* means the same as *polar bear*; and so on. These intuitions corroborate the need for a UFK, or something like it, which is independent of the words in any natural language.

Again, it is widely recognized that, if machine translation of natural languages is ever to reach the standard of good human translators, it will be necessary to provide some kind of *interlingua*—an abstract language-independent representation—to express the meaning of the source language and to serve as a bridge between the source language and the target language.* Any such interlingua is likely to be similar to or the same as a UFK.

---

*See, for example, "Interlingual machine translation," Wikipedia, bit.ly/1mCDTs3, retrieved January 24, 2014.

6.3.1.5   *The* semantic web, *the* internet of things, *and the* web of entities

The need for standardization in the representation of knowledge is recognized in writings about the *semantic web* (e.g., [3]), the *internet of things* (e.g., [6]), and in the Okkam project, aiming to create unique identifiers for a global *web of entities*.*

6.3.1.6   *Long-term preservation of data*

The continual creation of new formalisms and new formats for information and their subsequent obsolescence can mean that old data, which may include data of great value, may become unreadable or otherwise unusable. A UFK would help to reduce or eliminate this problem.

6.3.1.7   *Knowledge, brains, and autonomous robots*

In keeping with the long tradition in engineering of borrowing ideas from biology, the structure and functioning of brains provide reasons for trying to develop a UFK:

- Since brains are composed largely of neural tissue, it appears that neurons and their inter-connections, with glial cells, provide a universal framework for the representation and processing of all kinds of sensory data and all other kinds of knowledge.

- In support of that view is evidence that one part of the brain can take over the functions of another part (see, e.g., [1,2]). This implies that there are some general principles operating across several parts of the brain, perhaps all of them.

- Most concepts are an amalgam of several different kinds of data or knowledge. For example, the concept of a *picnic* combines the sights, sounds, tactile and gustatory sensations, and the social and logistical knowledge associated with such things as a light meal in pleasant rural surroundings. To achieve that kind of seamless integration of different kinds of knowledge, it seems necessary for the human brain to be or to contain a UFK.

- In a similar way, human versatility in intelligence seems to demand the seamless integration of diverse aspects of intelligence—unsupervised learning, pattern recognition, natural language processing, different kinds of reasoning, and so on—and that seems to demand a UFK as the vehicle for the representation and processing of diverse kinds of knowledge.

Those last two points also apply to autonomous robots that aspire to human-like versatility in intelligence, as discussed in [22, Section IV-A].

6.3.2   Potential of the SP system as a universal framework for the representation and processing of knowledge

In the SP program, the aim has been to create a system that, in accordance with Occam's Razor, combines conceptual *simplicity* with descriptive or explanatory *power* [17, Section 1.3; 24, Section 2]. Although the SP computer model is relatively simple—its *exec* file requires less than 500 KB of storage space—and despite the great simplicity of SP patterns as a vehicle for

---

*See "Okkam: Enabling the Web of Entities. A scalable and sustainable solution for systematic and global identifier reuse in decentralized information environments," project reference: 215032, completed: 2010-06-30, bit.ly/OSjc1b, retrieved March 24, 2014.

knowledge (Section 6.2), the SP system, without additional programming, may serve in the representation and processing of several different kinds of knowledge, as listed below:

■ *Syntax and semantics of natural languages*: The system provides for the representation of syntactic rules, including discontinuous dependencies in syntax, and for the parsing and production of language [17, Chapter 5; 20, Section 8]. It has potential to represent non-syntactic 'meanings' via such things as class hierarchies and part-whole hierarchies (next), and it has potential in the understanding of natural language and in the production of sentences from meanings [17, Section 5.7].

■ *Class hierarchies and part-whole hierarchies*: The system lends itself to the representation of class hierarchies (species, genus, family, etc.), heterarchies (class hierarchies with cross-classification), and part-whole hierarchies (e.g., [[head [eyes, nose, mouth, ...]], [body ...], [legs ...]]) and their processing in pattern recognition, reasoning, and more [17, Sections 6.4.1 and 6.4.2; 20, Section 9.1].

■ *Networks and trees*: The SP system supports the representation and processing of such things as hierarchical and network models for databases [18, Section 5], and probabilistic decision networks and decision trees [17, Section 7.5]. And it has advantages as an alternative to Bayesian networks [17, Section 7.8] (reproduced in [20, Sections 10.2 through 10.4]).

■ *Relational knowledge*: The system supports the representation of knowledge with relational tuples, and retrieval of information in the manner of query-by-example [18, Section 3], and it has some apparent advantages compared with the relational model [18, Section 4.2.3].

■ *Rules and reasoning*: The system supports several kinds of reasoning, with the representation of associated knowledge. Examples include one-step 'deductive' reasoning, abductive reasoning, chains of reasoning, reasoning with rules, nonmonotonic reasoning, and causal diagnosis [17, Chapter 7].

■ *Patterns and pattern recognition*: The SP system has strengths in the representation and processing of one-dimensional patterns [17, Chapter 6; 24, Section 9], and it may be applied to medical diagnosis, viewed as a pattern recognition problem [16].

■ *Images*: Although the SP computer model has not yet been generalized to work with patterns in two dimensions, there is clear potential for the SP system to be applied to the representation and processing of images and other kinds of information with a 2D form. This is discussed in [17, Section 13.2.1] and also in [21].

■ *Structures in three dimensions*: It appears that the multiple alignment framework may be applied to the representation and processing of 3D structures via the stitching together of overlapping 2D views [21, Section 7.1], in much the same way that 3D models may be created from overlapping 2D photos,* or a panoramic photo may be created from overlapping shots.

■ *Procedural knowledge*: The SP system can represent simple procedures (actions that need to be performed in a particular sequence); it can model such things as 'variables',

---

*See, for example, "Big Object Base" (bit.ly/1gwuIfa), "Camera 3D" (bit.ly/1iSEqZu), or "PhotoModeler" (bit.ly/MDj70X).

'values', 'types', 'function with parameters', repetition of operations, and more [24, Section 6.6.1]; and it has potential to represent sets of procedures that may be performed in parallel [24, Section 6.6.3]. These representations may serve to control real-world operations in sequence and in parallel.

As a candidate for the role of UFK, the SP system has other strengths:

■ Because of the generality of the concept of information compression via the matching and unification of patterns, there is reason to believe that the system may be applied to the representation and processing of all kinds of knowledge, not just those listed above.

■ Because all kinds of knowledge are represented in one simple format (arrays of atomic symbols in one or two dimensions), and because all kinds of knowledge are processed in the same way (via the creation of multiple alignments), the system provides for the seamless integration of diverse kinds of knowledge, in any combination [24, Section 7].

■ Because of the system's existing and potential capabilities in learning and discovery (Section 6.4), it has potential for the automatic structuring of knowledge, reducing or eliminating the need for handcrafting, with corresponding benefits in terms of speed, cost, and reducing errors.

■ For reasons given in Section 6.11, the SP system may facilitate the visualization of structures and processes via static and moving images.

In summary, the relative simplicity of the SP system, its versatility in the representation and processing of diverse kinds of knowledge, its provision for seamless integration of different kinds of knowledge in any combination, the system's potential for automatic structuring of knowledge, and for the visualization of structures and processes, makes it a good candidate for development into a UFK.

6.3.3   Standardization and translation

The SP system, or any other UFK, may be used in two distinct ways:

■ *Standardization in the representation of knowledge*: There is potential, on relatively long timescales, to standardize the representation and processing of many kinds of knowledge, cutting out much of the current jumble of formalisms and formats. But for the kinds of reasons mentioned in Section 6.3, it is likely that some of those formalisms or formats will never be replaced or will co-exist with representation and processing via the UFK.

■ *Translation into the universal framework*: Where a body of information is expressed in one or more non-standard forms but is needed in the standard form, it may be translated. This may be done via the SP system, as outlined in Section 6.5. Or it may be done using conventional technologies, in much the same way that the source code for a computer program may, using a compiler, be translated into object code. The translation of natural languages is likely to prove more challenging than the translation of artificial formalisms and formats.

Either way, any body of big data may be expressed in a standard form that facilitates the unsupervised learning or discovery of structures and associations within those data (Section 6.4), and facilitates forms of interpretation as outlined in Section 6.5.

6.4   Learning and Discovery

> While traditional computers must be programmed by humans to perform specific
> tasks, cognitive systems will learn from their interactions with data and humans
> and be able to, in a sense, program themselves to perform new tasks. [9, p. 7]

In broad terms, unsupervised learning in the SP system means lossless compression of a body
of information, **I**, via the matching and unification of patterns (Section 6.4.1).

The SP computer model (SP70, [17, Chapter 9; 20, Section 5]) has already demonstrated an
ability to discover generative grammars from unsegmented samples of English-like artificial
languages, including segmental structures, classes of structure, and abstract patterns. As it
is now, it has shortcomings, outlined in [20, Section 3.3]. But I believe these problems are
soluble, and that their solution will clear the path to the unsupervised learning of other kinds of
structures, such as class hierarchies, part-whole hierarchies, and discontinuous dependencies in
data. In what follows, we shall assume that these and other problems have been solved and that
the system is relatively robust and mature. A strength of the SP system is that it can discover
structures in data, not merely statistical associations between pre-established structures.

As noted in Section 6.2, evidence to date suggests that the system conforms to the
DONSVIC principle [20, Section 5.2].

6.4.1   The product of learning

The product of learning from a body of data, **I**, may be seen to comprise a *grammar* (**G**) and an
encoding (**E**) of **I** in terms of the grammar. Here, the term 'grammar' has a broad meaning that
includes grammars for natural languages, grammars for static and moving images, grammars
for business procedures, and so on.

As with all other kinds of data in the SP system, **G** and **E** are both represented using SP patterns. In accordance with the principles of *minimum length encoding* [14], the SP system aims
to minimize the overall size of **G** and **E**.* Together, **G** and **E** achieve lossless compression of **I**.

G is largely about *redundancies* within **I**, while **E** is mainly a record of the non-redundant
aspects of **I**. Here, any symbol or sequence of symbols represents redundancy within **I** if it
repeats more often than one would expect by chance. To reach that threshold, small patterns
need to occur more frequently than large patterns.†

G may be regarded as a distillation of the 'essence' of **I**. Normally, **G** would be more
interesting than **E**, and more useful in the kinds of applications described in Sections 6.5 and
6.10.2.

With data that is received or processed as a stream rather than a static body of data of fixed
size (Section 6.6), **G** may be grown incrementally. And, with many instances of **I**, there is
likely to be a case for merging **G**s from different sources, with unification of patterns that are
the same. In principle, there could be a single 'super' **G**, expressing the essentials of the world's
knowledge in a compressed form. Similar remarks apply to **E**s—if they are needed.

---

*The similarity with research on grammatical inference is not accidental: the SP program of research has grown out of
earlier research developing computer models of language learning (see [15] and other publications that may be downloaded
via bit.ly/JCd6jm). But in developing the SP system, a radical reorganization has been needed to meet the goal of simplifying
and integrating concepts across artificial intelligence, mainstream computing, and human perception and cognition. Unlike
the earlier models and other research on grammatical inference, multiple alignment is central in the workings of the SP
computer model, including unsupervised learning. A bonus of the new structure is potential for the unsupervised learning of
such things as class hierarchies, part-whole hierarchies, and discontinuous dependencies in data.

†**G** may contain some patterns that do not, in themselves, represent redundancy but are included in **G** because of their
supporting role [17, Section 3.6.2].

6.4.2 Unsupervised learning and the problem of variety in big data

Systems for unsupervised learning may be applied most simply and directly when the data for learning come in a uniform style as, for example, in DNA data: simple sequences of the letters A, T, G, and C. But as outlined in Section 6.3.1.1, it may be difficult to discover recurrent associations or structures when there is a variety of formalisms and formats.

The discussion here focuses on the relatively challenging area of natural languages, because the variety of natural languages is a significant part of the problem of variety in big data, because the SP system has strengths in that area, and because it seems likely that solutions with natural languages will generalize relatively easily to other areas.

With natural languages, learning processes in a mature version of the SP system may be applied in four distinct but inter-related ways, discussed in the following subsections.

6.4.2.1 Learning the surface forms of language

If the data for learning are text in a natural language, then the product of learning (Section 6.4.1) will be about words and parts of words; about phrases, clauses, and sentences; and about grammatical categories at all levels. Likewise with speech.

Even with human-like capabilities in learning, a **G** that is derived without the benefit of meanings is likely to differ in some respects from a grammar created by a linguist who can understand what the text means. This is because there are subtle interdependencies between syntax and semantics [24, Section 6.2] which cannot be captured from text on its own, without information about meanings.

6.4.2.2 Learning non-syntactic knowledge

The SP system may be applied to learning about the non-syntactic world: objects and their interactions, scenery, music, games, and so on. These have an intrinsic interest that is independent of natural language, but they are also the things that people talk and write about: the non-syntactic meanings or semantics of natural language. Some aspects of this area of learning are discussed in [17, Section 13.2.1; 21].

6.4.2.3 Connecting syntax with semantics

Of course, for any natural language to be effective, syntax must connect with semantics. Examples that show how syntax and semantics may work together in the multiple alignment framework, in both the analysis and production of language, are presented in [17, Section 5.7]. As noted in Section 6.3.2, seamless integration of different kinds of knowledge is facilitated by the use of one simple format for all kinds of knowledge and a single framework—multiple alignment—for processing diverse kinds of knowledge.

In broad terms, making the connection between syntax and semantics means associational learning, no different in principle from learning the association between lightning and thunder (Section 6.3.1.1), between smoke and fire, between a savory aroma and a delicious meal, and so on.

For the SP system to learn the connections between syntax and semantics, it will need speech or text to be presented alongside the non-syntactic information that it relates to, in much the same way that, normally, children have many opportunities to hear people talking and see what they are talking about at the same time.

6.4.2.4 *Learning via the interpretation of surface forms*

Since speech and text in natural languages are an important part of big data, it is clear that if the SP system, or any other learning system, is to get full value from big data, it will need to learn from the meanings of speech or text as well as from their surface forms.

For any given body of text or speech, **I**, the first step, of course, will be to derive its meanings. This can be done via processes of interpretation, as described in Section 6.5. The set of SP patterns that represent the meanings of **I** may then be processed as if it was new information, searching for redundancies in the data, unifying patterns that match each other, and creating a compressed representation of the data. Then it should be possible to discover such things as the association between lightning and thunder (Section 6.3.1.1), regardless of how the data was originally presented.

6.5 Interpretation of Data

By contrast with unsupervised learning, which compresses a body of information (**I**) to create **G** and **E**, the concept of *interpretation* in this chapter means processing **I** in conjunction with a pre-established grammar (**G**) to create a relatively compact encoding (**E**) of **I**.

Depending on the nature of **I** and **G**, the process of interpretation may be seen to achieve such things as pattern recognition, information retrieval, parsing or production of natural language, translation from one representation to another, several kinds of reasoning, planning, and problem solving. Some of these were touched on briefly in Section 6.3.2. Here is a little more detail:

- ■ *Pattern recognition*: With the SP system, pattern recognition may be achieved: at multiple levels of abstraction; with *family resemblance* or *polythetic* categories; in the face of errors of omission, commission, or substitution in data; with the calculation of a probability for any given identification, classification, or associated inference; with sensitivity to context in recognition; and with the seamless integration of pattern recognition with other aspects of intelligence—reasoning, learning, problem solving, and so on [20, Section 9; 17, Chapter 6]. As previously mentioned, the system may be applied in computer vision [21] and in medical diagnosis [16], viewed as pattern recognition.

- ■ *Information retrieval*: The SP system lends itself to information retrieval in the manner of query-by-example and, with the provision of SP patterns representing relevant rules, there is potential to create the facilities of a query language like SQL [18].

- ■ *Parsing and production of natural language*: As can be seen in Figure 6.2, the creation of a multiple alignment in the SP system may achieve the effect of parsing a sentence in a natural language (see also [17, Section 3.4.3 and Chapter 5]). It may also function in the production of sentences [17, Section 3.8].

- ■ *Translation from one representation to another*: There is potential with the SP system for the integration of syntax and semantics in both the understanding and production of natural language [17, Section 5.7], with corresponding potential for the translation of any one language into an SP-style interlingua and further translation into any other natural language [24, Section 6.2.1]. Probably less challenging, as mentioned earlier, would be the translation of artificial formalisms and formats—JPEG, MP3, and so on— into an SP-style representation.

■ *Several kinds of reasoning*: The SP system can perform several kinds of reasoning, including one-step 'deductive' reasoning, abductive reasoning, reasoning with probabilistic networks and trees, reasoning with 'rules', nonmonotonic reasoning, Bayesian reasoning and *explaining away*, causal diagnosis, and reasoning that is not supported by evidence [17, Chapter 7].

■ *Planning*: With SP patterns representing direct flights between cities, the SP system can normally work out one or more routes between any two cities that are not connected directly, if such a route exists [17, Section 8.2].

■ *Problem solving*: The system can also solve textual versions of geometric analogy problems, like those found in puzzle books and IQ tests [17, Section 8.3].

6.6 Velocity: Analysis of Streaming Data

Most of today's computing tasks involve data that have been gathered and stored in databases. The data make a stationary target. But, increasingly, vitally important insights can be gained from analyzing information that's on the move. ... This approach is called streams analytics. Rather than placing data in a database first, the computer analyses it as it comes from a variety of sources, continually refining its understanding of the data as conditions change. This is the way humans process information. [9, pp. 49–50]

Although, in its unsupervised learning, the SP system may process information in batches, it lends itself most naturally to an incremental style. In the spirit of the quotation above, the SP system is designed to assimilate *New* information to a steadily-growing body of relatively-compressed *Old* information, as shown schematically in Figure 6.1.

Likewise, in interpretive processes such as pattern recognition, processing of natural language, and reasoning, the SP system may be applied to streams of data as well as the processing of data in batches.

6.7 Volume: Making Big Data Smaller

Very-large-scale data sets introduce many data management challenges. [12, p. 41]

In addition to reducing computation time, proper data representations can also reduce the amount of required storage (which translates into reduced communication if the data are transmitted over a network). [12, p. 68]

Because information compression is central in how the SP system works, it has potential to reduce problems of volume in big data by making it smaller. Although comparative studies have not yet been attempted, the SP system has potential to achieve relatively high levels of lossless compression for two main reasons:

■ It is designed so that, if required, it can perform a relatively thorough search for redundancies in data, in accordance with the 'DONSVIC' principle—the discovery of 'natural' structures via information compression [20, Section 5.2].

- There is potential to tap into discontinuous dependencies in data, an aspect of redundancy that appears to be outside the scope of other systems for compression of information [24, Section 6.7].

More generally, information compression in the SP system can yield *direct benefits* in the storage, management, and transmission of data, and *indirect benefits* as described elsewhere in this chapter: unsupervised learning (Section 6.4), processes of interpretation such as pattern recognition and reasoning (Section 6.5), economies in the transmission of data via the separation of grammar and encoding (Section 6.8), gains in computational efficiency (Section 6.9), and assistance in the management of errors and uncertainties in data (Section 6.10).

6.8   Additional Economies in the Transmission of Data

> One roadblock to using cloud services for massive data analysis is the problem of transferring the large data sets. Maintaining a high-capacity and wide-scale communications network is very expensive and only marginally profitable. [12, p. 55]

> To control costs, designers of the [DOME] computing system have to figure out how to minimize the amount of energy used for processing data. At the same time, since so much of the energy in computing is required to move data around, they have to discover ways to move the data as little as possible. [9, p. 65]

Although the second of these two quotes may refer in part to movements of data such as those between the CPU and the memory of a computer, the discussion here is about transmission of data over longer distances such as, for example, via the internet.

As we have seen (Section 6.7), the SP system may promote the efficient transmission of data by making it smaller. But there is potential with the SP system for additional economies in the transmission of data and these may be very substantial [24, Section 6.7.1].

What is described here is essentially an analysis/synthesis scheme as described in [13, Chapter 18]. But, in this connection, the SP system has some potential advantages as outlined below.

Any body of data, **I**, may be compressed by encoding it in terms a grammar (**G**), provided that **G** contains the kinds of structures that are found in **I** (Section 6.4.1). Then **I** may be sent from A to B by sending only the encoding (**E**). Provided that B has a copy of **G**, **I** may be recreated with complete fidelity by means of the SP system [20, Section 4.5; 17, Section 3.8]. Since **E** would normally be very much smaller than the **I** from which it was derived, it seems likely that there would be a net gain in efficiency, allowing for the computational costs of encoding and decoding.

Since a copy of **G** must be transmitted to B, any savings will be relatively small if it is used only for the decoding of a single instance of **E**. But significant savings are likely if, as would normally be the case, one copy of **G** may be used for the decoding of many different instances of **E**, representing many different **I**s.

In this kind of application, it would probably make sense for there to be a division of labor between creating a grammar and using it in the encoding and decoding of data. For example, the computational heavy lifting required to build a grammar for video images may be done by a high-performance computer. But that grammar, once constructed, may serve in relatively low-powered devices—smartphones, tablet computers, and the like—for the much less demanding processes of encoding and decoding video transmissions.

As part of an analysis/synthesis scheme, the SP system offers these potential benefits:

■ As noted in Section 6.7, the system has potential to achieve relatively high levels of information compression: (1) because it can, when required, perform a relatively thorough search for redundancies in data and (2) because, unlike most other systems for compression of information, it has potential to encode discontinuous dependencies in data.

■ When the SP system has been generalized to work with 2D patterns [17, Section 13.2.1], and the creation of 3D digital models [21, Sections 6.1 and 6.2], it may, potentially, be applied to model-based encoding of video data, as outlined in [13, Section 19.6].

6.9 Energy, Speed, and Bulk

... we're reaching the limits of our ability to make [gains in the capabilities of CPUs] at a time when we need even more computing power to deal with complexity and big data. And that's putting unbearable demands on today's computing technologies—mainly because today's computers require so much energy to perform their work. [9, p. 9]

The human brain is a marvel. A mere 20 W of energy are required to power the 22 billion neurons in a brain that's roughly the size of a grapefruit. To field a conventional computer with comparable cognitive capacity would require gigawatts of electricity and a machine the size of a football field. So, clearly, something has to change fundamentally in computing for sensing machines to help us make use of the millions of hours of video, billions of photographs, and countless sensory signals that surround us. ... Unless we can make computers many orders of magnitude more energy efficient, we're not going to be able to use them extensively as our intelligent assistants. [9, pp. 75, 88]

Supercomputers are notorious for consuming a significant amount of electricity, but a less-known fact is that supercomputers are also extremely 'thirsty' and consume a huge amount of water to cool down servers through cooling towers ....*

It is now clear that, if we are to do meaningful analyses of more than a small fraction of present and future floods of big data, substantial gains will be needed in the computational efficiency of computers, with associated benefits such as the following:

■ Cutting demands for energy, with corresponding cuts in the need for cooling of computers.

■ Speeding up processing with a given computational resource.

■ Consequent reductions in the size and weight of computers.

With the possible exception of the need for cooling, these things are particularly relevant to mobile devices, including autonomous robots.

---

*From "How can supercomputers survive a drought?," *HPC Wire*, bit.ly/LruEPS, retrieved January 26, 2014.

Subsections 6.9.1 and 6.9.2 describe how the SP system may contribute to computational efficiency, via information compression and probabilities, and via a synergy with 'data-centric' computing.

6.9.1 Computational efficiency via information compression and probabilities

In the light of evidence that the SP system is Turing-equivalent [17, Chapter 4], and since information processing in the SP system means compression of information via the matching and unification of patterns (Section 6.2), *anything that increases the efficiency of searching for good full and partial matches between patterns will also increase the efficiency of information processing.*

It appears that information compression and associated probabilities can themselves be a means of increasing the efficiency of searching, as described in Subsections 6.9.1.1 and 6.9.1.2.

6.9.1.1 Reducing the sizes of data to be searched and of search terms

As described in [24, Section 6.7.2], if we wish to search a body of information, **I**, for instances of a pattern like *Treaty on the Functioning of the European Union*, the efficiency of searching may be increased:

- By reducing the size of **I** so that there is less to be searched. The size of **I** may be reduced by replacing all but one of the instances of *Treaty on the Functioning of the European Union* with a relatively short code or identifier like *TFEU*, and likewise with other recurrent patterns. More generally, the size of **I** may be reduced via unsupervised learning in the SP system. It is true that the compression of **I** would give rise to a computational cost, but this investment is likely to pay off in later processing.

- By searching with a short code like *TFEU* instead of a relatively large pattern like *Treaty on the Functioning of the European Union*. Other things being equal, a smaller search pattern means faster searching.

- With regard to the previous point, there is potential to cut out some searching altogether by creating direct connections between each instance of a code (*TFEU* in this example) and the thing that it represents (*Treaty on the Functioning of the European Union*). In *SP-neural* (Section 6.9.2.1), there are connections of that kind between *pattern assemblies*, as shown schematically in Figure 6.3 (see later in the chapter [22, Section III-C.2]).

6.9.1.2 Concentrating search where good results are most likely to be found

If we want to find some strawberry jam, our search is more likely to be successful in a supermarket than it would be in a hardware shop or a car-sales showroom. This may seem too simple and obvious to deserve comment but it illustrates the extraordinary knowledge that most people have of an informal 'statistics' of the world that we inhabit, and how that knowledge may help us to minimize effort.*

---

*See also G. K. Zipf [25].

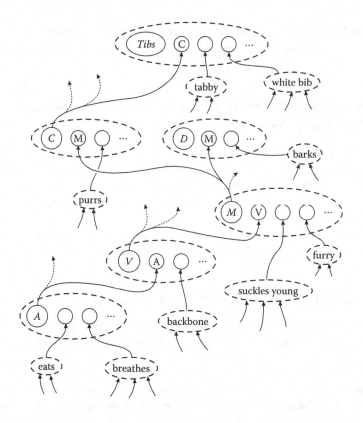

Figure 6.3: An example showing schematically how SP-neural may represent class-inclusion relations, part-whole relations, and their integration. *Key*: 'C' = cat, 'D' = dog, 'M' = mammal, 'V' = vertebrate, 'A' = animal, '...' = further structure that would be shown in a more comprehensive example. Pattern assemblies are surrounded by broken lines and each neuron is represented by an unbroken circle or ellipse. Lines with arrows show connections between pattern assemblies and the flow of sensory signals in the process of recognizing something (there may also be connections in the opposite direction to support the production of patterns). Connections between neurons within each pattern assembly are not marked. (Reproduced from Wolff, J.G., *Unifying Computing and Cognition: The SP Theory and Its Applications*, CognitionResearch.org, Menai Bridge, 2006, Figure 11.6, with permission.)

Where does that statistical knowledge come from? In the SP theory, it flows directly from the central role of information compression in our perceptions, learning and thinking, and from the intimate relationship between information compression and concepts of prediction and probability [11].

Although the SP computer model may calculate probabilities associated with multiple alignments (Section 6.2), it actually uses levels of information compression as a guide to search. Those levels are used, with heuristic search methods (including escape from 'local peaks'), to ensure that searching is concentrated in areas where it is most likely to be fruitful [17, Sections 3.9, 3.10, and 9.2]. This not only speeds up processing but yields Big-O values for computational complexity that are within acceptable limits [17, Sections 3.10.6, 9.3.1, and A.4].

6.9.1.3   *Potential gains in computational efficiency*

No attempt has yet been made to quantify potential gains in computational efficiency from the compression of information, as described in Sections 6.9.1.1, 6.9.1.2, and 6.8, but they could be very substantial:

- Since information compression is fundamental in the workings of the SP system, there is potential for corresponding savings in all parts and levels in the system.

- The entire structure of knowledge that the system creates for itself is intrinsically statistical, with potential on many fronts for corresponding savings in computational costs and associated demands for energy.

It may be argued that, since object-oriented programming already provides for compression of information via class hierarchies and inheritance of attributes, the benefits of information compression are already available in conventional computing systems. In response, it may be said that, while there are undoubted benefits from object-oriented programming, existing object-oriented systems run on conventional computers and suffer from the associated inefficiencies.

Realizing the full potential of information compression as a means of improving computational efficiency will probably mean new thinking about computer architectures, probably in conjunction with the development of data-centric computing (next).

6.9.2   *A potential synergy with data-centric computing*

What's needed is a new architecture for computing, one that takes more inspiration from the human brain. Data processing should be distributed throughout the computing system rather than concentrated in a CPU. The processing and the memory should be closely integrated to reduce the shuttling of data and instructions back and forth. [9, p. 9]

Unless we can make computers many orders of magnitude more energy efficient, we're not going to be able to use them extensively as our intelligent assistants. Computing intelligence will be too costly to be practical. Scientists at IBM Research believe that to make computing sustainable in the era of big data, we will need a different kind of machine—the data-centric computer. ... Machines will perform computations faster, make sense of large amounts of data, and be more energy efficient. [9, p. 88]

The SP concepts may help to integrate processing and memory, as described in the next two subsections.

6.9.2.1   *SP-neural*

Although the main emphasis in the SP program has been on developing an abstract framework for the representation and processing of knowledge, the theory includes proposals—called *SP-neural*—for how those abstract concepts may be realized with neurons [17, Chapter 11].

Figure 6.3 shows in outline how an SP-style conceptual structure would appear in SP-neural. It is envisaged that SP patterns would be realized with *pattern assemblies*—groupings of neurons like those shown in the figure within broken-line envelopes.

The whole scheme is quite different from 'artificial neural networks' as they are commonly conceived in computer science.* It may be seen as a development of Donald Hebb's [8] concept of a 'cell assembly', with more precision about how structures may be shared, and other differences.†

In SP-neural, what is essentially a statistical model of the world is reflected directly in groupings of neurons and their interconnections, as shown in Figure 6.3. It is envisaged that such things as pattern recognition would be achieved via the transmission of impulses between pattern assemblies, and via the transmission of impulses between neurons within each pattern assembly. In keeping with what is known about the workings of brains and nervous systems, it is likely that there would be important roles for both excitatory and inhibitory signals.

In short, neurons in SP-neural serve for both the representation and processing of knowledge, with close integration of the two—in accordance with the concept of data-centric computing. One architecture may promote computational efficiency by combining the benefits of information compression and probabilistic knowledge with the benefits of data-centric computing (see also [22, Section III-C.3]).

6.9.2.2   Computing with light or chemicals

The SP concepts appear to lend themselves to computing with light or chemicals, perhaps bypassing such things as transistors or logic gates that have been prominent in the development of electronic computers [24, Section 6.10.6].‡

At the heart of the SP system is a process of finding good full and partial matches between patterns. This may be done with light, with the potential advantage that light beams may cross each other without interference. Another potential advantage is that, with collimated light, there may be relatively small losses over distance—although distances should probably be minimized to save on transmission times and to minimize the sizes of computing devices. There appears to be potential to create an optical or optical/electronic version of SP-neural.

Finding good full and partial matches between patterns may also, potentially, be done with chemicals such as DNA,§ with potential for high levels of parallelism, and with the attraction that DNA can be a means of storing information in a very compact form, and for very long periods [7].

With both light and chemicals, the SP system may help realize data-centric integration of knowledge and processing. As before, there is potential for gains in computational efficiency via one architecture that combines the benefits of information compression and probabilistic knowledge with the benefits of data-centric computing.

---

*See, for example, "Artificial neural network," Wikipedia, en.wikipedia.org/wiki/Artificial_neural_network, retrieved December 23, 2013.

†In particular, unsupervised learning in the SP system [20, Section 5; 17, Chapter 9] is radically different from the *Hebbian* concept of learning (see, e.g. *Hebbian theory*, Wikipedia, en.wikipedia.org/wiki/Hebbian_learning, retrieved December 23, 2013), described by Hebb [8] and adopted as the mechanism for learning in most artificial neural networks. By contrast with Hebbian learning, the SP system, like a person, may learn from a single exposure to some situation or event. And, by contrast with Hebbian learning, it takes time to learn a language in the SP system because of the complexity of the search space, not because of any kind of gradual strengthening or *weighting* of links between neurons [17, Section 11.4.4].

‡"The most promising means of moving data faster is by harnessing photonics, the generation, transmission, and processing of light waves." [9, p. 93].

§See, for example, "DNA computing," Wikipedia, bit.ly/1gfEP4p, retrieved December 30, 2013.

6.10 Veracity: Managing Errors and Uncertainties in Data

> In building a statistical model from any data source, one must often deal with the fact that data are imperfect. Real-world data are corrupted with noise. Such noise can be either systematic (i.e., having a bias) or random (stochastic). Measurement processes are inherently noisy, data can be recorded with error, and parts of the data may be missing. [12, p. 99]

> Organizations face huge challenges as they attempt to get their arms around the complex interactions between natural and human-made systems. The enemy is uncertainty. In the past, since computing systems didn't handle uncertainty well, the tendency was to pretend that it didn't exist. Today, it is clear that approach won't work anymore. So rather than trying to eliminate uncertainty, people have to embrace it. [9, pp. 50–51]

The SP system has potential in the management of errors and uncertainties in data as described in the following subsections.

6.10.1 Parsing or pattern recognition that is robust in the face of errors

As mentioned in Section 6.2, the SP system is inherently probabilistic. Every SP pattern has an associated frequency of occurrence, and probabilities may be derived for each multiple alignment [20, Section 4.4; 17, Section 3.7 and Chapter 7].

The probabilistic nature of the system means that, in operations such as parsing natural language or pattern recognition, it is robust in the face of errors of omission, of commission, or of substitution [20, Section 4.2.2; 17, Section 6.2.1]. In the same way that we can recognize things visually despite disturbances such as falling leaves or snow (and likewise for other senses), the SP system can, within limits, produce what we intuitively judge to be 'correct' analyses of inputs that are not entirely accurate.

Figure 6.4 shows how the SP system may achieve a 'correct' parsing of the same sentence as in Figure 6.2 but with errors: the addition of 'x' within 't h e', the omission of 'l' from

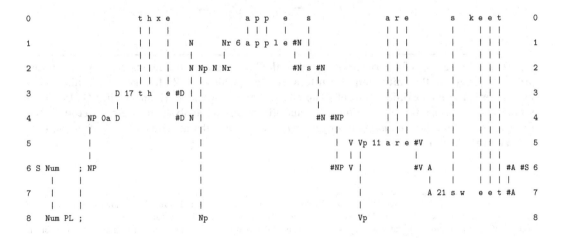

Figure 6.4: A parsing via multiple alignment created by the SP computer model, like the one shown in Figure 6.2, with the same sentence as before but with errors of omission, commission, and substitution as described in the text.

'a p p l e s', and the substitution of 'k' for 'w' in 's w e e t'. In effect, the parsing identifies errors in the sentence and suggests corrections for them: 't h x e' should be 't h e', 'a p p e s' should be 'a p p l e s', and 's k e e t' should be 's w e e t'.

The system's ability to fill in gaps—such as the missing 'l' in 'a p p l e s'—is closely related to the system's ability to make probabilistic inferences—going beyond the information given—discussed in some detail in [17, Chapter 7] and more briefly in [20, Section 10].

6.10.2 Unsupervised learning with errors and uncertainties in data

Insights that have been achieved in research on language learning and grammatical inference [20, Section 5.3; 15; 17, Sections 2.2.12 and 12.6] may help to illuminate the problem of managing errors and uncertainties in big data.

The way we learn a first language has some key features:

- We learn from a finite sample of the language, normally quite large.* This is represented by the smallest of the envelopes shown in Figure 6.5.

- It is clear that mature knowledge of a given language, **L**, includes an ability to interpret and, normally, to produce an infinite number of utterances in **L**.[†] It also includes an ability to distinguish sharply between utterances that belong in **L**—represented by the middle-sized envelope in Figure 6.5—and those that don't—represented by the area between the middle-sized envelope and the outer-most envelope in the figure.

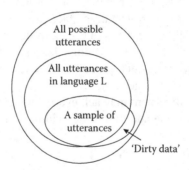

Figure 6.5: Categories of utterances involved in the learning of a first language, L. In ascending order size, they are the finite sample of utterances from which a child learns; the (infinite) set of utterances in L; and the (infinite) set of all possible utterances. (Adapted from Wolff, J.G., Learning syntax and meanings through optimization and distributional analysis, in: Levy, Y. et al. (eds.), *Categories and Processes in Language Acquisition*, Lawrence Erlbaum, Hillsdale, NJ, 1988, pp. 179–215, Figure 7.1, with permission.)

---

*An alternative view, promoted most notably by Noam Chomsky, is that we are born with a knowledge of 'universal grammar'—structures that appear in all the world's languages. But despite decades of research, there is still no satisfactory account of what that universal grammar may be or how it may function in the learning of a first language. Notice that the concept of a universal grammar is different from that of a UFK because the former means linguistic structures hypothesized to exist in all the world's languages, while the latter means a framework for the representation and processing of diverse kinds of knowledge.

[†]Exceptions in the latter case are people who can understand language but, because of physical handicap or other reason, may not be able to produce language (more below).

■ The finite sample of language from which we learn includes many utterances which are not correct, meaning that they do not belong in **L**. These include false starts, incomplete sentences, garbled words, and so on. These utterances are marked *dirty data* in the figure.

From these key features, two main questions arise, described here with putative answers provided by unsupervised learning via information compression:

■ *Learning with dirty data*: How is it that we can develop a keen sense of what does or does not belong in our native language or languages, despite the fact that much of the speech that we hear as children contains the kinds of haphazard errors mentioned above, and in the face of evidence that language learning may be achieved without the benefit of error correction by a teacher, or anything equivalent.*

It appears that the principle of minimum length encoding (Section 6.4.1) provides an answer. In a learning system that seeks to minimize the overall size of the grammar (**G**) and the encoding (**E**), most of the haphazard errors that people make in speaking—rare individually but collectively quite common—would be recorded largely in **E**, leaving **G** as a relatively clean expression of the language.

Anything that is comparatively rare but exceeds the threshold for redundancy (Section 6.4.1) may appear in **G**, perhaps seen as a linguistic irregularity—such as 'bought' (not 'buyed') as the past tense of 'buy'—or as a dialect form.

■ *Generalization without overgeneralization*: How is it that, in learning a first language, **L**, we can generalize from the finite sample of language which is the basis for learning to the infinite set of utterances that belongs in **L**, without overgeneralizing into the region between the middle-sized envelope and the outer-most envelope in Figure 6.5. As before, there is evidence, discussed in the sources referenced above, that language learning does not depend on error correction by a teacher or anything equivalent.

As with learning with dirty data, it appears that generalization without overgeneralization may be understood in terms of the principle of minimum length encoding. It appears that a learning process that seeks to minimize the overall size of **G** and **E** normally results in a grammar that generalizes beyond the data in **I** but does not overgeneralize. Both undergeneralization and overgeneralization result in a greater overall size for **G** and **E**.

These principles apply to any kind of data, not just linguistic data. With unsupervised learning from a body of big data, **I**, the SP system provides two broad options:

■ Users may focus on both **G** and **E**, taking advantage of the system's capabilities in lossless information compression, and ignoring the system's potential with dirty data and the formation of generalizations without overgeneralization. This would be the best option in areas of application where the precise form of the data is important, including any 'errors'.

---

*In brief, the evidence is that people with a physical handicap that prevents them producing intelligible speech can still learn to understand their native language [4,10]. If such a child is saying nothing that is intelligible, there is nothing for adults to correct. Christy Brown [4] went on to become a successful author, using his left foot for typing, and drawing on the knowledge of language that he learned by listening.

■ By focusing on **G** and ignoring **E**, users may see the redundant features in **I** and exclude everything else. As a rough generalization, redundant features are likely to be 'important'. They are likely to exclude most of the haphazard errors in **I** such as typos, misprints, and other rarities that users may wish to ignore (but see Section 6.10.3). And **G** is likely to generalize beyond what is in **I**—filling in apparent gaps in the data— and to do so with generalizations that are sensitive to the statistical structure of **I**, and excluding overgeneralizations without that statistical support.

These two options are not mutually exclusive. Both would be available at all times, and users may adopt either or both of them according to need.

6.10.3 Rarity, probabilities, and errors

Some issues relating to what has been said in Sections 6.10.1 and 6.10.2 are considered briefly here.

6.10.3.1 Rarity and interest

It may seem odd to suggest that we might choose to ignore things that are rare, since antiques that are rare may attract great interest and command high prices, and conservationists often have a keen interest in animals or plants that are rare.

The key point here is that there is an important difference between a body of information to be mined for its recurrent structures and things like antiques, animals, or plants. The latter may be seen as information objects that are themselves the products of learning processes designed to extract redundancy from sensory data. Like other real-world objects, an antique chair is a persistent, recurrent feature of the world, and it is the recurrence of such an entity in different contexts that allows us to identify it as an object.

6.10.3.2 The flip side of probabilities

As we have seen (Sections 6.10.1 and 6.10.2), a probabilistic machine can help to identify probable errors in big data. But contradictory as it may seem, a consequence of working with probabilities—for both people and machines—is that mistakes may be made. We may bet on *Desert King* but find that *Midnight Lady* is the winner. And in the same way that people can be misled by a frequently repeated lie, probabilistic machines are likely to be vulnerable to systematic distortions in data.

These observations may suggest that we should stick with computers in their traditional form, delivering precise, all-or-nothing answers. But also consider the following:

■ There are reasons to believe that computing and mathematics are fundamentally probabilistic:

> I have recently been able to take a further step along the path laid out by Gödel and Turing. By translating a particular computer program into an algebraic equation of a type that was familiar even to the ancient Greeks, I have shown that there is randomness in the branch of pure mathematics known as number theory. My work indicates that—to borrow Einstein's metaphor—God sometimes plays dice with whole numbers. [5, p. 80].

■ As noted in Section 6.2, the SP system can be constrained to deliver all-or-nothing results in the manner of conventional computers. But *constraint* is the key word here:

it appears that the comforting certainties of conventional computers come at the cost of restrictions in how they work, restrictions that may have been motivated originally by the low power of early computers [17, p. 28].

6.10.4   Deception

A third aspect of veracity, to be considered briefly here, is how to detect deliberate attempts at deception—including both the concealment of information and distortions in information—and what to do when such an attempt has been detected.

In broad terms, it seems that the detection of inconsistencies is the main and perhaps the only means of discovering attempts at deception. And the SP system may have a role to play in the detection of inconsistencies because of its strengths in pattern recognition [17, Chapter 6], parsing [17, Chapter 5], and several kinds of reasoning [17, Chapter 7].

It seems that the detection of inconsistencies is quite closely related to the detection or recognition of ambiguities. How the SP system may recognize syntactic ambiguity is illustrated in [17, Figure 5.1] which shows how, via multiple alignment, the system may create two alternative parsings of the ambiguous sentence *Fruit flies like a banana.**

If, for example, there is ambiguity about whether or not, at the time that a crime was committed, a suspect was present at the scene of the crime, that ambiguity is an indication that more information is needed. Further information may be gathered until the ambiguity is resolved.

6.11   Visualization

> ... methods for visualization and exploration of complex and vast data constitute a crucial component of an analytics infrastructure. [12, p. 133]

> [An area] that requires attention is the integration of visualization with statistical methods and other analytic techniques in order to support discovery and analysis. [12, p. 142]

In the analysis of big data, it is likely to be helpful if the results of analysis, and analytic processes, can be displayed with static or moving images.

In this connection, the SP system has three main strengths:

■ *Transparency in the representation of knowledge*: By contrast with subsymbolic approaches to artificial intelligence, there is transparency in the representation of knowledge with SP patterns and their assembly into multiple alignments. Both SP patterns and multiple alignments may be displayed as they are or, where appropriate, translated into other graphical forms such as tree structures, networks, tables, plans, or chains of inference.

■ *Transparency in processing*: In building multiple alignments and deriving grammars and encodings, the SP system creates audit trails. These allow the processes to be inspected and could, with advantage, be displayed with moving images to show how knowledge structures are created.

---

*This sentence is the second part of *Time flies like an arrow. Fruit flies like a banana*, attributed to Groucho Marx.

■ *The DONSVIC principle*: As previously noted, the SP system aims to realize the DONSVIC principle [20, Section 5.2] and is proving successful in that regard. This means that structures created or discovered by the system—entities, classes of entity, and so on—should be ones that people regard as natural. Those kinds of structures are also likely to be ones that are well suited to representation with static or moving images.

6.12   A Road Map

As mentioned in Section 6.2, it is envisaged that the SP computer model will provide the basis for the development of a new version of the SP machine. How things may develop is shown schematically in Figure 6.6. It is envisaged that this new version will be realized as a software virtual machine, hosted on an existing high-performance computer, that it will employ high levels of parallelism, that it will be accessible via a user-friendly interface from anywhere in the world, that all software will be open source, and that users will be able to create new versions of the system. This high-parallel, open source version of the SP machine will be a means for researchers everywhere to explore what can be done with the system and to create new versions of it.

As argued persuasively in [9, Chapters 5 and 6], and echoed in this chapter in Sections 6.9.1.3 and 6.9.2, getting a proper grip on the problem of big data will probably require the development of new architectures for computers.

But there is plenty that can be done with existing computers. Most of the developments proposed in this chapter may be pursued without waiting for the development of new kinds of computer. Likewise, many of the potential benefits and applications of the SP system, described in [24] and including such things as intelligent databases [18] and new approaches to medical diagnosis [16], may be realized with existing kinds of computer.

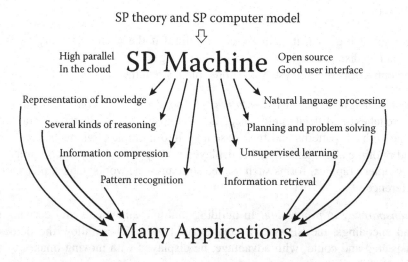

Figure 6.6: Schematic representation of the development and application of the SP machine. (Reproduced from Wolff, J.G., *Information*, 4, 283, 2013, Figure 2, with permission.)

6.13 Conclusion

The SP system, designed to simplify and integrate concepts across artificial intelligence, mainstream computing, and human perception and cognition, has potential in the management and analysis of big data.

The SP system has potential as a *universal framework for the representation and processing of diverse kinds of knowledge* (UFK), helping to reduce the problem of variety in big data: the great diversity of formalisms and formats for knowledge, and how they are processed. The system may discover 'natural' structures in big data, and it has strengths in the interpretation of data, including such things as pattern recognition, natural language processing, several kinds of reasoning, and more. It lends itself to the analysis of streaming data, helping to overcome the problem of velocity in big data.

Apart from several indirect benefits described in this chapter, information compression in the SP system is likely to yield direct benefits in the storage, management, and transmission of big data by making it smaller. The system has potential for substantial additional economies in the transmission of data (via the separation of encoding from grammar), and for substantial gains in computational efficiency (via information compression and probabilities, and via a synergy with data-centric computing), with consequent benefits in energy efficiency, greater speed of processing with a given computational resource, and reductions in the size and weight of computers. The system provides a handle on the problem of veracity in big data, with potential to assist in the management of errors and uncertainties in data. It may help, via static and moving images, in the visualization of knowledge structures created by the system and in the visualization of processes of discovery and interpretation.

The creation of a high-parallel, open source version of the SP machine, as outlined in Section 6.12, would be a means for researchers everywhere to explore what can be done with the system and to create new versions of it.

Acknowledgments

I am grateful to Daniel J. Wolff for drawing my attention to big data as an area where the SP system may make a contribution.

References

1. P. Bach-y-Rita. Theoretical basis for brain plasticity after a TBI. *Brain Injury*, 17(8):643–651, 2003.

2. P. Bach-y-Rita and S. W. Kercel. Sensory substitution and the human–machine interface. *Trends in Cognitive Science*, 7(12):541–546, 2003.

3. T. Berners-Lee, J. Hendler, and O. Lassila. The semantic web. *Scientific American*, 284(5):35–43, May 2001.

4. C. Brown. *My Left Foot*. Mandarin, London, 1989.

5. G. J. Chaitin. Randomness in arithmetic. *Scientific American*, 259(1):80–85, 1988.

6. N. Gershenfeld, R. Krikorian, and D. Cohen. The internet of things. *Scientific American*, 291(4):76–81, 2004.

7. N. Goldman, P. Bertone, S. Chen, C. Dessimoz, E. M. LeProust, B. Sipos, and E. Birney. Towards practical, high-capacity, low-maintenance information storage in synthesized DNA. *Nature*, 494(7435):77–80, 2013.

8. D. O. Hebb. *The Organization of Behaviour*. John Wiley & Sons, New York, 1949.

9. J. E. Kelly and S. Hamm. *Smart Machines: IBM's Watson and the Era of Cognitive Computing*. Columbia University Press, New York, 2013.

10. E. H. Lenneberg. Understanding language without the ability to speak: A case report. *Journal of Abnormal and Social Psychology*, 65:419–425, 1962.

11. M. Li and P. Vitányi. *An Introduction to Kolmogorov Complexity and Its Applications*. Springer, New York, 2009.

12. National Research Council. *Frontiers in Massive Data Analysis*. The National Academies Press, Washington, DC, 2013. Online edition: bit.ly/14A0eyo.

13. K. Sayood. *Introduction to Data Compression*. Morgan Kaufmann, Amsterdam, the Netherlands, 2012.

14. R. J. Solomonoff. A formal theory of inductive inference. Parts I and II. *Information and Control*, 7:1–22 and 224–254, 1964.

15. J. G. Wolff. Learning syntax and meanings through optimization and distributional analysis. In Y. Levy, I. M. Schlesinger, and M. D. S. Braine, editors, *Categories and Processes in Language Acquisition*, pp. 179–215. Lawrence Erlbaum, Hillsdale, NJ, 1988. See bit.ly/ZIGjyc.

16. J. G. Wolff. Medical diagnosis as pattern recognition in a framework of information compression by multiple alignment, unification and search. *Decision Support Systems*, 42:608–625, 2006. See bit.ly/XE7pRG.

17. J. G. Wolff. *Unifying Computing and Cognition: The SP Theory and Its Applications*. CognitionResearch.org, Menai Bridge, 2006. Detailed on bit.ly/WmB1rs.

18. J. G. Wolff. Towards an intelligent database system founded on the SP theory of computing and cognition. *Data & Knowledge Engineering*, 60:596–624, 2007. See bit.ly/Yg2onp.

19. J. G. Wolff. The SP theory of intelligence: An introduction. Technical report, CognitionResearch.org, 2013. Unpublished document. See bit.ly/1cFYTfw.

20. J. G. Wolff. The SP theory of intelligence: An overview. *Information*, 4(3):283–341, 2013. See bit.ly/1hz0lFE.

21. J. G. Wolff. Application of the SP theory of intelligence to the understanding of natural vision and the development of computer vision. *SpringerPlus*, 3(1):552–570, 2014. See bit.ly/1scmpV9.

22. J. G. Wolff. Autonomous robots and the SP theory of intelligence. *IEEE Access*, 2(1):1629–1651, 2014. See bit.ly/1zrSemu.

23. J. G. Wolff. Information compression, intelligence, computing, and mathematics. Technical report, CognitionResearch.org, 2014. See bit.ly/1jEoECH.

24. J. G. Wolff. The SP theory of intelligence: Benefits and applications. *Information*, 5(1):1–27, 2014. See bit.ly/1lcquWF.

25. G. K. Zipf. *Human Behaviour and the Principle of Least Effort*. Hafner, New York, 1949. Republished by Martino Publishing, Mansfield Centre, CT, 2012.

Chapter 7

A Qualitatively Different Principle for the Organization of Big Data Processing

Duoduo Liao

Maryam Yammahi

Adi Alhudhaif

Faisal Alsaby

Usamah AlGemili

Simon Y. Berkoich

CONTENTS

Big Data is not just a technological challenge, but it is a philosophical outlook. This chapter relates to the nowadays-prevailing idea of the Big Data revealing its most intriguing aspect— transition from quantity to quality. The proposed approach in this chapter may lead to an extraordinary form of computational intelligence.

Dealing with the "Big Data" problems requires a radical change in the philosophy of the organization of information processing. Primarily, the Big Data approach has to modify the underlying computational model in order to manage the uncertainty in the access to information items in a huge nebulous environment. As a result, the produced outcomes are directly influenced only by some active part of all information items, whereas the rest of the available information items just indirectly affect the choice of the active part. The rest of the information serves implicitly as a selection context playing a crucial role of the "unconsciousness." That is to say, when zillions of data processing is done haphazardly, the necessary result may be unexpected. An analogous functionality exhibits the organization of the brain featuring the unconsciousness, and a characteristic similarity shows the retrieval process.

In this chapter, we propose a completely different computer organization to describe our brain—a huge memory of loosely structured data and a small amount of relatively random data that change in the processing. In such a computational brain model, some effective stream algorithms can be employed to work a huge memory and some meaningful, but haphazard cache memory.

The initial idea of the computational brain model was first presented in the paper "On Clusterization of 'Big Data' Streams" [1] at COM.Geo 2012 published by the Institute of

Electrical and Electronics Engineers. This paper has been already recognized and raised substantial interest. This is a foundation for many future developments.

The following sections will be discussed in detail to cover different aspects of this side and show how this can be applied to the design of the brain to copy with Big Data problems.

1. Organization of the brain in light of the Big Data processing philosophy (Section 7.1)

2. Realization of intelligent software-defined storage (SDS) (filed a patent) (Section 7.2)

3. Realization of Golay code clustering (Section 7.3)

4. Realization of stream algorithms (filed a patent) (Section 7.4)

5. Realization of pipelining with field programmable gate arrays (FPGAs) (Section 7.5)

7.1 Organization of the Brain in Light of the Big Data Philosophy

This section presents a computational scheme for the brain using cloud computing and stream processing in the holographic universe. The surmised construction captures all the basic operational characteristics of the brain in health and disease. It is especially suitable for the Big Data computational model as it materializes the requirement of purposeful selection of information items in the unsteady framework of cloud computing and stream processing.

7.1.1 Introduction

The appearance of intractable complexity of the brain merely shows that its operational principle is not fully understood. Mass production of zillions of human and animal brains on the Earth suggests that the design of the brain must be robust. The brain encounters an overabundance of information, the so-called Big Data situation, which is qualitatively different from conventional computing.

With the Big Data methodology, the developed design of the brain captures all the characteristics the brain in norm and pathology. The Big Data computational model explicitly operates only with a small portion of the available information. The rest of the information serves implicitly as a selection context playing a crucial role of the unconsciousness.

7.1.2 Transition from quantity to quality

The Big Data processing features of the brain have been considered in the aspect of their implementation by the emerging computer technology. This may lead to a new approach to artificial intelligence (AI). Just as mentioned in the recent article from *New Scientist*, the traditional AI stalls after several decades "AI funding dried up." The article claims a breakthrough—"A soon as we gave up the attempt to produce mental, psychological qualities we started finding success." Douglas Heaven, the author of the article, states: "We have created a completely new form of intelligence, although no human can understand it" [2]. The explanation of the indicated situation in the article has been addressed in this chapter. It is related to the nowadays-prevailing idea of the Big Data revealing its most intriguing aspect—transition from quantity to quality.

The Big Data is not just a technological challenge, but it is a philosophical outlook. The proposed approach presented in this chapter may lead to an extraordinary form of computational intelligence. The obtained results are of far-reaching consequences showing a unique approach to monumental problems of information technology and fundamental science. These results offer a universal framework for effective handling of Big Data. Ultimately, they show an actual operational solution for the design of the brain.

7.1.3   Nature is Big Data enterprise

Nature is Big Data enterprise. This implies that information processing by the brain must be organized in accordance with the specifics of the Big Data computational model.

First of all, this processing cannot directly handle all available information items, but nonetheless every information item must be able to contribute to the output. Therefore, any computational concept of the brain must explicitly include the concept of unconsciousness. The major procedure for manipulation of large information repositories is associative access with some intelligent resolution of multiple responses. In Big Data systems, this resolution of multiple responses should be performed on data streams, excluding most data as irrelevant. As a result, the response of the associative memory would be determined, in some way, by the whole contents of the memory, creating what can be called context-addressable access. In practice, the most successful organization of the context-addressable access has been presented by Google on account of the PageRank algorithm [3].

Second, the brain performs a great variety of unrelated tasks in a real-time multiplexing mode. Seemingly, for a regular arrangement, the execution of these tasks would require very complex software programming with laborious machine learning. Apparently, this is not the case, and information processing in the brain is simply accomplished by some robust operations, merely through the amassment of the given data.

Third, realization of the brain definitely shows the prevalent property of fault tolerance. This designates that the organization of the brain must rest upon on cloud computing.

7.1.4   Holographic associative memory for brain construction

Due to the immensity of information, filling brain's memory stepwise, even in a parallel fashion, is not feasible—it would take thousands years. Instead, it is considered that formation of human mind starts in the womb by joining cloud-computing facilities in the holographic universe. A decisive question is how the brain gets instructions for performing its countless tasks. The suggested computational scheme for the brain is data driven; it runs by a transition diagram from memory as a Turing machine, not by special software instructions as a von Neumann computer. The notoriously slow processing of neural circuits is offset by fast memory access, and learning occurs just through the amassment of Big Data.

A new type of holographic associative memory for the brain construction is surmised. Incoming data are continuously recorded in a moving two-dimensional (2D) layer at the border of the memory. This construction provides a combined operation of "writing" and "marking–reading." Such a double functionality access results in a common illusion that brain has distinct short-term and long-term memories. Indicatively, structural distortions of the brain affect both operations uniformly, so different symptoms of Alzheimer's dementia—loss of recent memory and recall of early events—come out simultaneously.

The resolution of multiple responses in the introduced holographic associative memory is performed by stream selection of a prevalent element. The stream recall of information from earlier stages of life is easier to process than that from subsequent stages because the former case involves lesser samples of memory layers than the latter case. Roughly speaking, it is about 3 times easier to retain some information in 20 year of age than in 60 years of age. It is this limitation of the algorithm for stream resolution of multiple responses that imposes the upper bound on human life.

Functional specializations of the different regions of the brain are determined by their correspondence to the respective places in the 2D recording layers. Glia cells fix the states of the functional regions of the brain and thus constitute an interface between the neuronal circuitry and the memory. This interface is updated by asynchronous inputs from the environment and by regular information streams from the memory. Maintaining the given interface, an organism acquires self-awareness leading to the concept of consciousness.

7.1.5 Computer science innovation

The Big Data organization of the brain involves qualitatively new computer science solutions that can be basically described as follows:

1. A new construction of holographic associative memory. It combines recording and searching and thus leads to an illusion of having short- and long-term memory in one-level storage.

2. A new method for resolution multiple responses by a rational extraction of a predominant element. This produces "unconsciousness" and "intuition," so the brain works analogously to the theoretical "oracle" machine.

3. A new simple way to avoid complexity of programming by data-driven control. The brain operates using "mosaic" transformations of different cortical regions. Thus, brain maintains continuous self-awareness through integration of inputs

4. A new principle for fault tolerance with graceful degradation and progressive enhancement. Such adaptation is incorrectly attributed to neuroplasticity. Actually, this quality elucidates the most mysterious property of "mind over matter."

7.1.6 Concluding remarks

The conceptual scheme of the brain for handling the Big Data situation is simple as shown in Figure 7.1. Major publications associated with this work are presented in [1,4–7] and the remaining sections in this chapter.

Most vividly, the suggested organization of the brain can be verified considering variations from the usual population patterns of birth dates for a number of neuropsychiatric disorders, such as observable excess of autism births in March [8]. According to the presented concept, the etiological factor behind these effects is the sensitivity of the biological processes to the positioning of the Earth on the solar orbit as it is related to the holographic mechanism of the universe.

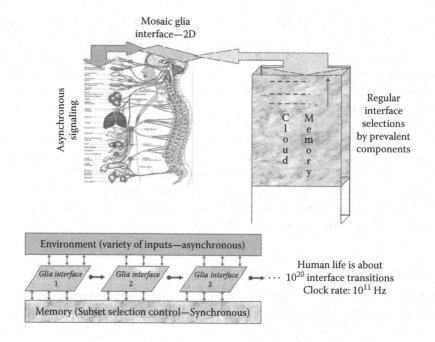

Figure 7.1: The summarized outlook of the brain: Slow Turing machine with fast component-wise mixed transitions for 2D states.

7.2 Realization of Intelligent SDS

In this section, we will introduce a special repository for diversity of information items as a practical realization of Big Data systems. It is a very large SDS for the construction of intelligent system. The suggested system emulates the basic features of a suggested memory organization of the brain that is based on a new type of computational model for processing Big Data. This new SDS will enable the access to very large data of diversified files. Thus, it enhances the speed and efficiency to the storage for various data. It uses different and unique operational techniques to allow efficient access to the stored data items in the storage.

7.2.1 Introduction

How efficiently Big Data can be processed? Big Data is the new term of the exponential growth of data in the Internet. The importance of Big Data is not about how large it is, but about what information you can get from analyzing these data. Such analysis would help many businesses on making smarter decisions and provide time and cost reduction. Therefore, to make such analysis, you will definitely need to store, analyze, search, and process the terabytes files on Big Data. Utilization of Big Data in contrast to the traditional information retrieval employs explorations akin to the "scientific method," in order to gain knowledge about the data that we have. In general, the Big Data situation requires a qualitatively different type of information processing. This problem brings in a new type of a computational model that explicitly works only with a relatively small portion of the available data, whereas the rest of the data just implicitly affects the selection of the given working portion [1]. To implement such kind of Big Data processing facilities in practice, we introduce a particular construction of intelligent SDS. This construction emulates the basic features of the suggested memory organization of the brain in [9], in which it has multiattribute cortical map, content-addressable access, and stream resolution of multiple responses. The unavoidable restrictions on the operations with overabundant data translate into the design of the brain in accordance with the fundamental Freud's idea of unconsciousness [9]. As a matter of fact, the intelligence facilities of the brain can be considered as a necessary condition to deal with the Big Data challenge [9].

The envisioned SDS incorporates two developments: memory device for multiattribute items that can be accessed by any combinations of attributes using "FuzzyFind" procedures with "Pigeonhole access" in [10] and massive distributed streaming for resolution of multiple responses [11].

7.2.2 Big Data computational model

Information processing begins with the idea of a computational model. Computational model is an abstract scheme for transforming information. It operates in the following cycling: extracting an item of data from the memory—transforming the given item of data—returning the transformed item to the memory [1]. John von Neumann had introduced a practical computational model using random access memory for the realization of first computers. The famous Church–Turing thesis conveys an informal statement that all reasonable computational models are in fact equivalent in their algorithmic expressiveness. In simple words, any calculation that can be done on one computer can be done on another computer; the difference is only in performance. This immediately raises the question about the brain. Thus, on the one hand, the facilities of the brain have to be equated with those of a conventional computer; on the other hand, this does not seem likely. A different computational model specifically suitable for a Big Data environment was introduced in [1]. In this computational model, the extraction of a data item from memory is determined by the whole bulk of data. Thus, only a relatively small part of the given Big Data explicitly contributes to actual computations; the vast majority of these Big Data contributes to the computations implicitly by determining what data items should be extracted for definite usage. Therefore, access to specific data items in this computational model is determined by the context of all data. This context-addressable access is different from that in traditional associative memories. It is similar to what is provided by Google's PageRank algorithm. The traditional associative memory, or the content-addressable memory, handles subsets rather than individual elements. Taking out the information from the content-addressable memory incurs resolution of multiple responses. A more intelligent form of content-addressable memory (which is the context-addressable memory), would entail extraction of the elements not on the basis of some arbitrary ordering of their locations, rather on a certain collective property of the retrieved subset. An example of that is Google search engine, which is based on PageRank algorithm. Access to holographic model of the brain encounters a particular problem of multiple responses resolution, because it supports content-addressable access. For the given environment, we employ a digital–analog adjustment of a streaming algorithm for finding a predominant element. This construction markedly exhibits all well-recognized properties of human memory, including better remembrance of repeated items and earlier events [9]. The resolution of multiple responses in the introduced holographic associative memory is performed by stream selection of a prevalent element. In general, this computational model encounters two specific fundamental procedures: (1) Every element is touched in stream processing and (2) appropriate selection is done through clustering. We introduced specific methods and algorithms in dealing with the last mentioned two operations, which introduced in Sections 7.2.2 and 7.2.3.

7.2.3 Basic embodiment to the suggested SDS

The suggested approach presents a very large SDS with component-wise access to multiattribute data for the construction of intelligent systems. Software-defined storage is a term for computer data storage technologies that separate storage hardware from the software that manages the storage infrastructure. This way the storage infrastructure resources can be automatically and efficiently allocated and managed to fulfill the enterprise's need. The general scheme of our suggested storage mimics the design for the brain suggested in [9]. The suggested storage combines two operational techniques for access: "Pigeonhole access" in [10] and "Stream extraction of items" in [11]. The suggested model works as follow: The storage in the Figure 7.2 represents the hardware storage, which is represented in the physical layer. The storage could contain a different storage repository from multivendor hardware, including Flash Disk, clouds, and Hard Drive. Each computer from the application layer sends its data to be stored in the storage. The storage accumulates amassment of unstructured information items—around 1 terabyte (10^{12}), from outside at arbitrary rates. Each item contains number of attributes (m) (A0, A1, A2,..., Am), where m about 100–1000. Attributes of information items are characterized by 23-bit metaknowledge templates [1]; thus, the organized items will have less number of attributes. When a computer from the application layer sends a request, the software appliance that gets the request will communicate with other software appliances, and thus, all the software appliances will work in synchronization. The enterprise's request is a set of some number of specified attributes (which is less than m), because the requester will not have all the attributes of that item. Access to "Big Data" systems is typically aimed at fault-tolerant keyword-based algorithms. Because the request has less number of attributes than the given item, this would enable mismatches in the target item, and this requires fuzzy search procedure. Many applications require approximate match rather than the exact match of the searching pattern, including gene search and other biomedical applications. Thus, the suggested device in

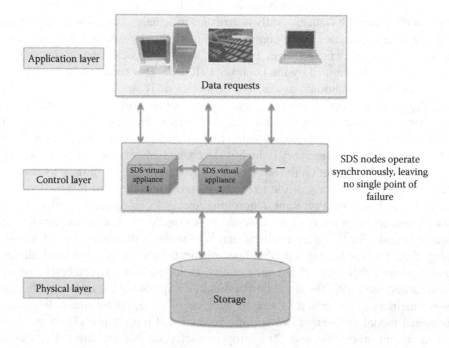

Figure 7.2: An SDS basic scheme.

this research is a whole system that stores, indexes, searches, and retrieves desired items, especially if these items are characterized by attributes and could be processed in a stream fashion. Thus, accessing the storage issues a request from a different component. Content-addressable access is arranged by inverted files for each type of attribute using the "FuzzyFind Dictionary" (FFD) with "Pigeonhole search" algorithm introduced in [10]. The result of this request marks a large subset of various data items. Accessing these items with individual attributes, we will get a lot of responses (multiple responses), especially if we access approximately with the suggested Pigeonhole method using FFDs in [10]. A stream algorithm is used for the resolution of multiple responses. The result of this request marks a large subset of various data items, and selection of appropriate items is done by either a "cyber-physical stream" algorithm (CPS) or "multi-buffer based" algorithm in [11]. Thus, we will formulate certain common criterion for retrieval and perform the extraction of an element in a stream fashion.

Generally speaking, we have a lot of attributes, and to make inverted FFDs for each attribute is expensive. For these attributes, search is rarer and can be done by sequential streaming of the whole storage. Therefore, we have used parts of the attributes only to access in our design, and this is based on the newly discovered switch in the brain consciousness [12]. This will substantially simplify the construction. In our design, we select primary attributes on which we will perform access search and here we use the FFD for access. We select the secondary attributes, at which only the selection of the retrieved attributes is to be performed, and here we use the stream algorithm.

7.2.4 Summary

The suggested SDS will enable the access to very large data of diversified files. Thus, it enhances the speed and effieciency to the storage for various data. This invention could get numerous applications in Big Data intelligent systems. Biomedical applications and searching for genetic disorders in genome databases are one of the important applications that could benefit from this system, as well as many other applications that deal with large data files that are beyond terabaytes (Big Data). Clustering to gain more knowledge about the data we are dealing with will be included in the suggested system in which we will use the method suggested in [13].

7.3 Realization of Golay Code Clustering

In this section, unsupervised clustering has been primarily tackled in various contexts and by researchers in different fields ranging from computer vision, computational biology, to social science. Hence, this reflects its broad demand and significance as an essential step in the process of analyzing very large collections of information. Therefore, in this section, we are presenting a novel clustering technique that employs the error correction Golay code to cluster Big Data streams. This clustering methodology can be effectively applied to various computational intelligence problems in the Big Data situation. It outperforms all other conventional techniques because it has linear time complexity, that is, $O(n)$ with one passage through the file.

7.3.1 Introduction

Clustering techniques partition the data items into groups or clusters in which the objects within a cluster are similar to one another and dissimilar to objects in other clusters. Similarity is determined in terms of how close the objects are in the space based on a distance measurement function. The quality of a cluster is evaluated according to the maximum distance between any two objects in that cluster. The emerging of the new forms of technology and media allows data to be increasingly generated and gathered; datasets grow dramatically in size. Mobile devices, radiofrequency identification chips, genome sequencing, finance (such as stocks) logs, Internet search, and wireless sensor networks are examples of those data or technologies that are contributing in forming Big Data streams [14]. Organizing and analyzing these massive data would produce unpredictable knowledge and help the decision maker to make smarter decisions. Thus, unsupervised categorization of this large-scale data is an extreme desire [15–18]. The significance of processing large-scale datasets lies in insights that can be achieved by analyzing and investigating this data. Based on this analysis, smarter decision can be made, which allows businesses to be proactive in a time and cost-effective manner.

7.3.2 Related work and contribution of this work

This approach extends a previously developed method of clustering Big Data streams by utilizing the Golay code encoding mechanism. In practical, grouping large datasets is a rather complicated computer science problem. It can be computationally expensive in terms of time and space complexity if conventional clustering methods are considered. In addition, further expense may be incurred by the need to repeat data clustering. Hence, developing a new method of clustering large datasets is an increasing demand. Different clustering algorithms have been developed to address data clustering problems such as hierarchical and partitional algorithms.

Table 7.1 lists the time and space complexities of several well-known algorithms. In the table, n is the number of patterns to be clustered, k is the number of clusters, and l is the number of iterations [19].

7.3.3 Golay code clustering algorithm

7.3.3.1 23-bit meta-knowledge template

Clusters may correspond to a hidden pattern that can represent different concepts. Realization of the Golay code clustering algorithm requires using templates of yes/no questions to examine the properties of each data item. Template should contain 23 questions that can be used to produce a 23-bit binary attribute vector. Each of these questions investigates the presence or

Table 7.1   Computational complexities of clustering algorithms

Cluster Algorithm	Complexity	Capability of Addressing Large-Scale Data
k-Means	$O(NKd)$	No
Fuzzy c-means	Near $O(N)$	No
Hierarchical clustering	$O(N^2)$	No
BIRCH	$O(N)$	No
CURE	$O(N^2 \log N)$	Yes
ROCK	$O(n^2 + nm_m m_a + n^2 \log n)$	No
CLICK	$O(m \log 3\ n)$	No
Golay code	$O(N)$	Yes

Note: BIRCH, balanced iterative reducing and clustering using hierarchies; CURE, clustering using representatives; ROCK, robust clustering using links; CLICK, cluster identification via connectivity kernels.

absence of a property, for instance, symptom. The 23-bit meta-Knowledge template is inspired by the popular 20 questions game [1]. The output of this investigation is a 23-bit binary attribute vector that represents the data item. This procedure facilitates mapping similar data items into the same cluster as long as the distance between them does not exceed a certain threshold [21]. After that, the generated 23-bit binary attribute vector is transferred into a 12-bit index by reversing the error-correction Golay code. This 12-bit index is then used to map the item into the associated cluster [20].

7.3.3.2   Structure of clusters and cluster validity

Clusters are core components of our classification and prediction methodology due to its ability to discover the connected components and the underlying relationships [22]. Because fuzziness is one of the most salient features of the "Big Data" concept, underlying relationships can be detected by using the Golay code clustering technique. Furthermore, clusters assist in reducing the influence of vectors that have little or no similarity, that is, common features. When applying the Golay code clustering algorithm to the possible 23-bit vectors (8,388,608), a total of 1,267,712 nonempty clusters were created. The generated scheme has two cluster size, a large cluster which can hold up to 139 data items and a small cluster which can hold at most 70 data items. Both the small and the large cluster assure that the Hamming distance between their members does not exceed 8.

7.3.4   Pattern recognition

Semi-supervised training is a means for reducing the effort needed to prepare the training set by training the model with a small number of fully labeled examples and an additional set of unlabeled or weakly labeled examples [22]. Our pattern recognition procedure starts by providing the system with a fully labeled training dataset (we call them centers). Specifically, a dataset is a collection of vectors that represent the identity of corresponding objects, for instance, cats, dogs, and rabbits. These vectors will be employed to label objects that already have been clustered. We sequence through clusters and find the nearest center to each codeword within clusters in terms of Hamming distance. The label of the codeword is the same label of the nearest center. When classifying all objects, labeling clusters becomes trivial.

A simple approach to label a cluster is based on the object frequency. The label of the cluster depends solely on the majority weight within this cluster, that is, prevalent element. Some clusters have different types in which one type dominates that cluster or has more weight. Therefore, the weight of each object within a cluster is simply its frequency in that cluster. In other words, the class label is the majority voting of its members. Clusters are not labeled until they contain a certain number of data items. Assume the threshold equals 10. Clearly, cluster (5 and 7) do not meet this threshold condition, therefore, they are not labeled.

Hence, the cluster label is basically the same label of its prevalent element. Table 7.2 presents an example of the labeling process.

7.3.5   Summary

New computing technologies such as cloud computing, sensors, mobile phones, geospatial computing, medical imaging, climate informatics, and social networks have raised the scale of the generated data. This digital data is stored in different formats, including texts, geometries, images, videos, sounds, or their combination. If the data to be collected is processed and clustered, this would be extremely effective and significantly advance our understanding

Table 7.2   Illustration of cluster labeling based on the majority vote

Cluster Number	Object Frequency		Cluster Size	Label
	Cats	Dogs		
1	20	3	23	Cat
2	12	2	14	Cat
3	104	10	114	Cat
4	14	1	15	Cat
5	1	2	3	<threshold
6	65	6	71	Cat
7	2	1	3	<threshold
8	15	7	22	Cat
–	–	–	–	–
14	78	3	81	Cat
15	30	2	32	Cat
The new object is:				**Cat**

of the given dataset. In this contribution, we addressed two Big Data problems: clustering massive data on the stream mode and semi-supervised pattern recognition. We investigated the methodology of clustering Big Data stream based on reversing of conventional usage of error correction code. We employed the Golay code in this technique. Interestingly, this clustering system can be used in order to perform pattern recognition operations. We tested our system with a synthetic dataset.

7.4   Realization of Stream Algorithms

This section is about data streaming. Data streaming systems have their foundations and applications in many domains, such as database systems, network monitoring, data mining, algorithms, sensor networks, theory, and statistics. Therefore, there has been a substantial interest in the design of algorithms that process data streams using single-pass (or on-the-fly) algorithms to provide up to the moment analysis and statistics on current arrival streams. The amount of data stream is extremely large that can hardly be stored in the main memory for online processing. The challenge of stream processing algorithm is to extract valuable knowledge from large volume using very limited space requirements. Many applications rely directly or indirectly on knowledge extraction from Big Data streams, and implementations are in use in large-scale industrial systems.

7.4.1   Introduction

"Big Data" situation relates to the problem of dealing with very large amounts of data [23]. It presents a qualitatively different state of affairs for the organization of information processing, namely, this organization cannot utilize all the data explicitly. Data stream processing is one of the important challenges, and many studies have been made on this topic [1]. Stream processing uses different methods; compared to traditional dataset computing, it requires relatively smaller response time to deal with huge amount of data. Because vast amounts of data need to be operated continuously regardless of storage and access distribution, and respond quickly to new information, Big Data community is beneficial to study stream algorithms. In the field of computer science, the streaming algorithms are designing for processing data streams in a way of limited time and limited memory. It was first introduced in 1999 [24,25], and then used in most domains of computer science such as networking [26], machine learning [32], database [27], information security [28], privacy [29], web application, manufacturing, financial applications, and telecommunications data management. Traditional dataset computing does not seem to be the best practice of dealing with large amount of data in small respond time. It is not feasible to simply insert the incoming stream data into traditional database management system and process them locally. Traditional database management systems are not structured for continuous and frequent loading of individual data streams, and they do not directly support the continuous queries that are typical of data stream applications [30]. Now the Big Data society comes to study stream algorithms when large amounts of data can be operated continuously regardless of storage and access distribution, meanwhile respond quickly to new information. In reality, stock market data is a typical stream data. The data contains real-time price, transaction, and other financial information. Traders usually receive and analyze data streams to make decisions by advanced systems [31]. Under the stream processing framework, the space is also constrained to specific requirements. It also contributes to more advanced computations such as detection [32] and prediction. It is critical to apply stream processing for prediction from the aspect of solving frequency problems.

In this section, we investigate previous efforts for finding most frequent item and introduce novel algorithms that have better performance for Big Data processing. First, we investigate the majority algorithm that extracts single item with occurrence of more than 50% overall processed stream. Next, we introduce our novel CPS algorithm that retrieves the most frequent item of frequency as low as 2% in time complexity of linear time, and show how CPS performs better compared to previous efforts, and that the efficiency of CPS increases by increasing the volume of the stream. Moreover, we propose a design of an analog device that holds CPS algorithm to extract items from large data streams.

7.4.2 Previous work

Finding the most frequent item is one of the classic problems on a data stream [33,34]. It has been considered one of the most studied problems since the1980s [35,36]. The problem is popular due to its clarity to state, and its intuitive interest and value. The frequency problem is an important branch in stream algorithms both in itself and as a subroutine within other complex data stream computations. The problem is usually to find the item occurring most frequently. Back in 1981, Boyer and Moore [35] invented the majority algorithm that guarantees the extraction of the predominant item of a frequency of more than 50%. However, in order to optimally solve this problem of finding guaranteed item of frequency of more than $N/2$. Work [37] proposed a two-pass algorithm that works as follow: First, it finds a single item candidate using the majority algorithm; second, it scans the stream again to count the number of occurrences of the candidate item to see whether it occurs more than $N/2$.

7.4.3 CPS algorithm

CPS algorithm is a novel algorithm that extracts the item of most frequent occurrence of frequencies as low as 2% in very high probabilities and space requirements efficiency. The CPS algorithm was inspired by the possibility of its usage in the model of the brain as considered in [9]. The algorithm is structured as a physical process that can be illustrated in Figure 7.3. This

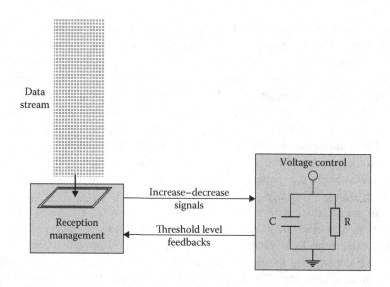

Figure 7.3: Illustration of the scheme for extracting the prevalent item from a stream.

scheme presents a capacitor and a resistor; the level of the voltage serves as a control factor for the developing process.

The algorithm can be stated as follows: initially assign the value 1 to weight (T) and the value 0 to voltage control (v). For the next coming stream, do the following: store the new arrival of the stream slice and set v to 1. Keep comparing the incoming stream slice with previously stored item. If they are equal, and v is greater than or equal to 1, increase v by the result value of weight (T) divided by the current v, and increase the T by 1. However, if not equal, and v is greater than 1, decrease v by a decrement rate known by α. Otherwise, next new arrival of stream slice will replace the previous stored item and set v to 1. By the end of this stream, output the stored item. The algorithm can be described as follows:

```
T := 1;
v := 0;
REPEAT
get next item
IF (v >= 1 and item = top) THEN (v = v + T/v; T = T + 1 )
ELSE_IF (v > 1 and item ≠ top) THEN v = v * α
ELSE (v := 1; top := item)
UNTIL more items;
```

Moreover, we introduced a design of an analog device that holds CPS algorithm to extract predominant item from Big Data streams (Figure 7.3). At reception management, new item comes, and it is necessary to send a signal to voltage management. If the incoming item is the same as that at the reception register, voltage control increases the voltage. However, voltage control does nothing, and voltage decreases exponentially; in a short time, it is about $v*\alpha$.

7.4.4 Experiments

Experiments are divided into two main sections: (1) performing cyber-physical algorithm using streams of pseudorandom numbers and (2) performing cyber-physical algorithm using streams of Zipf's law distribution.

7.4.4.1 Pseudorandom numbers

We performed cyber-physical algorithm using different iterations of streams of pseudorandom numbers. Each stream of size 1000 contains a determined element frequency. Using $\alpha = 0.99$, the probabilities of selecting the prevalent element at a certain frequency can be illustrated in Figure 7.4. Moreover, we tested various stream sizes with smaller frequencies: 1%, 2%, 3%, 4%, and 5%. Figure 7.5 shows the behavior of selecting the predominant element of smaller frequencies with variable streams sizes: 1,000, 10,000, and 100,000 at $\alpha = 0.99$ and a range of 1,000.

7.4.4.2 Zipf's law distribution

In this section, we examined cyber-physical algorithm using streams that follow Zipf's law distribution [38]. For a stream of size 10,000 and $\alpha = 0.8$, Figure 7.6 shows the probabilities of retrieving the most frequent element (first element) and the probabilities of retrieving the second most frequent element (second element). Figure 7.7 shows the factor of stream size in processing Zipf's law distribution using cyber-physical algorithm. Stream sizes used are 10,000 and 200,000 elements.

7.4.5 CPS analysis

The performance of the algorithm is determined by the relationship between increment and decrement rates. For random distribution determining a certain increment rate, the optimal

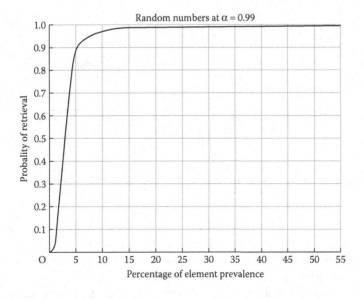

Figure 7.4: Probabilities of retrieving the predominant element.

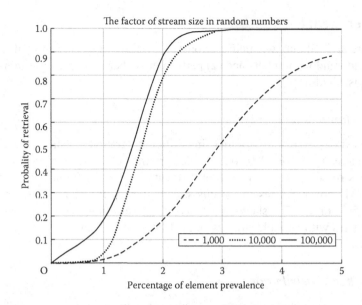

Figure 7.5: Probabilities of retrieving the predominant element in various stream sizes.

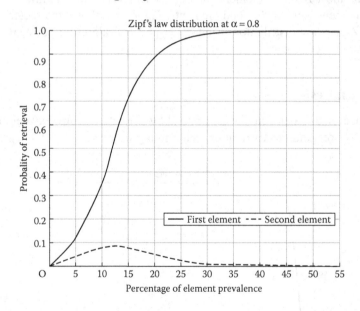

Figure 7.6: Probabilities of retrieving the first and second most frequent items.

decrement rate turns out to be presented by $\alpha = 0.99$. By comparing the majority algorithm (single buffer) in the work [35] and the new CPS algorithm using uniform random data (Figure 7.6), we observed that the new CPS algorithm performs better at providing high probabilities of retrieving the prevalent item of frequencies as low as 2% as shown in Figure 7.5.

In Figure 7.5, we show that the factor of stream size has an impact on the performance. Under the same distribution, the larger file input has better performance. The input size that has 100,000 numbers has better performance than the one that has 10,000 elements. And the stream

Figure 7.7: Probabilities of retrieving the prevalent element in different stream sizes.

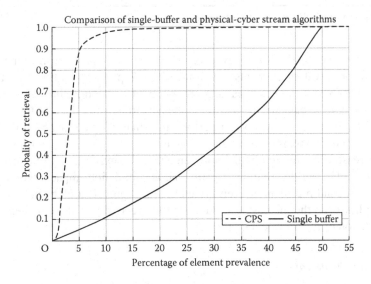

Figure 7.8: Majority algorithm vs. CPS algorithm with uniform random data.

size that has 10,000 numbers has better performance than the one that has 1,000 elements. Consider the stream having the following input: the prevalent element is in the beginning of the stream (Figure 7.8).

By comparing the majority algorithm and CPS algorithm using data distribution that follows Zipf's law (Figure 7.9), we found out that the new CPS algorithm has better performance of providing high probabilities of retrieving the prevalent item of low frequencies. For Zipf's law, the decrement rate that is optimal for a selection of a single element is $\alpha = 0.8$. However, Zipf's law is intended for selection of a group of elements. Thus, selecting two most frequent

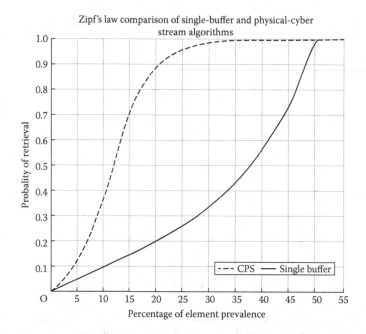

Figure 7.9: Majority algorithm vs. CPS algorithm with Zipf's law.

elements in accordance with Zipf's law using $\alpha = 0.99$ may provide more better results if applied to pairs of prevalent elements. Anyhow, single Zipf's distribution selection does not have to be optimal for $\alpha = 0.99$ as does random distribution selection. In Figure 7.5, the results show that the factor of stream size also impacts the performance of the new CPS, and the input size that has 200,000 numbers has better performance than the one that has 10,000 elements. However, compared to the improvement in random distribution, the improvement here is minimal. The reason for this minimal improvement is the existence of the second prevalent element in the Zipf's law distribution stream. The accumulation impacts would increase the influence of the second prevalent element. Even the most prevalent element would have more accumulation; however, the increasing influence of the second prevalent element would offset some accumulation of the prevalent element.

7.4.5.1 Future direction

In the given computer program, we do not worry about high values of the counter. In actual physical implementation of the algorithm, the voltage cannot go too high; there should be some practical limits. Apparently, if the voltage limit corresponds to a value that is somewhat higher then the average of the counter; this would be sufficient for the physical implementation of the algorithm. As we consider hardware realization of the suggested method, we will have to take this condition into account. At this moment, the obtained results just show the feasibility of physical design for future works. In case of the possible broad usage of the presented algorithm, it might be advantageously realized in hardware as shown in Figure 7.1. This imposes some limitation factors. The performance of CPS algorithm will extremely increase in the case of preprocessing the input stream by the following steps: (1) Split the input stream into smaller data chunks, (2) sort all items within the smaller data chunks, and (3) feed this sorted smaller

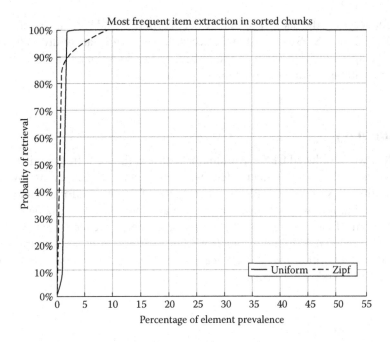

Figure 7.10: Probability of extracting the frequent item using the multicore pipelined architecture.

data chunks to the CPS algorithm. Figure 7.10 illustrates the performance of CPS algorithm using the previous steps. The sorting task is done by the multicore pipelined architecture. Such kind of preprocessing can be efficiently implemented by a new pipelining architecture [39]. This multicore pipelined architecture provides a simple and effective solution to the on-the-fly computations by transferring the operating states from core to core [40,41]. The multiprocessor pipeline allows an arbitrary algorithm to be performed on the fly on a data chunk, given a sufficient number of processors. The major condition for continual mode of stream operations—equal durations of time intervals for computations at each section of the pipeline—is realized by forced interrupts at each processing stage. It was observed that the chunk size of 100 items would achieve the most probability chances for extracting the most frequent item among all of the chunks.

7.4.6   Stream algorithms in applications

In this section, we introduce applications that take advantages of the use of the previously introduced algorithms. We discus the major role of CPS algorithm in the newly introduced structure of SDS and how the novel CPS algorithm contributes to the suggested construction of SDS.

7.4.6.1   Intelligent SDS

The suggested invention presents a very large SDS with component-wise access to multiattribute data for the construction of intelligent systems. The general scheme of such storage mimics the design for the brain suggested in [1]. The storage combines two operational techniques for access to the data: "Pigeonhole access" in [24] and "Stream extraction of items" in [11,42,43]. This new SDS will enable the access to very large data of diversified files. Thus, it enhances the speed and efficiency to the storage for various data. This invention could get numerous applications in Big Data intelligent systems.

7.4.7 Summary

In this section, we introduced CPS algorithm that extracts the most frequent item using single pass. CPS extracts the most frequent item with appearance frequency as low as 2%. We presented results of performing CPS over two types of inputs: uniform random distributed items and items that follow Zipf's distribution. We also introduced an analog physical design model that holds the functionality of CPS. During experiments, we identify the best values of α for both uniform random distribution and Zipf's law distribution. Also, we explored the major role of introduced algorithm into the applications of SDS and the clusterization of Big Data streams using Golay code transformation.

7.5 Realization of Pipelining with FPGAs

7.5.1 Introduction

Conventional architectures overloaded by stream-based applications. The huge demand of stream data stimulated the industry to respond by introducing a joint of traditional stream processing engines coupled with emerged technologies. The ongoing paradigm embraces parallel computing as the most suitable proposition. However, Big Data situation needs a reconfigurable pipelined architecture that aims at Big Data clustering. Pipelining with FPGAs overpasses internal memory constraints of multicore architectures by applying forced interrupts and crossbar switching. This reduces memory operations, and consequently reduces complexity, data dependency, high latency, and cost overhead of parallel computing [44]. In this model, program slicing uses dynamic resource management that allows organizing on-the-fly processing of arbitrary complexity without parallel programming, and it adopts symmetric multiprocessing (SMP) technology along with crossbar switch and forced interrupt. The main goal of this promising architecture is to efficiently process Big Data streams on the fly, although it can process sequential programs on the pipelined model. The new architecture is very suitable for handling Big Data systems. The experimental results and performance comparison with existing multicore architectures demonstrate the effectiveness, flexibility, and diversity of the new architecture, in particular, for large data parallel processing [45]. However, we noticed that it is very rare to find a shared memory application SMP that scales to more than 64 cores practically. The considered pipelining processing is of especial significance for applications of the presented technique of Golay code clustering as it involves very diverse and rather sophisticated computations for realization of multiple data cross sections with sophisticated "meta knowledge" templates.

This section discusses stream processing requirements, followed by general outlook over the current limitations of parallel systems. It suggests the FPGA hardware model that is especially intended to process Big Data clustering on the fly, although this model can process variable lengths of data using parallel pipelined multicore design. Finally, it proposes the same model based on SMP and forced interrupts.

7.5.2 Why pipelining with FPGAs

Big Data introduces unconventional pressure on time and memory performance. Consequently, new computation models are significantly required to cope up with Big Data situation. Researchers introduced "on-the-fly" clusterization of amorphous data. On-the-fly processing deals with a continuous stream of data, and it must maintain a certain throughput of information flow. In this pattern, hardware design should not tolerate any postponement of oncoming stream.

There are two obvious reasons to exclude graphics processing units as a suitable architecture for text processing algorithms: First, this type of algorithms creates multiple "thread divergent" points; second, the use of malloc() inside the kernel further slows down performance [46].

In contrast, multicore pipelined architecture provides a simple yet effective solution to the on-the-fly computation by transferring the operating states from core to core down the pipeline [41]. Pipelined architectures consist of a sequence of processing elements in which the output of one processor is the input of the next one. "By pipelining, processing may proceed concurrently with input and output, and consequently overall execution time is minimized. Pipelining plus multiprocessing at each stage of a pipeline should lead to the best-possible performance" [44].

7.5.3 Multiprocessor pipelined architecture by forced interrupt

Multiprocessor pipelined architecture uses forced interrupts in order to automatically slice each program into fixed durations. Each processing cycle starts with (L) loading, (P) processing, and (U) unloading. Because each cycle has a fixed duration, this design performs efficiently on large volume of data with relatively small algorithms. The initial design is limited by algorithm size that requires additional round of processing [47].

The challenges of algorithm size as well as the overhead of memory data transformation are well addressed as follows: The multicore system uses program slicing and forced interrupts. Theoretically, it receives data, and then it generates blocks of different sizes based on slice function.

Figure 7.11 shows the ideal situation in which the blocks are in equal size, and each processor would execute one block, respectively. In this case, each processor executes the block with the same color, and timing/slicing barriers excluded as conflicting issue. In contrast, Figure 7.12 illustrates variable length processing. This situation presents data blocks that are not equal in size or processing time. Hence, we need special handling by forced interrupts and program slicing. Each processor may process a certain amount of data; then it can be stopped.

The number of processors required for one block execution may vary according to data length and processing time. The multiprocessor pipeline allows an arbitrary algorithm to be performed on the fly on a data chunk, given a sufficient number of processors.

7.5.4 Multicore memory and cache architecture based on crossbar switching

This technology does not relocate memory data down the pipeline; instead, it uses a crossbar switch in order to assign memory data blocks to the corresponding processor. Memory data may include program status information that allows the next processor to resume the work starting from the last state. This approach has been applied on a multi-memory/multi-cache design.

This design does not assign the corresponding memory for each processor; it reduces the cost of data relocation. Cache and memory can be grouped into one set, and the crossbar switch

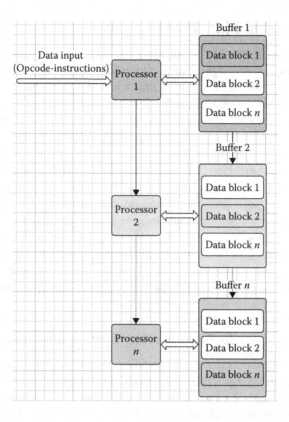

Figure 7.11: Blocks in equal size.

would assign each processor to one group. The number of groups should be equal to processing elements.

This organization eliminates the cost overhead behind memory relocation as well as additional round setup as it is described by the forced interrupts technique. This organization replaces loading/unloading operations by a switching mechanism. Theoretically, we can add as much processing elements as we need to reasonably process any given input of Big Data clusters.

Figure 7.13 shows the processing steps of the multicore memory and cache architecture based on crossbar switching. Once the data are loaded in G1, the corresponding core P1 starts processing G1 data for a fixed time; then the crossbar switch assigns the remaining data to P2. At this stage, P1 starts processing the new incoming data assigned by the crossbar switch [41].

7.5.5 Multicore architectures based on SMP

In the context of processing Big Data clusters, memory management plays an important role, a modern technology of memory and cache management may provide a very simple, yet effective solution by switching the state of processors among data blocks, and it can eliminate the overhead of internal data transformation. As a result, the performance of this pipeline architecture increases significantly.

This technology follows the SMP architecture, whereas many processors can share single main memory. This approach solves the issue of algorithm size, one main memory can receive

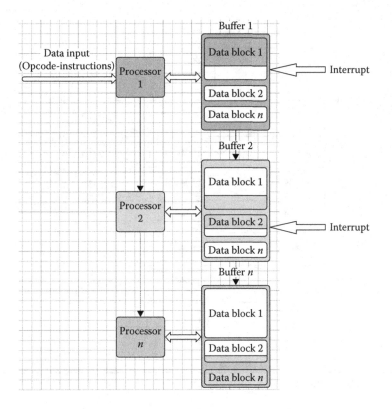

Figure 7.12: Variable length processing.

the input data, and then it can assign each memory chunk to the processor that would process the instruction set in a fixed time. Then, it shifts to the next block of data, respectively; this allows the next processor proceed operations in which the previous processor stopped keeping in mind that the number of pipeline processors has direct relationship with system complexity.

Computer cluster systems have proved the efficiency of this technique on intensive amounts of data. However, this organization works at the processing level. In stream processing, data storage can add high-cost operation within processing path; hence, the system must minimize unnecessary storage operations to archive low latency. Figure 7.14 shows an architecture that decreases time-intensive operations by processing messages on the fly.

In real-time situations, we try to avoid dependencies; the program must process messages in a given time by interrupts during which this architecture can proceed with partial data.

The system promotes multithreading by allowing data partition among processing blocks without having the developer write low-level code. This would prevent blocking external data, thereby minimizing latency.

The objective of this system is to be able to efficiently process external data that can arrive in either variable lengths or high volumes or both. In order to achieve better performance, the system should optimize execution path, and it must minimize boundary-crossing overhead. The desired length of the pipeline must be tested with performance in mind to insure sufficient processing path while decreasing the additional cost of multi-processing passage.

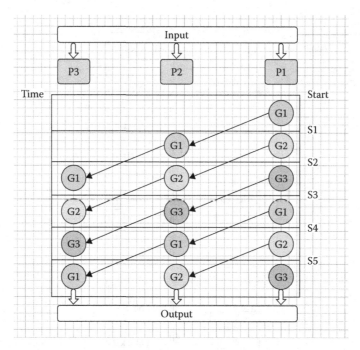

Figure 7.13: Processing steps.

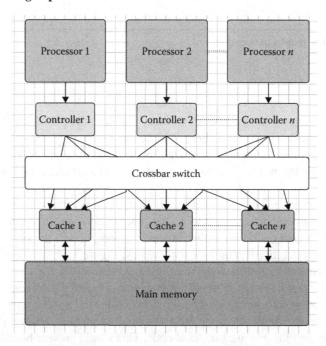

Figure 7.14: The SMP architecture.

Figure 7.4 illustrates the desired communication between each processor and main memory. This architecture provides more flexibility to add more processing units into the pipeline sharing the same data source [45].

Raman and Clarkson [46] carried out an interesting project that proves the efficiency of this specific type of architectures. Their project recognizes parallelism with nonidentical processing units. These units can work simultaneously with one shared memory.

Conventionally, the design of multiprocessor pipeline moves data chunks when the processor loading state changes. In contrast, this new architecture offers two advantages over conventional models: First, the data blocks do not relocate when switch operation occurs; second, it allows other processors to load data as long as it is in the ideal state in order to utilize the pipeline.

References

1. Berkovich, S. and Liao, D., On clusterization of "Big Data" streams, COM.Geo '12, *Proceedings of the 3rd International Conference for Geospatial Research and Applications*, ACM, New York, 2012.

2. Heaven, D., Not like us: Artificial minds we can't understand, *New Scientist Magazine*, Vol. 2929, pp. 32–35, 2013.

3. Google Inc., PageRank, US Patent 6,285,999, Mountain View, CA.

4. Berkovich, S., An algorithm to access human memory showing alzheimer symptoms when distorted, Bio-IT '14, Boston, Massachusetts, 2014. http://www.bio-itworldexpo.com/Bioinformatics/.

5. Berkovich, S., Formation of artificial and natural intelligence in big data environment, R.E. Pino (ed.), *Network Science and Cybersecurity*, Advances in Information Security, Volume 55, Springer, New York, pp. 189–203, 2014.

6. Berkovich, S., On the information processing capabilities of the brain: Shifting the paradigm, *Nanobiology*, Vol. 2, pp. 99–107, 1993.

7. Fordyce, D. R., Long-term memory: Scaling of information to brain size, *Frontiers in Human Neuroscience*, 2014. http://journal.frontiersin.org/Journal/10.3389/fnhum.2014.00397/full.

8. Stevens, M. C., Fein, D., and Watrhouse, L., Season birth effects in autism, *Journal of Clinical and Experimental Neuropsychology*, Vol. 22, No. 3, pp. 399–407, 2000.

9. Berkovich, S., Organization of the brain in light of the big data philosophy, COM. BigData '14, *Proceedings of the 1st International Summits on Big Data Computing*, IEEE, Washington, DC, 2014.

10. Yammahi, M., Kowsari, K., Shen, C., and Berkovich, S., An efficient technique for searching very large files with fuzzy criteria using the Pigeonhole Principle, COM. BigData '14, *Proceedings of the 1st International Summits on Big Data Computing*, IEEE, Washington, DC, 2014.

11. Alhudhaif, A., Yan, T., and Berkovich, S., A cyber-physical algorithm for selecting a prevalent element from big data streams, *GSTF Journal on Computing*, Vol. 4, No. 1, pp. 83–88, 2014.

12. Koubeissi, M., Consciousness on-off switch discovered deep in brain, *New Scientists Magazine*, July 2014.

13. Alsaby, F. and Berkovich, S., Realization of clustering with Golay code transformations, *GSTF Journal on Computing*, Vol. 4, No. 1, pp. 89–94, 2014.

14. Feldman, D., Schmidt, M., and Sohler, C., Turning Big data into tiny data: Constant-size coresets for k-means, PCA and projective clustering, *Proceedings of the 24th Annual ACM-SIAM Symposium on Discrete Algorithms*, pp. 1434–1453, 2013.

15. Biswas, A. and Jacobs, D., Active image clustering: Seeking constraints from humans to complement algorithms, *Proceedings of the IEEE Computer Society Conference on Computer Vision and Pattern Recognition*, Providence, RI, pp. 2152–2159, 2012.

16. Cai, X., Nie, F., Huang, H., and Kamangar, F., Heterogeneous image features integration via multi-modal spectral clustering. *Proceedings of the IEEE Conference on Computer Vision and Pattern Recognition*, pp. 1977–1984, 2011.

17. Dueck, D. and Frey, B. J., Non-metric affinity propagation for unsupervised image categorization, *Proceedings of the 11th IEEE International Conference on Computer Vision*, pp. 1–8, 2007.

18. Lee, Y. J. and Grauman, K., Foreground focus: Unsupervised learning from partially matching images, *International Journal of Computer Vision*, Vol. 85, pp. 143–166, 2009.

19. Wunsch, D. and Rui, X., Survey of clustering algorithms. *IEEE Transactions on Neural Networks,* Vol. 16, No. 3, pp. 645–678, 2005.

20. Strauss, M. J., Procedural enhancements to some approximate searching techniques, Doctoral dissertation, George Washington University, Washington, DC, 2008.

21. Bari, N., Liao, D., and Berkovich, S., Organization of meta-knowledge in the form of 23-bit templates for big data processing, *Computing for Geospatial Research and Application*, IEEE, Washington, DC, pp. 87–90, 2014.

22. Davis, D., Chawla, N., Blumm, N., Christakis, N., and Barabasi, A., *Predicting Individual Disease Risk Based on Medical History*, ACM, New York, 2008.

23. Berkovich, S., Physical world as an internet of things, COM.Geo '11, *Proceedings of the 2nd International Conference on Computing for Geospatial Research & Applications*, New York, p. 66, May 2011.

24. Alon, N., Matias, Y., and Szegedy, M., The space complexity of approximating the frequency moments, *Proceedings of the 28th Annual ACM Symposium on Theory of Computing*, ACM, New York, 1996.

25. Babcock, B., Babu, S., Datar, M., Motwani, R., and Widom, J., Models and issues in data stream systems, *Proceedings of the 21st ACM SIGMODSIGACT-SIGART Symposium on Principles of Database Systems*, New York, pp. 1–16, 2002, doi:10.1145/543613.543615.

26. Abadi, D. J. et al., The design of the borealis stream processing engine, *Proceedings of the CIDR*, Vol. 5, pp. 277–289, 2005.

27. Chandrasekaran, S. et al., Telegraph CQ: Continuous dataflow processing, *Proceedings of the 2003 ACM SIGMOD International Conference on Management of Data*, ACM, New York, pp. 668–668, June 2003.

28. Gruschka, N., Jensen, M., Iacono, L., and Luttenberger, N., Server-side streaming processing of WS-Security, *IEEE Transactions on Services Computing*, Vol. 4, No. 4, pp. 272–285, 2011.

29. Oliveira, S. R. M. and Zaiane, O. R., Privacy preserving frequent itemset mining. *Proceedings of the IEEE International Conference on Privacy, Security and Data Mining*, Vol. 14, Australian Computer Society, Darlinghurst, Australia, pp. 43–54, 2002.

30. Terry, D., Goldberg, D., Nichols, D., and Oki, B., Continuous queries over append-only databases, *Proceedings of the ACM SIGMOD International Conference on Management of Data*, New York, pp. 321–330, June 1992.

31. Babcock, B., Babu, S., Datar, M., Motwani, R., and Widom, J., *Models and Issues in Data Stream Systems*, Stanford University, Stanford, CA, Report, 2003.

32. Rosten, E. and Drummond, T., Machine learning for high-speed corner detection. Computer Vision–ECCV 2006, Springer, Berlin, Germany, pp. 430–443, 2006; *International Conference on Database Theory*, Edinburgh, pp. 398–412.

33. Alon, N., Matias, Y., and Szegedy, M., The space complexity of approximating the frequency moments, *Journal of Computer and System Sciences*, Vol. 58, No. 1, pp. 137–147, 1999.

34. Henzinger, M., Raghavan, P., and Rajagopalan, S., Computing on data streams, Technical Report TR 1998-011, Compaq Systems Research Center, Palo Alto, CA, December 1998.

35. Boyer, R. S. and Moore, J.S., A fast majority vote algorithm, Technical Report ICSCA-CMP-32, University of Texas, Austin, TX, February 1981.

36. Boyer, R. S. and Strother Moore, J., A fast string searching algorithm. *Communications of the ACM*, Vol. 20, No. 10, pp. 762–772, 1977.

37. Misra, J. and Gries, D., Finding repeated elements, *Science of Computer Programming*, Vol. 2, pp. 143–152, 1982.

38. Adamic, L A. Zipf, power-laws, and pareto-a ranking tutorial. Xerox Palo Alto Research Center, Palo Alto, CA, http://ginger.hpl.hp.com/shl/papers/ranking/ranking. html 2000.

39. Berkovich, S., Berkovich, E., and Loew, M., Multi-layer multi-processor information conveyor with periodic transferring of processor's states for on-the-fly transformation of continuous information flows and operating method therefor, US Patent No. 6,145,071, issued November 7, 2000.

40. Liao, D., Real-time solid voxelization using multi-core pipelining, dissertation, The George Washington University, February 2009, http://gradworks.umi.com/3344878.pdf.

41. Liao, D. and Berkovich, S. Y., A new multi-core pipelined architecture for executing sequential programs for parallel geospatial computing, *Proceedings of the 1st International Conference on Computing for Geospatial Research & Application*, Washington, DC, COM.Geo '10, ACM, New York, June 21–23, 2010.

42. Alhudhaif, A., Yan, T., and Berkovich, S., On the organization of cluster voting with massive distributed streams, *Proceedings of the 5th International Conference on Computing for Geospatial Research and Application*, IEEE, Washington, DC, 2014.

43. Alhudhaif, A., Yammahi, M., Yan, T., and Berkovich, S., A cyber-physical stream algorithm for intelligent software defined storage, *International Journal of Computer Applications*, Vol. 109, No. 5, pp. 21–25, January 2015.

44. Dewdney, A. K., *The (New) Turing Omnibus*. Henry Holt and Company, New York, 1993.

45. Algemili, U. and Berkovich, S., A design of pipelined architecture for on-the-fly processing of big data streams, *International Journal of Advanced Computer Science and Applications*, Vol. 6, No. 1, 2015. http://dx.doi.org/10.14569/IJACSA.2015.060104.

46. Raman, S. and Clarkson, T., Parallel image processing system—A modular architecture using dedicated image processing modules and a graphics processor, *Proceedings of the IEEE, Conference on Computer and Communication Systems*, Hong Kong, China, pp. 319–323, September 1990.

47. Berkovich, S., Kitov, Z., and Meltzer, A., On-the-fly processing of continuous data streams with a pipeline of microprocessors, *Proceedings of the International Conference on Databases, Parallel Architectures, and Their Applications*, IEEE Computer Society, Miami Beach, FL, pp. 85–97, March 1990.

BIG DATA SECURITY: SECURITY, PRIVACY, AND ACCOUNTABILITY

Chapter 8

Integration with Cloud Computing Security

Ibrahim A. Gomaa

Emad Abd-Elrahman

CONTENTS

Abstract

With the movement of Big Data to cloud computing platforms, security issue has become sophisticated. The claim of risk duplication is introduced in this chapter due to the virtualization nature of the cloud environment. Cloud computing has become an appropriate infrastructure for Big Data processing and analytics due to its massive features. Nowadays, in data-centric world, Big Data processing and analytics have become critical to most enterprises and government applications. Thus, there is a need for novel approaches to secure Big Data analytics in the cloud. This chapter discusses the motivations and outcomes of the integration of Big Data with cloud computing security. Therefore, it studies the relationship between Big Data and cloud computing. Then, we explain general theoretical properties that motivate the evaluation of Big Data integration technology and capabilities with case studies. Moreover, we survey identity-based approaches, cryptographic approaches, and some techniques that are used to secure Big Data in the cloud. Finally, we will introduce a suitable paradigm for achieving high degree of security and efficiency in such sophisticated network access to many online services and applications introduced by the integrity of Big Data with cloud computing. In conclusion, the main objective of this chapter is to address the main risks arising from the integration of Big Data with the cloud. Moreover, it will propose some solutions to overcome the security weakness of Big Data access through cloud computing assets.

8.1 Introduction

Nowadays, we are experiencing an explosion in the volume of data exploration and analysis that are created by the large-scale distributed services, which usually work with large datasets. Consequently, the rapid development of Internet, Internet of things (IoT), and virtualization technologies appeared later. Therefore, the online applications and services in organizations, businesses, companies, and large-scale and small-scale industries are increasing greatly.

Most notably, cloud computing facilitates the on-demand management and control of computing, storage, and connectivity resources in the network, by automatically moving or scaling up or down the resources, required to distribute content and applications. It can also be rapidly provisioned and released with minimal management effort or service provider interaction [1]. Many international IT corporations now offer powerful public cloud services to users on a scale from individual to enterprise all over the world such as Amazon Web Services (AWS), Microsoft Azure, and IBM Smart Cloud. Data security/privacy is one of the major concerns in the adoption of cloud computing [2,3]. Compared to conventional systems, users will lose direct control over their data while migrating to the cloud environment.

Big Data has become the most critical trend that shines through cloud computing. Indeed, it is the core driver in cloud computing and will define the future of IT. It is the high-volume, high-velocity, and high-variety information assets that demand cost-effective, innovative forms of information processing for enhanced insight and decision making [4]. In addition, it brings about the 5V (volume, variety, velocity, variability, and veracity) and complexity of the information disclosure, hence highlights security and privacy concerns.

In this chapter, we will investigate the problem of secure integrity for Big Data storage in cloud computing. Also, to manage the complex data and to identify the data patterns, it is important to securely handle this data. As known, cloud computing comes with explicit security challenges. In addition, Big Data have its own security challenges; thus, in this chapter, we have to find a framework, a model, or a method to satisfy the integration of Big Data with cloud computing security.

8.1.1 State of the art

Cloud computing has become a fashion, a prevalent solution for data manipulation (processing, storage, and distribution). Within the two years (2010–2012), AWS has gone from 262 billion objects stored in its S3 cloud storage, to over 1 trillion. However, enterprises and corporations that work with Big Data have been unable to realize the full potential of the cloud, because of the inherent bottlenecks of moving Big Data in, out, and across cloud infrastructures. Consequently, integrity verification for outsourced data storage has attracted spacious importance and research interests for enterprises, corporations, universities, and institutes.

Aspera On Demand (AOD) enables secure, efficient, large-scale workflows in the cloud. Aspera introduces many applications such as BGI, Netflix, Zencoder, and SendToNews to their customers and partners who are using the cloud for Big Data. In addition, AOD enables them to get their Big Data onto the cloud infrastructure in order to meet their IT infrastructure needs [5].

Hadoop is an open source project of Apache developed by Nutch. Hadoop's initial design was completed by Doug Cutting and Mike Cafarella. Now, Hadoop is widely deployed by Yahoo!, Facebook, and other Internet enterprises, because it forms a powerful Big Data systematic solution through system management, data processing, data storage, and integration of other modules. Many security issues that need to be addressed because Hadoop is still new and is being developed to add more features. Many researchers have identified some of the security issues related to Hadoop and started working on this.

SPARQL is used for diverse data sources. The World Wide Web Consortium identifies the importance of SPARQL in dealing with Big Data. Later on, to increase the security and privacy of queried data, the idea of secured query will be proposed. Kevin Hamlen et al. proposed that they could store the encrypted data in a database. Therefore, even though an attacker can get into the database, he/she cannot get the actual data. Nevertheless, the disadvantage is that encryption may be requiring a lot of overhead. Most of the operation will take place in cryptographic form, instead of processing the plaintext. Hence the approach of processing in cryptographic form added an extra security layer [6]. In Section 8.5, we will identify and propose identity-based and cryptographic approaches to secure the accessing of Big Data in the cloud.

IBM researchers have also ensured the need of a secure environment for query processing. There are many methodologies used to prepare these environments. One of them is Kerberos, which is considered effective to prepare the secured environment. It uses an encryption technology along with a trusted third party, an arbitrator, to be able to perform a secure authentication in an open network. Moreover, Kerberos uses cryptographic tickets to avoid transmitting plaintext passwords over the wire [6]. Therefore, it can mitigate a man-in-the-middle (MITM) attack, which may be happening in these environments.

Airavat [7] has shown some significant enhancement of security in the MapReduce environment. In this work, Roy et al. have used the access control mechanism along with differential privacy. They have worked upon mathematical bound potential privacy violation, which prevents information leak beyond the data provider's policy.

Jules et al. [8] proposed the first model of *proofs-of-retrievability* (PoR) concept. But their scheme is not applicable for dynamic data storage; hence, it is applicable only for static data storage such as an archive or library. Ateniese et al. proposed a model (similar to PoR) named *provable data possession* (PDP) [9]. PDP introduces *blockless verification*, in which the verifier can confirm the integrity of a proportion of the outsourced file, through ensuring a combination of precomputed file tags, called homomorphic verifiable tags (HVTs) or homomorphic linear authenticators (HLAs). Erway et al. [10] proposed the first PDP mechanism based on skip list, despite public auditability and file blocks of various sizes not being supported in this scheme. It can support full dynamic data updates.

Shacham et al. [11] introduced an enhancement for PoR model with stateless verification. Moreover, they also proposed a MAC-based private verification scheme and a public verification scheme based on the BLS (Boneh–Lynn–Shacham) signature [12]. In the public verification scheme, integrity generation and verification proofs are similar to signing and verifying of BLS signatures.

Wang et al. [13] proposed public data auditing scheme that supports data dynamics; this is considered one of the latest works with dynamics support based on BLS signature. It can support public auditing from a third-party auditor (TPA). In another work [14], Wang's scheme has been changed by adding a random masking technology to ensure that the TPA cannot derive the raw data file from a series of integrity proofs. In their scheme, they also integrated a strategy first proposed in [10] to divide file blocks into many sectors.

Ateniese et al. [15] proposed how to transform a mutual identification protocol to a PDP scheme, whereas Zhu et al. [16] introduced a scheme that can allow different service providers in a hybrid cloud to prove data integrity to the data owner.

Curtmola et al. [17] proposed an MR-PDP scheme based on PDP [9] with which they can efficiently prove the integrity of multiple replicas along with the original data file. Moreover, Meng and others [18] have explained the basic framework of Big Data processing and analyzed the effect of cloud computing technology on data management. Then, Shen Derong [19] summarized the relevant research of No-SQL systems systematically, including access method, replication policy, data consistency policy, and data processing policies based on MapReduce. Finally, Zhao et al. [20] have consolidated scientific workflow systems with cloud platforms as a service to deal with the growing amount of data and analysis complexity.

8.1.2 Sample research problems

In this section, we identify some of the major open research problems that must be addressed to ensure the success of secure Big Data Cloud (B-DC). In brief, a single secure Big Data management solution after integrating with the cloud platform is yet to be designed. Different systems target different aspects in the design space, and multiple open research problems still remain. The key-value stores is one of them, these systems are supporting only very simple functionality. Therefore, they needed to generalize these proposals to support different classes of applications and different key-value stores. In addition, they needed to extend the key-value stores for supporting richer set of services and applications. Moreover, in the domain of relational database management, an important open problem is how to protect the systems from possible attacks by effectively utilizing the available resources and minimizing the risk. Designing secure, scalable, elastic, and autonomic multitenant database systems is another important challenge that must also be addressed. In addition, ensuring the security and privacy of the data outsourced to the cloud is also an important problem for ensuring the success of Big Data management systems in the cloud.

Traditional IT infrastructure for data management and security cannot adapt to the rapid growth of Big Data, especially, after integrating with the cloud. Therefore, we have to try to refine and enhance the integration of Big Data with cloud computing without affecting cloud security. More security problems of B-DC appeared, and we categorize some of these problems into the nine categories shown in Table 8.1.

8.1.2.1 Security and privacy problem

For structured data, relational database management systems (RDBMSs) have already formed a set of extensive methods for access and security after decades of development. Usually, Big Data is unstructured data; therefore, distributed file storage systems (DFSS), distributed No-SQL databases (D-NOSQL-DB), and others are derived to manage and secure the mentioned data. But such emerging systems required to be further enhanced, especially in areas such as data access privileges and security controls. To mitigate the risk we have to ensure the following:

■ Prevention of data loss

■ Protection of data from unauthorized access

Nowadays, we search for different storage and access mechanisms, a unified security access control mechanism for multisource and multitype data used to deal with large amounts of unstructured data. As known, Big Data contains sensitive data; therefore, it is more attractive to potential attackers. Some of these attacks resulted in 6.5 million LinkedIn user

Table 8.1   B-DC main security problems

Problems	Challenges
Security and privacy	Secure structured and non-structured data
Distributed nodes and distributed data	Ensure the security of the place where computation is done
Inter/intra-node communication	Anyone can tap and modify the node communication for breaking into systems
Data protection	Store the data without encryption
Access control	Uncontrolled access
Nodes authentication	No authentication
Logging	No activity is recorded
Traditional security tools	Scalability is not huge
Different technologies	Many interacting complex components

account passwords was leaked in 2012; 450,000 Yahoo! user IDs were stolen by network attacks; 6,000,000 user login names, passwords, and email addresses were leaked in December 2011 when Chinese Software Developer Network's security system got hacked [21].

One of the very important issues in Big Data world is the privacy problem. Because of the speedy development of Internet technologies and the IoT, all kinds of information related to our lives and jobs have been collected and stored. Therefore, our actions are always being monitored and analyzed, which drive more researchers to propose techniques, methods, and algorithms for anonymous communication. Moreover, more enterprises and multinational companies can serve customers better and make precision marketing possible by investigating the in-depth analysis and modeling of user behaviors.

A user was still identified and disclosed by the data going by the name *Anonymous*, although the data was carefully anonymized and hidden by the company. On Twitter, Facebook, and others, many users are accustomed to publishing their personal data, their locations, and activities at any time. Therefore, we can speculate the times when they are not at home, get their home addresses, and even find personal and house photos, just based on the information they published; many sites used to analyze these data, such as http://pleaserobme.com/ and http://ww1.weknowyourhouse.com/. Nowadays, many countries around the world are creating and improving laws related to data use, access, and privacy to protect secrete and personal information from getting hacked [21].

8.1.2.2 Distributed nodes and distributed data

Distributed nodes [22] are defined as an architectural problem. As known, the computation in Big Data is done in any set of nodes at anytime and anywhere across the clusters; therefore, it is difficult to know the exact location of computation at certain time. Hence, it is very difficult to secure the place where computation is done.

Moreover, in the cloud environment, we believe that it is extremely difficult to find exactly where pieces of a file are stored. By applying the concept of defense in depth in centralized data security system, the critical data (treasure) is wrapped around various security layers. The same method cannot be applied to cloud environments since not all related data are presented in one place and it changes. Therefore, we have to use other techniques and methods to secure data in the cloud.

8.1.2.3 Inter/Intra-node communication

Many Big Data distributions and applications use RPC over TCP/IP for data transfer between nodes. This behavior usually happens over a network, distributed around environments consisting of wireless and wired networks. Therefore, by using MITM attacks and its various techniques, anyone can intercept the inter-node communication [22] for breaking into systems.

8.1.2.4 Data protection

Many B-DC environments store the data as plain text without encryption to improve efficiency and neglect encryption and decryption steps. Therefore, if the attacker can access a set of workstations, it is easy for him to steal the critical data present in those workstations.

8.1.2.5 Access control

Each node in B-DC has administrative rights and privileges [22]. The uncontrolled access to any data is very dangerous and considered as vulnerability, as a malicious node can steal or manipulate critical user data.

8.1.2.6 Node authentication

Nodes can join the cloud to increase the parallel operations and availability. In case of no authentication, malicious nodes can join the cloud to steal user data or disrupt the operations of the cloud.

8.1.2.7 Logging

In B-DC environments, the absence of logging means that no activity is recorded and no information is stored. In the absence of these logs, it is difficult to find if someone has breached the cloud or if any malicious altering of data is done, which needs to be reverted. Moreover, internal users can perform malicious data manipulations without being caught. Moreover, we have to know that, one of the most critical steps the attacker uses is to clear tracks and logs for any operation and access. Therefore, logging is an urgent issue to track any activity on the systems.

8.1.2.8 Traditional security tools

Scalability issue must be taken into consideration when dealing with distributed systems. Traditional and normal security tools that have been developed and used over years cannot be directly applied to this distributed form of cloud environments, and these tools do not scale as well as cloud scales do.

8.1.2.9 Different technologies

The cloud environment consists of different technologies, which have many interacting complex components, such as databases, computing power, and networks. Small security vulnerabilities in one or small parts of a component can bring down the whole system. Therefore, maintaining security in the B-DC environment is considered an important challenge.

8.1.3 Chapter road map

This chapter discusses problems, motivations, and outcomes of integrating Big Data with cloud computing security. Therefore, the overview, related works, and sample research problems of Big Data and cloud computing integrity were introduced in Section 8.1. Then, Section 8.2 explains general theoretical properties that motivate the evaluation of Big Data integration technology and capabilities with case studies. Section 8.3 reviews the mechanisms used to secure data in cloud computing and its challenges. The relationship between cloud computing security and Big Data will be introduced in Section 8.4 in addition to how to secure Big Data in the cloud. Moreover, Section 8.5 surveys and proposes identity-based approaches, cryptographic approaches, and some techniques that are used to secure Big Data in the cloud. Section 8.6 introduces conclusion and future works to address the main risks arising from the integration of Big Data with the cloud.

8.2 Big Data Integration

With the huge variety in Big Data sources and types, the enterprises are obligated to move this huge amount of data to be stored in the cloud. As a result, it is mandatory to assure the security either for this kind of storage or in accessing this data. So, the integration of Big Data with

Figure 8.1: Big Data location in the cloud service reference model.

the cloud will lead to a complex security scenario. This means that, many challenges will arise because of this integration. Many enterprises are migrating their applications to the cloud in order to take the pros of cloud computing flexibility, expandability, and cost savings. Google, Facebook, Apple, and others are moving their Big Data to the cloud business model. They are taking into account the highest return of investments (ROIs) from this model while keeping the security measures as high as possible.

In cloud computing service models, storage as a service (Storage-a-a-S) is a sub-offering service of infrastructure as a service (IaaS). Moreover, if the root key factor of Big Data is the storage, then it belongs to the IaaS family. This means that, securing Big Data will be considered more as consumers' responsibility than as that of cloud service providers (CSPs), as shown in Figure 8.1. As a result, we consider that identity management in Big Data accessing behind cloud computing is a major security issue in this domain.

8.2.1 Big Data SWOT analysis

The strengths, weaknesses, opportunities, and threats (SWOT) analysis is an evaluation methodology used either in new project planning or in the running projects to have a clear vision about the erstwhile discussed four matrix elements involved in this project. The correct analysis and mapping between those elements will lead to a successful business and high investments revenue. Moreover, it will give good average revenue per user (ARPU) in case of client-based business chain. Therefore, the main scope of conducting SWOT analysis on Big Data integration with cloud is to offer a good planning for enterprises about the main issues released from this integration and how they can overcome them in a professional way. In this analysis, we will focus on the security aspects relevant to Big Data integration in cloud computing:

- The strengths of the Big Data integration with the cloud when faced with Big Data challenges.

- The weaknesses of this integration when faced with Big Data challenges.

- The opportunities offered to the integration of the Big Data with the cloud.

- The threats to the integration resulting from moving Big Data to the cloud environment.

To better understand the security challenges in Big Data integration with the cloud, we conduct the following SWOT analysis as shown Table 8.2.

Table 8.2 Big Data SWOT analysis

Strengths	Internal Factors	Weaknesses
Experienced classification for data by sources/formats will help in assuring data integration.	Data integration	Different sources of data. Different types of data format.
Dynamic policy-based access control for B-DC provides efficient access channels to the required fields.	Data access	Different way of access. Restricted types of securing access.
The high efficient orchestrators can organize the B-DC, which leads to secure distribution for those data in the cloud.	Data place	Unknown storage place. Out of control data allocation.
The adaptability and scalability methods applied in the cloud will assure the reliable security in Big Data.	Methods	Securing B-DC environment needs more developments and new methods.
Applying portability for virtual service security.	Standards	Need more adaptations for the current standards.
New appliances and integrated security solutions cope with the new standards of quality of resources.	Quality of resources	Lack of quality control of Big Data with the available resources.
Professional skills in virtual firewall appliances, security analysis, and high professional training for new technologies will lead to reliable security in Big Data reside in the cloud environment.	Skills	High staff skills. Different types of skills. Different types of technology. Difficult management.
Opportunities	**External Factors**	**Threats**
Applying new trend for multi-tenancy applications in the cloud environment through a distributed security solutions based active virtual instances.	Trends	Big Data in the cloud is a new trend in securing multi-tenancy links/channels/records, all behind virtual environment, which is a complex trend in security solutions.
Improving the culture of Big Data producers and analyzers by applying balanced security measures.	Society culture	Different actors behind Big Data producers will lead to unbalanced policies either in data management or in proposed security solutions.
Considering enterprises or operators as trusted third parties will help in the extremist investments in such type of application.	Funding	More investments by enterprises or operators for their Big Data externally is out of control and lead to a sophisticated business model.
Building new business models based on new security strategy will save losses in legacy systems and lead to fast security and access to Big Data in the cloud environment, which will encourage this integration.	Systems	The traditional systems need more and more investments in order to keep their continuity measures outside their local sites, that is, *data away in cloud.*

8.2.2 *Big Data capabilities for data integration*

Today, enterprises and developers are able to manage and store massive volumes of data, which is then ending up in Big Data containers. Analyzing and feeding a large amount of data types (structured, unstructured, complex, social, etc.) can ensure vendor's challenges for Big Data integration. Therefore, a new service called integration platform as a service (iPaaS) appeared and many vendors started to use and develop it. By using iPaaS, you will be able to quickly feed, read, analyze, and manage Big Data, regardless of its volume, velocity, and variety.

The Big Data capabilities for data integration platforms should support some features to be comprehensive:

1. The platform should support all key life cycle steps for Big Data integration. The life cycle steps for data integration are illustrated in Figure 8.2.

2. The platform should handle all data types or formats for all the life cycle steps including the following:

    a. Structured data

    b. Unstructured data

    c. All simple and complex data types

3. The platform should have the capability to integrate data regardless of whether it is located inside or outside the security devices (firewalls, IDS, IPS, etc.). As known, the data may be within the enterprise or it may be coming from business partners or being managed by software-as-a-service (SaaS) vendors *in the cloud*.

4. The platform should have the capability to manage (access, discover, cleanse, integrate, and deliver) data whenever the data is located and needed by applications or users.

5. The platform should provide role-specific tools for each person involved in Big Data integration (analysts, architects, administrators, developers, etc.).

6. The platform should support both analytical (reporting, analysis, etc.) and operational (business processes/operations execution) different use cases.

One of the most useful researches in Big Data integration is Ventana Research [23], which was performed in May 2014 to determine attitudes and utilization of Big Data integration and

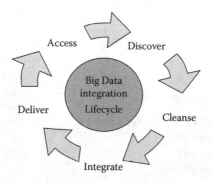

Figure 8.2: Big Data integration life cycle.

to investigate Big Data integration practices and potential benefits. Moreover, gaining the most benefit from Big Data integration requires an assessment of any organization's unique needs to identify gaps and priorities for improvement.

The research examined how organizations use Big Data integration. It also evaluated the aspects of Big Data integration and the requirements of technologies that Big Data needs. This research introduced valuable information about Big data integration. Therefore, we can use these data to investigate how to secure Big Data integration in cloud computing environments.

Moreover, this research showed that many organizations integrate Big Data technology with other environments in use in businesses such as intelligence systems, applications running in business processes, data warehouses, and business analytics managed in the lines of business. After this hot integration, how organizations plan to manage Big Data integration? The research introduces that 41% will do it across the lines of business and IT (according to data collected by this research, most large organizations will do this). Only 29% will leave integration management entirely by the IT organization, and just 12% assign the task to business analysts.

Nevertheless, from Figure 8.3, most organizations plan to manage Big Data integration across lines of business and IT. Therefore, the security and privacy role will be done by organization IT. Despite the need for IT leadership in this technical area, we conclude that Big Data is critical to business users and thus recommend that securing it be a critical activity.

Finally, from the analysis and investigation of the mentioned research, we can conclude that the data integration capabilities are most often called critical for Big Data. Enterprise capabilities are most often required at the system level, which are as follows:

- Load balancing

- Cross-platform support

- Development and testing environment

- Systems management

- Scalable execution of tasks

8.2.3 Evaluating Big Data integration technology

As mentioned before, the objective of Big Data integration is to collect data from numerous, different locations and sources, merge, cleanse it, and then deliver it to the end user in such a way that it appears to be a unified whole.

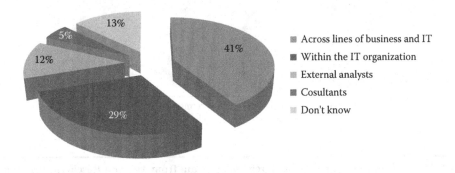

Figure 8.3: Managing Big Data integration.

Ventana Research [23] introduces the innovative technology trends that will be used in Big Data integration. The first one is cloud computing, which more than 50% of the organizations plan to utilize in managing Big Data. However, only 28% of the organizations prefer to use it to deploy Big Data integration software. However, almost 14% prefer to integrate Big Data on premises.

As known, many organizations most commonly integrate its Big Data between on-premises systems, while other organizations move cloud-based data on premises, and many move data in the reverse direction. In future, organizations plan to use data integration that involves cloud systems in some way than between on-premises systems.

After analysis and investigation, we can see that more organizations that were using cloud-to-cloud integration are confident in their abilities to process large volumes of Big Data.

Moreover, in these organizations, Big Data arrives at a high velocity than in those that rely on on-premises systems. Figure 8.4 introduces some types of Big Data integration processes and significant growth planned for cloud computing in many organizations.

The benefits of B-DC integration that organizations most often ranked most valuable are as follows:

■ Make information more available across enterprise

■ Meet analytic needs of business

■ Increase speed of integration

■ Load balance integration

■ Retain and analyze more data

■ Secure data and mitigate the risk in protected cloud

8.2.4   Case studies for Big Data integration

B-DC integration is accomplished by reported case studies on Big Data using cloud computing technology. We report two parts of the case studies here. The first one describes case studies

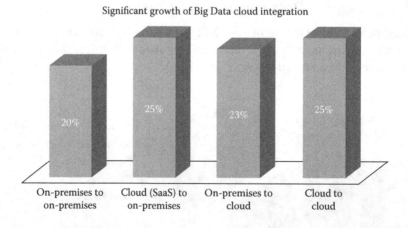

Significant growth of Big Data cloud integration

| On-premises to on-premises | Cloud (SaaS) to on-premises | On-premises to cloud | Cloud to cloud |
| 20% | 25% | 23% | 25% |

Figure 8.4: Significant growth of B-DC integration. (Data from Ventana Research, Pentaho, *Ventana Research Benchmark Research: Big Data Integration*, Sponsor Report, Prepared for Pentaho, May 2014. http://www.ventanaresearch.com/.)

Table 8.3   Case studies from vendors

Case	Business needs	Cloud service models	Big Data solution	Assessment
Swift-Key	Language technology	IaaS	Amazon Elastic MapReduce	Success
343 Industries	Video game developer	IaaS	Apache Hadoop	Success
Red-Bus	Online travel agency	IaaS, PaaS	Big Query	Success
Nokia Mobile	Communications	IaaS	Apache Hadoop, Enterprise Data Warehouse	Success
Alacer	Big Data solution	IaaS	Big Data algorithms	Success

Source: Hashem, I.A.T. et al., *Inform. Syst.*, 47, 98–115, 2015.

provided by different organizations that have completed the Big Data integration with their cloud environments as shown in Table 8.3.

The second part shows case studies that have been mentioned by academic sources [24]. Hence, the following case studies report recent cases of how researchers integrate the cloud with their Big Data projects. Table 8.4 will introduce the situation, objectives, approaches, and results for latest research case studies resulting from the integration of cloud computing with existing Big Data projects. We should investigate all mentioned cases to study how to mitigate the risks in these environments.

8.3   Cloud Computing Security

Despite continuous defending from new threats and discovering new vulnerabilities, Big Data and cloud security in general needs to be well understood. Therefore, B-DC security became a well-researched domain with a rich body of knowledge both in theory and in practice. In general, B-DC-based services can require more analyses of security architectures to address security requirements such as confidentiality, identity management, authentication, authorization, availability, integrity, audit, continuous monitoring, incident handling, incident response, and security policy management.

8.3.1   Cloud computing background

Cloud computing is a model for enabling convenient, on-demand management and control of computing, storage, and connectivity resources in the network, by automatically moving or scaling up or down the resources, required to distribute content and applications. It can also be rapidly provisioned and released with minimal management effort or service provider interaction [1]. It is composed of five essential characteristics (on-demand self-service, broad network access, resource pooling, rapid elasticity, and measured services), three service models (IaaS, platform-a as-a service [PaaS], and SaaS), and four deployment models (private, public, community, and hybrid).

Security and privacy were identified as cross-cutting concerns that are shared responsibilities for cloud vendor; therefore, the placement of security and privacy as a backplane to the

Table 8.4 Researchers case studies for B-DC integration

Case	Situation/Context	Objective	Approach	Result
1	Massively parallel DNA sequencing generates staggering amounts of data.	To provide accurate genomic results at a scale ranging from individuals to large cohorts.	Develop a Mercury analysis pipeline and deploy it in the AWS cloud via the DNA nexus platform.	Established a combination of a robust validated software pipeline that have been applied to more than 10,000 whole genome and whole exome samples.
2	Conducting analyses on large social networks such as Twitter requires considerable resources because of the large amounts of data involved.	To use cloud services as a possible solution for the analysis of large amounts of data.	Use PageRank algorithm on the Twitter user base to obtain user rankings.	Implemented a relatively cheap solution for data acquisition and analysis by using the Amazon Cloud infrastructure.
3	To study the complex molecular interactions that regulates biological systems.	To develop a Hadoop-based cloud computing application that processes sequences of microscopic images of live cells.	Use Hadoop cloud computing framework.	Allows users to submit data processing jobs in the cloud.
4	Applications running on cloud computing likely may fail.	Design a failure scenario.	Create a series of failure scenarios on an Amazon Cloud computing platform.	Help to identify failure vulnerabilities in Hadoop applications running in cloud.

cloud computing reference architecture is an appropriate change to the conceptual reference model as shown in Figure 8.5.

8.3.2   How to secure Big Data in the cloud

A secure B-DC is a complex issue and the integration is tricky to control. The more complex B-DC security environment, the more difficult it will be to handle and manipulate data in this environment.

Where security is concerned, the penetration testers in B-DC may have more to do with the unfamiliarity of the platforms. Platform heterogeneity is potentially B-DC security vulnerability. Many B-DC vendors involve deploying novel platforms (Hadoop, No-SQL,

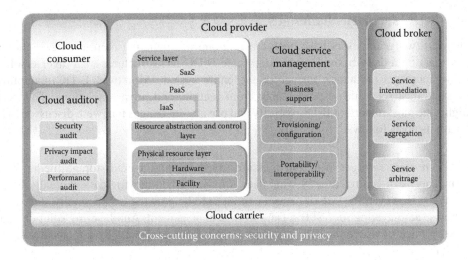

Figure 8.5: Security and privacy in cloud conceptual reference model. (Data from National Institute of Standards and Technology [NIST], Gaithersburg, MD, *US Government Cloud Computing Technology Roadmap*, Special Publication 500–293, vol. II, Release 2.0, October 2014. http://dx.doi.org/10.6028/NIST.SP.500-293.)

and in-memory databases) that you have never used before; therefore, security issue is a big challenge. Hence, if the B-DC model is new to the market, you may have difficulty finding a sufficient and suitable range of commercial security tools.

Anyway, we can control a B-DC in a coherent manner, just as we oversee data and analytics investments at smaller scales. Doing so requires that you completely assess your B-DC infrastructure, practices, staffing, and skillsets in the key areas of security controls such as encryption, time-stamping, non-repudiation, privacy, and monitoring, enforcing policies of authentication, entitlement, encryption, time-stamping, non-repudiation, privacy, monitoring, auditing, and protection on all access.

8.3.3 Risk-mitigation considerations for Big Data in the cloud

Any deployment model for B-DC often depends on a risk analysis. To study the risk analysis we have to review and consider the following parameters [26], which are listed in Table 8.5.

Table 8.5   Risk analysis for Big Data in the cloud

Risk	Approach
Scaling	Allow the solution to scale elastically in volume, velocity, and variety as B-DC analytics requirements.
Performance	Enable the solution to achieve performance objectives for all workloads, current and anticipated.
Total cost of ownership	Reduce the cost of deploying B-DC analytics and maximize the productivity and efficiency of IT operations.
Availability and reliability	Enable to meet requirements for 24 × 7 high availability and reliability in the provisioned B-DC analytics service.
Security and compliance	Meet requirements for security and compliance in B-DC analytics.
Manageability	Meet requirements for administration, monitoring, and optimization of the B-DC platforms.

8.3.4   B-DC security challenges

For the past two years, users have been creating and manipulating more than 2.5 quintillion bytes of data on a daily basis. Use of large-scale B-DC infrastructures with diverse software platforms increases the attack surface of the entire system. Moreover, B-DC manipulation requires ultrafast send and response times from security and privacy solutions. This section highlights the most important B-DC security and privacy challenges that were discussed in [27], which are listed as follows:

1. *Secure data storage*: Data and logs are saved in multilayered storage media. The IT manager can control exactly what data is moved when moving data manually between layers. However, the size of dataset continues to be growing exponentially, and we have to use autolayering for fast Big Data storage management. Autolayering solutions do not keep track of where the data is stored, which poses new challenges to secure Big Data storage.

2. *Input validation*: Many cases of Big Data use require data collection from many sources (security information and event management system collect event logs from millions of hardware devices and software applications in an enterprise network). Input validation is considered as a key challenge in the data collection process. How can we trust and validate that a source of input data is not malicious and how can we filter malicious input from our collection?

3. *Real-time security monitoring*: In all cases, real-time security monitoring is a continuous challenge; intrusion prevention system (IPS) and intrusion detection system (IDS) generate a number of alerts, which lead to many false positives. This problem might even increase with B-DC technologies. However, Big Data technologies do allow for fast processing and analytics. Therefore, it is used to provide real-time anomaly detection based on scalable security analytics.

4. *Secure communication and access control*: Data should be encrypted based on access control rules to ensure that private data is end-to-end secure and only accessible to the authorized entities. Attribute-based encryption (ABE) is one of the most specific researches in this area. To ensure security and privacy among the distributed entities in B-DC, a cryptographically secure communication framework has to be implemented and used. Section 8.5 explains some cryptographic approaches to secure B-DC.

8.4   Big Data and Cloud Computing Relationship

Along with the increasing popularity of Big Data cloud environments, the security issues introduced through integration and adaptation of this technology are also increasing.

Despite B-DC offering many benefits and many researches proposed approaches to secure B-DC environment, it is vulnerable to attacks. Attackers are consistently trying to find holes or vulnerabilities to attack the B-DC environment to get the treasure. The ability to control and inspect all network links and ports is required to ensure security and privacy. Hence, there is a need to invest in understanding the challenges, investigate vulnerabilities and components prone to attacks with respect to the integration of Big Data with cloud computing, and come up with a platform and infrastructure that is less vulnerable to attacks.

8.4.1 Big Data and cloud computing

As mentioned before, cloud comes with different security issues and problems because it includes many technologies such as networks, databases, operating systems, and virtualization. After B-DC integration, the need for privacy and security of Big Data causes the implementers, developers, and research communities to re-think and re-design solid and secured solutions and methodologies. Hence, security issues of these systems and technologies are solved as much as possible and became applicable to B-DC. First, the network that is used to interconnect the systems in this environment must be secured to mitigate risks and network attacks. Second, mapping of the guests (virtual machines) to the hosts (physical machines) have to be performed securely. Moreover, we have to ensure that appropriate policies are enforced for data sharing, as security alone does not involve the encryption of the data. In addition, memory management and resource allocation methodologies have to be secure. Finally, we can consider what befits of data mining techniques and analyze the malware detection in B-DCs.

8.4.2 Modern DPI and SDN in Big Data security

In today's exponentially evolving threat landscape, enterprises, operators, and content providers can gain an advantage through the implementation of an intrusion detection and prevention system (IDPS), which offers an additional security layer to protect against recent attacks. But, this belief could be changed when the data is moved to the cloud. This means that, the traditional security ideology will not be efficient with such type of security, especially for auditing compliance problem and new deep packet inspection (DPI) systems. Therefore, the software-defined networks (SDNs) can build, create, and update security policies in a dynamic way.

Figure 8.6 shows the global architecture of securing a B-DC-based SDN solution. Through this test, SDN controls the consumer traffic generated from the cloud by the OpenFlow Switches (OFS). Moreover, the rules build by the SDN controller can be generated/modified/updated/deleted using the OpenFlow protocol nature between the OFS and SDN and irrelevant to the internet service providers.

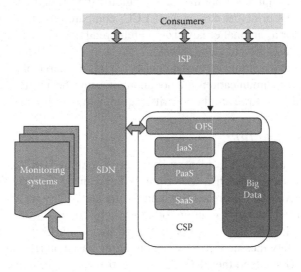

Figure 8.6: SDN architecture for securing Big Data in the cloud.

8.5 Identity-Based and Cryptographic Approaches to Secure Big Data Accessing in the Cloud

Cloud computing has become the tool of choice for Big Data processing and analytics due to its reduced cost, broad network access, elasticity, resource pooling, and measured service. B-DC integration enables consumers to store and analyze their data using shared computing resources while easily handling fluctuations in the volume and velocity of the data. However, cloud computing comes with risks. The shared computing infrastructure introduces many security concerns not present in more traditional computing architectures. The cloud provider and tenants may be untrusted entities who try to tamper with data storage or computation. These concerns motivate the need for a novel framework for analyzing cloud computing security, as well as for the use of cryptographic tools to address cloud computing security goals. In this section, we propose identity-based approaches for B-DC applications, and survey existing cryptographic tools used to secure Big Data in the cloud.

8.5.1 Identity-based approaches

This section aims at proposing new approaches for virtual identity (V_{ID}) for anonymous communication in B-DC. The objective is to create an identity that could help in preventing the reverse of access chain in the B-DC through hiding the main user identity. This means that, instead of executing a service for a known identity in an unknown place (B-DC), we hope to execute a service with an unknown identity (V_{ID}) in an unknown environment (B-DC). To achieve this, we propose a new anonymous access using virtual identity. This identity is created either based on identity-based encryption (IBE) or pseudonym-based encryption (PBE) mechanisms.

The two main secure mechanisms for creating V_{ID} are public-key cryptography for encryption and digital signatures. The key length for frequently used public-key cryptography algorithms has increased over recent years, and this has put a heavier processing load on applications using these algorithms. This burden has ramifications, especially for social and commercial sites that conduct large numbers of secure transactions. Elliptic curve cryptography (ECC) is showing up in standardization efforts, including the IEEE P1363 Standard for Public-Key Cryptography. The principal attraction of ECC, compared to others, is that it appears to offer equal security for a far smaller key size, thereby reducing processing overhead. The two solutions commonly use private-key generator (PKG) to calculate the V_{ID}. However, these approaches assume that a centralized trust authority (TA) is in charge of private-key generation. Thus, the anonymous communications are not anonymous to the TA. But, they are different in the encryption technique. However, the security requirements for both are the same.

8.5.1.1 First approach: IBE

Public-key-based solution, such as identity-based cryptography (IBC), is an asymmetric key cryptographic technique, in which a user's public key can be an identifier of the user, and the corresponding private key is created by binding the identifier with a system master secret. Therefore, they are used later as a solution for anonymous communications.

IBE is based on the IBC and was first proposed by Boneh [30]. We build a platform to provide security to safety messages in a cloud computing based on IBC and ECC. As shown in Figure 8.7, the user U sends to the PKG his/her identity U_{ID} (e.g., user@homeoperator.com) and the requested service name Ser. Hence, the PKG calculates the user V_{ID} and the key pair (public UP/private UD) by choosing an ECC with a point P as a generator of it. Then the PKG

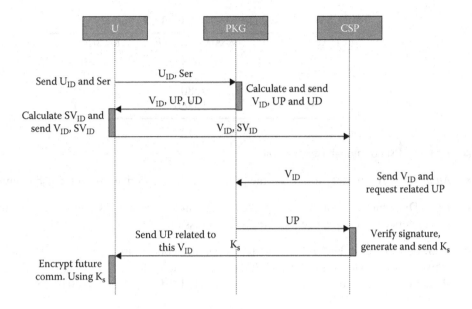

Figure 8.7: Proposed IBE message exchanges.

uses V_{ID} to generate the user's public and private keys and sends V_{ID}, UP, and UD to the user. The user wants to be authenticated by SP; therefore, he/she uses an identity-based signature (IBS) to calculate SV_{ID} and sends it with V_{ID} to the SP. Then the CSP sends V_{ID} to the PKG and asks for public key corresponding to the V_{ID}. The CSP verifies the SV_{ID} by decrypting it using the UP. If it retrieves the V_{ID}, then the authentication succeeds. At the end, the CSP generates and sends a shared secret key (K_s) to the user to encrypt the future communications between them.

We are motivated to propose a security system that considers the given conflicting goals of privacy and traceability, and the challenges in designing a privacy-preserving defense scheme for new V_{ID} approach for anonymous communication in cloud networking. Specifically, our new identity-based system includes

- New entities distribution

- New algorithms to encrypt, decrypt, sign, and verify messages

- Pseudonym-based scheme to assure device user privacy

- New pseudonym generation and update of the private key

- Assurance of non-repudiation of the device's owners in case of dispute

- Avoidance of attacks

Our solution assures:

- Non-repudiation signature (using ECDSA protocol).

- High level of security through the use of ECC protocol (see Table 8.6 to compare with RSA).

Table 8.6   Equivalent Key sizes for ECC and RSA

ECC Key size (bits)	RSA Key size (bits)	Key size ratio
163	1024	1: 6
256	3072	1: 12
384	7680	1: 20
512	15360	1: 30

- *Privacy*: Pseudonym-based scheme.

- *Anonymity*: Anonymity to PKG—the private key is self-derived from the pseudonym.

- *PFS*: Dynamicity through the updated phase in which the PKG updates its public parameters or through a pseudonym for each session.

- *PBS*: Elliptic curve Diffie–Hellman (ECDH) protocols assure a secure communication between users equipments (UEs). Through this amended EC protocol, each device should not have access to data that are transmitted before it joins the discovery phase.

8.5.1.2   Second approach: PBE

The second approach is based on PBE, which was proposed for key management for anonymous communication in mobile ad hoc networks [28,29]. In this approach, a user uses PBE to calculate its own VID. The PKG will just compute the user's private key, which depends on its secret master key. The PKG will act as an authority which certifies that the user has the private key corresponding to his/her public key. The following scheme shows the PBE algorithm with the communication process details in Figure 8.8:

Players: *U: User, PKG: Virtual ID (VID) generator and CSP: Cloud Service Provider*

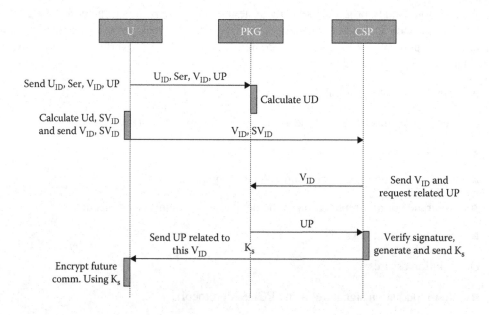

Figure 8.8: Proposed PBE message exchanges.

System setup: *UID: User ID, Ser: Requested Services, VID: Virtual ID (Pseudonym), UP: User Public Key, UD: User Private Key, SVID: Signature of Virtual ID (Pseudonym) and K_s: Shared Secret Key*

Key extraction: *Given UID, Ser, and VID (Pseudonym), user chooses random value k to calculate UP=K*P. PKG calculates UD=S*UP (P is a point on elliptic curve, S is the master secret key of PKG)*

Signature generation: *SVID=Encrypt VID using UD. Authenticate using VID and SVID*

Verification: *SP sends VID to PKG and request UP. SP decrypt SVID using UP if it retrieves UD, therefore the authentication succeeds*

Our proposed work uses (IBC+ECC) for master-key and private-key generation. Therefore, our approach increases the level of security in order to prevent any form of attack and guarantees the following:

- Non-repudiation/integrity and protection

- Privacy/anonymity

- Dynamicity with ECC

- Identity disclosure

Our two solutions based on (IBE and PBE) have two new contributions:

In the first solution, *dynamicity* is assured by regenerating virtual identity for each service.

In the second solution, *dynamicity* is assured by regenerating virtual identity for each service, whereas in *anonymity*, ECC is based on pseudonym instead of identity, which leads to assure privacy, anonymity, and solve identity disclosure problems.

Furthermore, in both approaches, the use of ECC achieves the highest speed of IBE's and PBE's functions.

8.5.2 Cryptographic approaches

In this section, we will highlight some approaches of cryptographic techniques that are particularly applicable to achieving secure B-DC [31]:

- Homomorphic encryption (HE)

- Verifiable computation (VC)

- Multiparty computation (MPC)

- Functional encryption (FE)

  - IBE (explained in Section 8.5.1)

  - ABE

8.5.2.1 Homomorphic encryption

HE is a cryptographic scheme that enables functions to be computed on encrypted data without decrypting it first.

From Figure 8.9, given only the encryption of a message, one can obtain an encryption of a function of that message by computing directly on the encryption.

The types include

1. Privacy homomorphic

   ■ Rivest, Adleman, Dertouzos (1978)

2. Partial homomorphic

   ■ RSA cryptosystem (1977)
   ■ El-Gamal cryptosystem [32] (1984)
   ■ Paillier cryptosystem (1999)
   ■ Boneh–Goh–Nissim cryptosystem (2005)

3. Fully homomorphic encryption (FHE)

   ■ Gentry's lattice-based cryptography (2009)

8.5.2.2 Verifiable computation

VC allows the data owner to check the integrity of the computation. In this scheme, the data owner (the prover) gives the data, along with a specification of the computation desired to some entity. We have to know that it must be easier to verify the proof than to perform the computation. There would be no reason for the data owner to outsource the computation to the prover in the first place if verification of the computation is as hard as the computation itself. Most of VC techniques are still much too slow for most B-DC analytics applications.

8.5.2.3 Multiparty computation

MPC is appropriate to take pros of the semi-trusted cloud setting. To ensure data and computation integrity and confidentiality, MPC leverages honest parties' presence. (It is not eligible to know which parties are honest.) In MPC, if many parties are corrupted by an adversary, they can break confidentiality, though no single party learns anything about the data. A threshold adversary model is used by most MPC schemes; this model limits the total number of nodes that can be corrupted by an adversary at any given time. Therefore, to obtain the actual output,

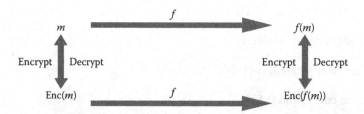

Figure 8.9: Homomorphic encryption.

the nodes can reproduce shares of the output that the receiver can reconstruct. The cloud node does not learn anything about the data or the computation output because it only sees individual shares of the data.

8.5.2.4   Functional encryption

FE is an asymmetric key algorithm and considers one of the new approaches in cryptography. Without revealing any other information, FE permits the secret key to decrypt only certain functions of the original texts. FE has great implication on the development of distributed computing and cloud computing. In 2005, FE was proposed by Amit Sahai and Brent Waters [33]. By 2012, researchers have developed FE schemes that support arbitrary functions. Later, some encryption techniques were introduced and used in distributed environments and cloud computing, including:

- IBE

- ABE

- Predicate-based encryption

In the previous techniques, the data owner predefines granular access privileges for the users to access data and encrypts it using public key. Users get secret keys from a private key generator (trusted key) server and then decrypt some parts of the encrypted data based on their assigned privileges.

Such generalization of techniques is an important step toward a unified theory for computational privacy in distributed environments. Generalization of the above-mentioned techniques has been formalized as FE [34]. The relationships between the above techniques and FHE have been studied, and interconnections have been established.

8.5.3   Other approaches

In this section, we will present various security measures and approaches that would enhance the security of B-DC environment. Since the B-DC environment is a mixture of many different technologies, we will introduce various solutions, which collectively will make the environment secure [6].

8.5.3.1   File encryption

All the data stored in B-DC environment should be encrypted to prevent the attacker from stealing collection of critical information from one node. Various encryption keys should be used on different nodes (machines), and the key information should be stored centrally behind strong security solution (firewalls–IPS–IDS).

8.5.3.2   Network encryption

To prevent MITM attacks and its various techniques, all the network communication should be encrypted. The RPC procedure calls and messages transferred through the network should happen over secure connection (i.e., SSL). Therefore, even if the attacker can get into network communication packets or sniff the connection, he or she still cannot extract useful information or manipulate packets.

8.5.3.3 Logging

To monitor all user activities, all these activities should be recorded as logs and audited regularly to find if any malicious actions are performed or any malicious intervention to manipulate the data in any machine.

8.5.3.4 Honeypot nodes

Honeypot nodes should be present in the B-DC and behave as a normal node or machine but it is a trap. These honeypots trap the attackers to get fake information.

8.5.3.5 Layered framework

As defense in depth strategy applied, we have to prepare a layered framework for assuring B-DC environment security. Several protection layers are used such as virtual machine layer, cloud storage layer, cloud data layer, and the virtual network monitor layer throughout the environment to secure it.

8.6 Summary

This chapter has set out a number of challenges and issues regarding Big Data security integration with the cloud computing environment. We proposed the main guidelines for Big Data integration through the SWOT analysis. Then, and based on the previous analysis, we addressed the weak points that output from this integration and potential solutions. After that, we highlighted cryptographic approaches that could be used in accessing different types of services in cloud computing. Moreover, we proposed new approaches (IBE and PBE) for V_{ID} for anonymous communication in B-DC. IBE and PBE are considered as a suitable paradigms for achieving high degree of security and efficiency in such sophisticated network access to many online services and applications introduced by the integration of Big Data with cloud computing.

References

1. National Institute of Standard and Technology (NIST), Gaithersburg, MD, *The NIST Definition of Cloud Computing*, Special Publication 800–145, September 2011. http://csrc.nist.gov/publications/nistpubs/800-145/SP800-145.pdf.

2. J. Yao, S. Chen, S. Nepal, D. Levy, and J. Zic, TrustStore: Making Amazon S3 trustworthy with services composition, in *Proceedings of the 10th IEEE/ACM International Symposium on Cluster, Cloud and Grid Computing*, 2010, pp. 600–605, IEEE, Melbourne, Australia.

3. D. Zissis and D. Lekkas, Addressing cloud computing security issues, *Future Generation Comput. Syst.*, vol. 28, no. 3, pp. 583–592, 2011.

4. Big Data. Garner. https://www.gartner.com/it-glossary/Big-Data/.

5. ASPERA, an IBM Company, Emeryville, CA, White paper, *Taking Big-Data to the Cloud.* http://cloud.asperasoft.com/fileadmin/media/Asperasoft.com/Resources/White_Papers/Taking_Big_Data_to_the_Cloud_AsperaWP.pdf.

6. V.N. Inukollu, S. Arsi, and S.R. Ravuri, Security issues associated with Big data in cloud computing, *Int. J. Network Secur. Appl.*, vol. 6, no. 3, pp. 45–56, 2014.

7. A. Kilzer, E. Witchel, I. Roy, V. Shmatikov, and S.T.V. Setty. *Airavat: Security and Privacy for MapReduce*, in *Proceedings of the 7th USENIX Symposium on Networked Systems Design and Implementation*, 2010, pp. 297–312, IEEE/ACM, San Jose, CA.

8. A. Juels and B.S. Kaliski Jr., PORs: Proofs of retrievability for large files, in *Proceedings of the 14th ACM Conference on Computer and Communications Security*, 2007, pp. 584–597, ACM, New York.

9. G. Ateniese, R.B. Johns, R. Curtmola, J. Herring, L. Kissner, Z. Peterson, and D. Song, Provable data possession at untrusted stores, in *Proceedings of the 14th ACM Conference on Computer and Communications Security*, 2007, pp. 598–609, ACM, New York.

10. C. Erway, A. Küpçü, C. Papamanthou, and R. Tamassia, Dynamic provable data possession, in *Proceedings of the 16th ACM Conference on Computer and Communications Security*, 2009, pp. 213–222, ACM, New York.

11. H. Shacham and B. Waters, Compact proofs of retrievability, in *Proceedings of the 14th International Conference on Theory and Application of Cryptology and Information Security*, 2008, pp. 90–107.

12. X. Zhang, C. Liu, S. Nepal, S. Panley, and J. Chen, A privacy leakage upper-bound constraint based approach for cost- effective privacy preserving of intermediate datasets in cloud, *IEEE Trans. Parallel Distrib. Syst.*, vol. 24, no. 6, pp. 1192–1202, 2013, IEEE, Australia.

13. Q. Wang, C.Wang, K. Ren, W. Lou, and J. Li, Enabling public auditability and data dynamics for storage security in cloud computing, *IEEE Trans. Parallel Distrib. Syst.*, vol. 22, no. 5, pp. 847–859, 2011.

14. C. Wang, Q. Wang, K. Ren, and W. Lou, Privacy-preserving public auditing for data storage security in cloud computing, in *Proceedings of the 30th IEEE Conference on Computers and Communications*, 2010, pp. 1–9, IEEE, Beijing, China.

15. G. Ateniese, S. Kamara, and J. Katz, Proofs of storage from homomorphic identification protocols, in *Proceedings of the 15th International Conference on Theory and Applications of Cryptology and Information Security*, 2009, pp. 319–333, Springer, Tokyo, Japan.

16. Y. Zhu, H. Hu, G.-J. Ahn, and M. Yu, Cooperative provable data possession for integrity verification in multi-cloud storage, *IEEE Trans. Parallel Distrib. Syst.*, vol. 23, no. 12, pp. 2231–2244, 2012.

17. R. Curtmola, O. Khan, R.C. Burns, and G. Ateniese, MR-PDP: Multiple-replica provable data possession, in *Proceedings of the 28th IEEE Conference on Distributed Computing Systems*, 2008, pp. 411–420, IEEE, Beijing, China.

18. X.F. Meng and X. Ci. Big data management: Concept, techniques and challenges, *J. Comput. Res. Dev.*, vol. 50, no. 1, pp. 146–169, 2013 (in Chinese).

19. D.R. Shen, G. Yu, X.T. Wang, T.Z. Nie, and Y. Kou, Survey on NoSQL for management of big data, *J. Softw.*, vol. 24, no. 8, pp. 1786–1803, 2013 (in Chinese).

20. Y. Zhao, Y. Li, W. Tian, and R. Xue, Scientific-workflow-management-as-a-service in the Cloud. in *Proceedings of the 2nd International Conference on Cloud and Green Computing*, 2012, pp. 97–104, IEEE, Beijing, China.

21. W. Tian and Y. Zhao, *Optimized Cloud Resource Management and Scheduling*, October 2014, Morgan Kaufmann, Waltham, MA.

22. Vormetric data security, Securosis, *Securing Big-Data: Security Recommendations for Hadoop and NoSQL Environments*, Securosis Blog, version 1.0, 2012. http://enterprise-encryption.vormetric.com/big-data-security-recommendations-for-hadoop-and-nosql-environments.html.

23. Ventana Research, Pentaho, *Ventana Research Benchmark Research: Big-Data Integration*, Sponsor Report, Prepared for Pentaho, May 2014. http://www.ventanaresearch.com/.

24. I.A.T. Hashem, I. Yaqoob, N.B.Anuar, S. Mokhtar, A. Gani, and S.U. Khan, The rise of "big data" on cloud computing: Review and open research issues, *Inform. Syst.*, vol. 47, pp. 98–115, Elsevier, 2015.

25. National Institute of Standards and Technology (NIST), Gaithersburg, MD, *US Government Cloud Computing Technology Roadmap*, Special Publication 500-293, vol. II, Release 2.0, October 2014. http://dx.doi.org/10.6028/NIST.SP.500-293.

26. Cloud Standards Customer Council, *Deploying Big-Data Analytics Applications to the Cloud: Roadmap for Success*, May 2014. http://www.cloud-council.org/CSCC_Deploying_Big_Data_Analytics_Applications_to_the_Cloud_FINAL.pdf.

27. Cloud Security Alliance, *Top Ten Big-Data Security and Privacy Challenges*, November 2012. https://downloads.cloudsecurityalliance.org/initiatives/bdwg/Big_Data_Top_Ten_v1.pdf.

28. L. Zhang and Y. Hu., An identity-based broadcast encryption protocol for ad hoc networks, in *International Conference for Young Computer Scientists*, 2008, pp. 1619–1623.

29. D. Huang, Pseudonym-based cryptography for anonymous communications in mobile ad hoc networks, *Int. J. Security Networks*, vol. 2, pp. 272–283, 2007.

30. D. Boneh and M. Franklin, Identity-based encryption from the weil pairing, in J. Kilian (ed.) *CRYPTO 2001, LNCS 2139*, 2001, pp. 213–229, Springer-Verlag, Berlin, Germany.

31. S. Yakoubov, V. Gadepally, N. Schear, E. Shen, and A. Yerukhimovich, *A Survey of Cryptographic Approaches to Securing Big-Data Analytics in the Cloud*, MIT Lincoln Laboratory, Lexington, MA, 2014, IEEE, Waltham, MA.

32. T. ElGamal, A public key cryptosystem and a signature scheme based on discrete logarithms, in G.R. Blakley and D. Chaum (eds.) *Advances in Cryptology*, Springer-Verlag, Berlin, Germany, 1985, pp. 10–18.

33. A. Sahai and B. Waters, Fuzzy identity based encryption, in *Proceedings of Eurocrypt*, 2005, University of Aarhus, Denmark.

34. D. Boneh, A. Sahai, and B. Waters, Functional encryption: Definitions and challenges, in *Theory of Cryptography*, Springer, 2011, pp. 253–273.

Chapter 9

Toward Reliable and Secure Data Access for Big Data Service

Fouad Amine Guenane

Michele Nogueira

Donghyun Kim

Ahmed Serhrouchni

CONTENTS

Abstract

Two innovative technologies have emerged on the last decade: *Big Data* and *Cloud Computing*. On the one hand, *Big Data* warrants innovative data analytics for a variety of new and existing data, providing real business benefits [4]. Digital traces from people and systems produce every day a massive volume of data collected from actions, sensors, algorithms, and social web. On the other hand, Cloud Computing has the potential to enhance business agility and productivity while enabling greater efficiencies and reducing costs. The convergence of Cloud Computing and Big Data technologies has envisioned to offer a cost-effective delivery model for Cloud-based Big Data analytics. However, this convergence imposes new challenges to the traditional security technologies that may need to be redesigned considering the current security models and tools [9]. In fact, Big Data-sets have made Big Data storage systems attractive targets for cyber attackers, whose goal lies in compromising the confidentiality, integrity, and availability of data and information. In order to address this new problem and further achieve a secure, available, and reliable Big Data Cloud-based service, we not only present in this chapter the state-of-the-art of Big Data Cloud-based services, but also a novel architecture to manage reliability, availability, and performance for accessing Big Data services running on the Cloud. The architecture is a service offered by the Cloud provider as an outsourcing of firewalling functions, characterized by a high availability, huge and dynamic resource provisioning. Those firewalling functions aim at ensuring a reliable and secure data access, high data availability, and cope with distributed denial-of-service (DDoS) attacks by managing their impacts in the Cloud and tilting the playing field in favor of the defender using network function virtualization technology [7] and Cloud benefits, such as flexibility, elasticity, and on-demand provisioning. This chapter follows several steps, such as the investigation of related works, the architecture design, software developing, testing, and results analysis. Hence, this chapter is organized as follows: introduction; the state-of-the-art; the description of the proposed reliable and secure data access for Big Data Cloud-based service; results from a case study applying the architecture; and conclusions and future work.

9.1   Introduction

By definition, *Big Data* relies on a high-volume, high-velocity, and high-variety information assets that demand cost-effective and innovative forms of information processing for enhancing the insight and decision making [4]. *High-volume* means the data resulting from volume transaction and other traditional data types, causing a massive analysis issue. *High-variety* is related to the existence of a vast amount of types of information (Rich-Data). Accordingly, there is a decision issue caused by the translation of this Rich-Data. *High-velocity* means how fast the data is being produced and how fast the data must be processed to meet demand. The advantages of Big Data are well established [11], such as enabling experimentation to discover needs by collecting more accurate and detailed data replacing or supporting the human decision making by unearthing valuable insights that would otherwise remain hidden; and innovating new business models, products, and services. Thus, it becomes difficult to process a Big Data using on-hand database management tools. An efficient solution is to store and process a high volume of data in the *Cloud*.

In fact, there is a growing interest in outsourcing services to a Cloud provider in order to reduce management and deployment [13]. On the one hand, we have the benefit of Cloud characteristics such as on-demand self service, rapid elasticity, flexibility, broad access network, and availability. On the other hand, the data security (confidentiality, availability, and integrity) becomes a major concern. In fact, if one centralizes the data in one place [14], it becomes a valuable target for attackers, which undermines trust in the organization and damages its reputation. Moreover, the reliability and availability of the Big Data service are important for Big Data model [6]. In fact, if it is not insured, the operation can cease to function and the downtime can lead to loss in productivity, in revenue, damaged customer relationships, bad publicity, and lawsuits.

Nowadays, the most critical threat impacting the reliability and availability of a Cloud service is undoubtedly the DDOS attacks. They are prepared and coordinated by multiple hosts and then directed to a single target making it weak faced to those attacks, particularly if the target is not prepared or equipped beforehand. With an average duration of 1.4 days of effective denied service and an average consumed bandwidth of 1.5 Gbps [10], all organizations are a susceptible target. For many companies even an hour of service unavailability is considered too long and many of them use firewalls as a DDOS mitigation tool, but unfortunately the limited physical resources of firewalls become overwhelmed and do not effectively respond to such attacks.

This chapter presents a reliable and secure data access for Big Data service, it is ensured by a novel Cloud-based DDOS mitigation architecture. The proposal is mainly based on virtual firewall (VF) instances characterized by a dynamic resource provisioning to deal with the traffic volume. Thus, it allows to cope with DDOS attacks by offering more resources to the defender.

The chapter proceeds as follows. Section 9.2 recapitulates the background and related works. Section 9.3 details the reliable and secure data access for Big Data Cloud-based service. Section 9.5 presents tests and results. Finally, Section 9.6 concludes the chapter and outlines future works.

9.2 Related Works

Denial-of-service attacks are characterized by an explicit attempt to prevent legitimate users of a service from using it [1]. This issue is important to the business model of the Cloud provider, simply because if customers get poor service, they will move their business elsewhere. Additionally, it affects the availability and integrity of the data and, consequently, the reputation of the Cloud provider. To the best of our knowledge, there has been no study or research focusing on DDOS attacks considering the Big Data context. First, this section presents an overview of DDOS attack types, categorizes mitigation tools, and discusses some prominent methods. Second, we discuss some access mechanisms in Big Data context.

The attacker uses several methods in order to consume the victim's resources and disrupt operation. All actions that cause server shutdown are called *denial of service* [12]. We are interested in the mitigation tools and we categorize them according to the methods employed to mitigate the DDOS [15]. There are four main approaches of mitigation tools [15]: bandwidth defense, rate filtering, signature filtering, and moving target. On the *bandwidth defense* approach, it uses a large distributed network characterized by sufficient capacity to cope with the attack. Note that services such as Akamai, a cloud computing service and content delivery network provider for media and software delivery, use this technique. In *rate filtering*, the majority of the DDOS attacks is carried by one provider then other providers filter out all traffic from that one. Its drawback is that only a partial service is preserved. The *signature filtering* approaches rely on recognizing signatures created for typical flood packets, such tools are efficient and do not suffer from performance problems, but they present high percentages of false positive. Finally, *moving target* techniques involve switching to a new IP address in case of attack. Such a method makes the attacker's behavior more difficult and forces it to retrieve the new IP address from the DNS server and resend the order to all the botnet, making the attack management more difficult.

Thus, tools exist and often represent an efficient solution against large-scale attacks [15]. However, the proposed techniques are not scalable with the dynamic network changes in the runtime environment, do not easily accommodate modifications, do not counter the critical issue of harnessing sufficient resource to overcome DDOS attacks, and are characterized by high financial and computational costs. Thus, these mechanisms are no more suitable for Big Data Cloud-based service because of the expensive cost.

Our work proposes an efficient Cloud-based service that makes use of the flexibility and availability of the Cloud to come as a reinforcement for traditional security infrastructures that cope with massive unanticipated volumes of traffic. Thus, the subscriber faces DDOS attacks by an innovative DDOS mitigation service characterized by self-scalabilty, high availability, and dynamic resource provisioning using a novel firewalling Cloud-based approach to purchase a high availability and reliability for a Big Data Cloud-based service.

9.3   Architecture for Reliability and Security on Data Access

From the client's side, a Cloud provider offers a Big Data service guarantee a high availability and reliability. However, the questions are: *How to ensure this high availability? How to cope with a huge flooding attack? How does it work technically?* Thus, our primary goal is to provide access to data by protecting the Big data service and the access control server from DDOS attacks, as well as avoiding the denial of the service. As shown in Figure 9.1, both services (Big Data and DDOS mitigation) are part of the operating environment of the Cloud provider. Hence, the main challenge for the provider is to manage the network operations such as routing, analysis, traffic balancing, and virtual component instances of the DDOS mitigation architecture.

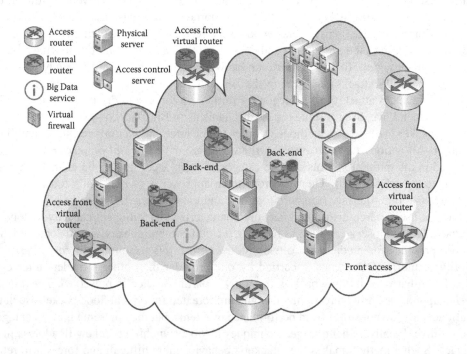

Figure 9.1: General architecture.

Therefore, the Cloud provider asks its customers to subscribe to the offer by guaranteeing them a 99% availability across a year. Additionally, it proposes to analyze a bandwidth capacity with functional traffic filtering rules and other proposed by the subscriber. To achieve this goal, it forwards the subscriber traffic across the DDOS mitigation architecture. Thereafter, the Cloud provider acts as an intermediate for the customer and forwards the legitimate traffic to Big Data service. This solution supplies the subscriber an additional processing capacity on-demand, if dynamically needed, without having to buy, deploy, or maintain new physical equipments. Thus, the Big Data service is more reliable and available to the end users and avoids denial of service.

Figure 9.2 represents a general overview of our proposal. The DDOS mitigation architecture is composed of several instances distributed all over the Cloud environment. It comprises three main virtual items: *access front virtual router*, *VF instance*, and *back-end virtual router*. They are respectively instantiated on the physical access router, physical server, and physical internal router. Each virtual item is characterized by huge resources, processing, and network capacity. We discuss and present these different virtual components in the next sections. In a nutshell, the traffic destined to Big Data service is first forwarded to the access front virtual router which is the point of entry for the service architecture. The access front virtual router balances the traffic across the VF instances that analyze the traffic and deliver the legitimate traffic to the back-end virtual router as suggested by Figure 9.2. Finally, the traffic is sent to Big Data service which includes storage and access control.

9.3.1 Access front virtual router

The access front virtual router is a virtual router characterized by a huge processing capacity and a large amount of resources. These resources are provided by the Cloud provider in order to be adaptable to the traffic volume. Its features include

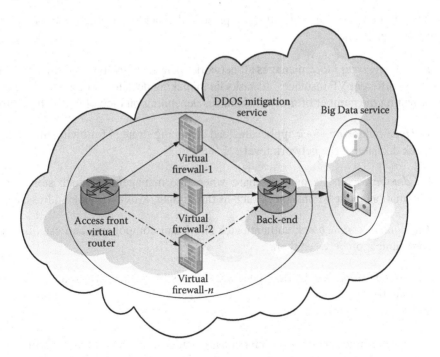

Figure 9.2: Cloud-based DDOS mitigation architecture: Overview.

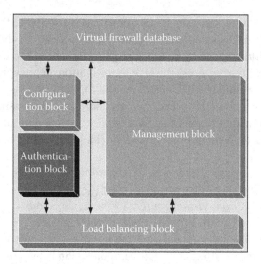

Figure 9.3: Access front virtual router framework.

- Authenticating all VF nodes and establishing a secure channel.

- Managing load balancing policies over different firewall nodes.

- Maintaining and updating VF resources database.

- Applying the proposed load balancing strategies decided by the management module.

As illustrated in Figure 9.3, each feature is represented by a block (module) on the access front virtual router framework:

- *The management block* manages all network operations involving traffic load balancing across different VF instances. Its tasks lie in making decision for the load balancing and providing the number of firewall instances depending on the volume of the traffic.

- *The load balancing block* applies the load balancing strategy following the management block decision at the network level.

- *VF database* consists of a database where the management block stores and reads information about VF instances such as throughput, availability, IP address, and so on.

- *The authentication block* authenticates all virtual components and establishes secure communication between them.

- *The configuration block* offers interface to the system's administrators, allowing them to instantiate or delete additional resources (firewall), modify the load balance policy, and manage the parameters.

The percentage of traffic sent to every VF node is calculated by a simplified algorithm as shown in Algorithm 9.1. In fact, this algorithm takes into account predefined parameters stored and updated in the VF database, such as latency, throughput, processing capacity, and other.

Algorithm 9.1: Load balancing management

Require: Firewall node parameters—local database
Ensure: Load balancing policy
  while true **do**
    $n = Number - of - firewall - instances$;
    Input= input traffic volumetry;
    Read Firewall database;
    if $Input \geq \sum_{i=1}^{n} Throughput_i$ **then**
      Create new firewall node
    else
      Execute session dynamic load balancing
    end if
  end while

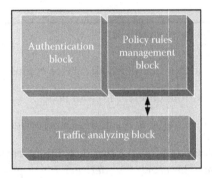

Figure 9.4: VF node framework.

9.3.2 VF node

Every firewall node (instance) acts as a traditional firewall. They analyze all incoming traffic from access front virtual router due to the *traffic analyzing block* as presented in Figure 9.4. The security policy rules are replicated in every instance and stored in *policy rules management block*. The *authentication block* coordinates and manages the authentication process with other authentication blocks.

For more reliability, availability, and efficiency, we deploy the VF, instances in parallel as shown in Figure 9.2. In [5], authors assume that this organization is more relevant in case of any trouble, such as system overloads, unreachable virtual machine, and massive attacks.

9.3.3 Back-end virtual router

The back-end virtual router is represented in Figure 9.5 and is characterized by a huge processing capacity and resources. It is the final checkpoint before reaching the Big Data service. Its features are

■ Broadcasting outcome requests from Big Data service to other services across the access front virtual router in order to manage the outcoming traffic.

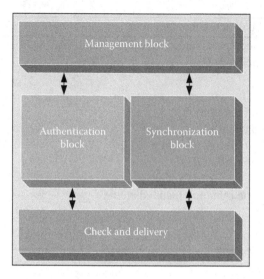

Figure 9.5: Back-end virtual router framework.

- Generating information to synchronize all service components.

- Achieving a last check from different firewall nodes.

- Delivering the legitimate traffic to reach Big Data service.

The *synchronization block* processes the broadcasting of outcome requests and generates synchronization information. The *check and delivery block* processes analysis and delivery tasks. The *management block* offers interface to the system's administrators, allowing them to manage the virtual router. The *authentication block*, as mentioned before, coordinates and manages the authentication process with other authentication blocks.

9.3.4 Network management

The network management (operations) features are supported by the proposed architecture. As presented in Figure 9.6, the proposed service executes and manages different network operations. Authentication and load balancing are located at the access front virtual router and the analysis is mainly and exclusively performed by the firewall nodes. Verification, delivery, and synchronization are performed by the back-end virtual router. We describe the operations as follows:

- *The authentication operation*: Its main feature lies in processing authentication and create a secure tunnel between all the components of the service. As previously mentioned, we develop the authentication operation in order to achieve this feature. It is based on radius server and provides different authentication protocols.

- *The forwarding or load balancing*: It assists in achieving the fluidity of the traffic across every network point, avoiding bottleneck, and improving availability. Moreover, our proposal is based on Khiyaita et al. study [8], we decided to adopt a centralized approach where a single node is responsible for managing the distribution of traffic across the network. In our case, we choose the access front virtual router to manage the forwarding and load balancing features because it is more relevant to dispatch the traffic

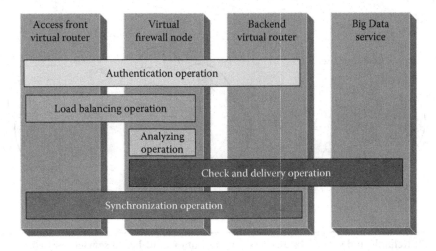

Figure 9.6: Network operations framework.

from the entry point of the service avoiding a bottleneck. (As reminder, the access front virtual router are characterized by a huge resources and is dynamically provided.)

■ *Analysis and monitoring*: In order to ensure the filtering function and according to our needs which are to store the traffic history when inspecting packets, we deploy a *Stateful* firewall. Indeed, this mechanism allows to ensure an application layer filtration and in the same time to not exceed the transport/session layer. Thus, every firewall node executes the Netfilter Firewall [3] software. Regarding parallel firewalling, our primary goal is to simultaneously treat huge and heterogeneous traffic. In fact, parallel designs are scalable since it is easier now to introduce additional firewall nodes and the result will improve reliability and availability [5]. Thus, a data-parallel firewall design is the mechanism that best fits our needs.

■ *The check and delivery operation*: As the name suggests, it is the last checkpoint of the legitimate flow, where the incoming traffic from every firewall nodes gather at a single point that simplifies delivery operation. Finally, the traffic is delivering to Big Data service.

■ *The synchronization operation*: It provides two features: (1) relies the outgoing requests of network subscriber and (2) supports messages exchanged between the access and back-end virtual router in order to ensure collaboration.

We detailed DDOS mitigation approach that provides a secure and available Big Data service. The proposal considers the network management for an optimal exploitation. The following section discusses the test scenarios to evaluate the effectiveness of the proposal.

9.4   Performance Analysis

We demonstrate the efficiency of the proposed service in terms of reliability, availability, and processing. We deploy a real testbed in local Cloud environment, illustrated by Figure 9.7. The environment consists of a virtual router that acts as access front virtual router. Additionally, virtual machines are instanciated and deployed in parallel such as firewall nodes. Every VF

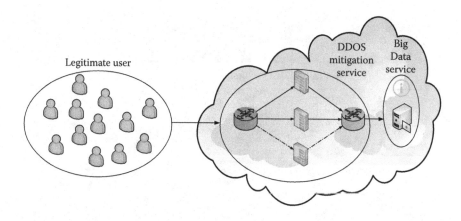

Figure 9.7: Reliable and secure data access for Big Data Cloud-based service testbed.

instance is a virtual machine equipped with a single processor, Intel (R) Xeon (R) CPU-E5335, with 2.00 GHz processing and 512 Mega Octet of RAM capacity. The operating system is Ubuntu 12.04 (x86-64-linux-gnu). In order to ensure the filtering function, as previously mentioned, we execute Netfilter instances.

9.4.1 Test scenario

A major challenge that faces the Big Data service is the congestion of the network bandwidth caused by various bottlenecks (stress point) that represent instances of firewalls, the gateway router, and many other. In order to simulate this congestion, we increase the bandwidth capacity by 10% and we monitor various network parameters such as latency and loss rates. The traffic is generated by the software Iperf [2]. Iperf is a tool characterized by two main functions. The first one is measuring maximum TCP bandwidth, allowing the tuning of various parameters, and UDP characteristics. The second one is generating and reporting network parameters such as bandwidth, delay, and packet loss rate. We apply the same test scenario to a *Basic* service.*

9.4.2 Baseline architecture

In order to analyze the improvement in performance that resulted from the proposed architecture, we deploy a *Basic* architecture based on a single firewall illustrated in Figure 9.8. The traffic is analyzed by a classical provider firewall. Note that we use this reference architecture because it is a basic topology commonly used by most small and medium companies. For comparison, we apply the test scenario to the *Basic* architecture and we monitor the network, mainly metrics such as end-to-end delay and packet loss rate. We use them as a reference to see the progress and the gain in terms of network performance by comparing them to the obtained results using our proposal.

We choose to employ for comparisons two main load balancing algorithms: round-robin and least connection. Round-robin algorithm maintains a list of servers and sends a new connection to the next server in the members list. Least connection algorithm (LCA) maintains a register of active server connections and transmits a new connection to the server with the fewest number of active connections.

---

*Basic Service: Big Data service provided without DDOS mitigation service.

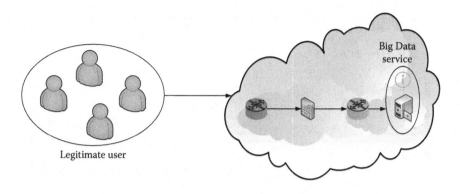

Figure 9.8: Basic architecture.

9.5   Results and Discussion

This solution is proposed to address resource limitations that characterize the physical firewall and represent a bottleneck. Our main goal is to improve fluidity of the traffic. In Figure 9.9, the *y*-axis represents the network latency (RTT: round-trip time delay) in milliseconds and the *x*-axis represents the percentage of bandwidth saturation. We vary the number of VF instances in 2 and 3 to see if it is impacting the solution performance.

Figure 9.9 shows the values of latency for the basic architecture (the line "Basic" with squares) and the two load balancing algorithms: round-robin (the line "RR-2-VF" with circles) and LCA (the line "LCA-2-VF" with diamonds). The results show a similar behavior with the architecture of reference but we observe that the latency decreases (5–10 ms) for 100% of bandwidth saturation. This decrease is confirmed in Figure 9.10, when we instantiate three VF instances with a maximum of 40%.

We can conclude that the DDOS mitigation mechanism provides a fluidity of the traffic by avoiding the bottleneck points in the network caused by the physical resource limitations.

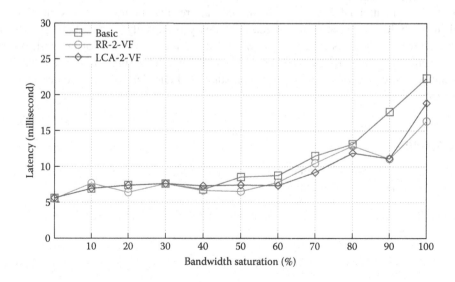

Figure 9.9: Packet delay over bandwidth saturation—two VF.

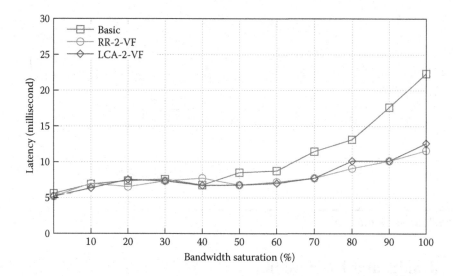

Figure 9.10: Packet delay over bandwidth saturation—three VF.

We observe that the latency is subject to the distance between the Cloud provider and the Client, this situation is a future study. Preliminary results are promising taking into account availability and reliability. However, further tests are ongoing.

9.6 Conclusion and Future Work

This chapter started from the idea of using large amount of computing resources offered by Cloud Computing associated to parallel firewall techniques to design a reliable and secure data access for Big Data Cloud-based service. This Cloud-based service is connected to DDOS mitigation service. This service is using security virtual machines characterized by significant resources to deal with DDOS attacks. The preliminary results obtained demonstrate that the service deals with flooding attacks and increase the analysis capacity by distributing traffic across multiple firewall nodes. As future work, we will treat the automation of network operations such as monitoring firewall instances, load balancing, and delivering. We are also interested in the dynamic resource allocation focused on instantiated and deleted VF instances *on demand* or *on need*.

References

1. CERT. Denial-of-service attack. https://www.cert.org/information-for/denial_of_service.cfm? (Online; accessed April 2015).

2. Iperf. The network bandwidth measurement tool. http://www.iperf.fr/. (Online; accessed April 2015).

3. Netfilter. The netfilter.org project. http://www.netfilter.org. (Online; accessed April 2015).

4. Mark A Beyer and Douglas Laney. *The Importance of 'Big Data': A Definition*. Stamford, CT: Gartner, 2012.

5. Errin W Fulp. Parallel firewall designs for high-speed networks. In *25th IEEE International Conference on Computer Communications INFOCOM Proceedings*, pp. 1–4, Barcelona, Spain, April 2006.

6. Vincent C Hu, Tim Grance, David F Ferraiolo, and D Rick Kuhn. An access control scheme for big data processing. In *International Conference on Collaborative Computing: Networking, Applications and Worksharing (CollaborateCom)*, pp. 1–7. IEEE, Miami Beach, FL, 2014.

7. Raj Jain and Subharthi Paul. Network virtualization and software defined networking for cloud computing: A survey. *IEEE Communications Magazine*, 51(11):24–31, 2013.

8. Abdelaziz Khiyaita, Mostapha Zbakh, Hanane El Bakkali, and Dafir El Kettani. Load balancing cloud computing: State of art. In *National Days of Network Security and Systems (JNS2)*, pp. 106–109. IEEE, Marrakech, Morocco, 2012.

9. Bharat B Madan and Manoj Banik. Attack tolerant architecture for big data file systems. *ACM SIGMETRICS Performance Evaluation Review*, 41(4):65–69, 2014.

10. Steve Mansfield-Devine. DDOS: Threats and mitigation. *Network Security*, (12):5–12, 2011.

11. James Manyika, Michael Chui, Brad Brown, Jacques Bughin, Richard Dobbs, Charles Roxburgh, and Angela H Byers. *Big Data: The Next Frontier for Innovation, Competition, and Productivity*. McKinsey Global Institute, 2011. http://www.mckinsey.com/insights/business_technology/big_data_the_next_frontier_for_innovation. (Online; accessed December 2015).

12. Jelena Mirkovic and Peter Reiher. A taxonomy of DDOS attack and DDOS defense mechanisms. *SIGCOMM Computer Communication Review*, 34(2):39–53, April 2004.

13. Hamid R Motahari-Nezhad, Bryan Stephenson, and Sharad Singhal. Outsourcing business to cloud computing services: Opportunities and challenges. *IEEE Internet Computing, Special Issue on Cloud Computing*, 6 February, 2009. http://www.hpl.hp.com/techreports/2009/HPL-2009-23.html.

14. Colin Tankard. Big data security. *Network Security*, 2012(7):5–8, 2012.

15. Saman T Zargar, James Joshi, and David Tipper. A survey of defense mechanisms against distributed denial of service (DDOS) flooding attacks. *IEEE Communications Surveys & Tutorials*, 15(4):2046–2069, 2013.

Chapter 10

Cryptography for Big Data Security

Ariel Hamlin

Nabil Schear

Emily Shen

Mayank Varia

Sophia Yakoubov

Arkady Yerukhimovich

CONTENTS

10.1 Introduction

With the amount of data generated, collected, and analyzed by computing systems growing at an amazing rate, big data processing has become crucial to most enterprise and government applications. For example, the rise of Internet of things (e.g., a connected refrigerator), smart phone location tracking to target advertising, and the growing adoption of health and wellness devices that collect personal statistics all represent new opportunities for novel analytics and services.

However, this proliferation of big data is not without its own dangers. The collected data often contain private information about individuals or corporate secrets that would cause great harm if they fell into the wrong hands. Criminal groups are creating underground markets where one can buy and sell stolen personal information [122]. Government intelligence services are targeting personal, corporate, and adversary government systems for espionage and competitive advantage [117]. This potential for harm is clearly demonstrated by many recent, highly publicized cyber attacks against commercial [156] and government targets [157], costing these organizations millions of dollars and causing serious damage to the individuals and institutions affected.

As these examples make clear, new and improved security tools are needed to protect systems collecting and handling big data to allow applications to reap the benefits of big data analysis without the risk of such catastrophic attacks. Fortunately, modern cryptography offers many powerful technologies that can help protect big data applications throughout the data life cycle, as it is being collected, stored in repositories, and processed by analysts. This chapter provides a brief survey of several of these technologies and explain how they can help big data security. We hope that this survey can provide big data practitioners with a better understanding of what protections are available to them as they design their applications and will also encourage the necessary discussion and collaboration to mature these tools for real-world consumption.

This chapter focuses on state-of-the-art provably secure cryptographic techniques for protecting big data applications. We do not focus on more established and commonly available cryptographic solutions. The goal is to inform practitioners of new techniques to consider as they develop new big data solutions rather than to summarize the current best practice for securing data.

We note that, in addition to the cryptographic techniques, there are also many non-cryptographic techniques, such as firewalls, data guards, mandatory access control, and data provenance, that are likely to be a part of any complete solution to secure big data applications. However, such security mechanisms are beyond the scope of this chapter and we refer readers to an excellent book on the topic by Anderson [5] for a good summary. Additionally, we do not cover the security of operation, monitoring, and maintenance of big data systems; refer to Limoncelli et al. for more details [110]. Finally, we focus on software-based solutions and do not discuss the work relying on trusted hardware components to achieve data security [8,9,152].

We also do not address the privacy implications of big data collection and processing. That is, our focus is on keeping the data and computations out of the hands of unintended parties and not on the questions of whether the data should be collected or the analyses performed in the first place or what the implications of such analyses are for the individuals and corporations supplying the data. These are very important considerations in any big data system, but they fall outside the scope of our discussion.

10.1.1 Data life cycle

We begin this chapter by presenting the cryptographer's view of the data life cycle and the stages that big data go through. To secure big data, it is necessary to understand the threats and protections available at each stage. For this reason, the cryptographic techniques presented in this chapter are organized according to the three stages of the data life cycle described below.

10.1.1.1 Data in transit

The first step in most big data processing architectures is to transmit the data from a user, sensor, or other collection source to a centralized repository where it can be stored and analyzed. To ensure that the data arrive at its destination unmodified and unstolen, it is necessary to guarantee that all data are transmitted in a protected form. This is the stage of the life cycle best served by existing tools with existing technology such as the Internet protocol security [89] and the transport layer security [90] protocol suites already standardized and widely deployed in many of the big data tools of interest to the reader. As this chapter focuses on newer technology, we do not discuss data in transit security further.

10.1.1.2 Data in storage

The next step in the big data life cycle is to store the data in a repository, where it will be stored until it is needed. It is critical that unauthorized parties not be able to read or modify the stored data. However, authorized parties should be able to efficiently access all or parts of these data as necessary. Thus, more advanced techniques are needed to enforce access control to stored data while allowing for efficient data retrieval and search.

10.1.1.3 Data in use

Finally, once the data have been collected and stored, it is necessary to run analytics over them to derive value from the collected information. It is necessary to guarantee that only authorized analytics are run on the data by authorized parties and that the computation is performed correctly. This is one area where current best practices are especially lacking, with all processing done on unprotected data. Modern cryptography offers several techniques to change this status quo and allows data to remain protected even while they are in use.

10.1.1.4 Chapter outline

The remainder of this chapter is organized according to the stages of the data life cycle. We begin in Section 10.2 with a description of the basic concepts of data security and an overview of the traditional cryptographic tools such as encryption and signatures. In Sections 10.3 and 10.4, we describe techniques for protecting and searching stored data. In Section 10.3, we present techniques such as *broadcast encryption* and *attribute-based encryption* for controlling access to data stored and retrieved via unique identifiers, such as in a file system. In Section 10.4, we describe *searchable encryption*, which allows complex queries (e.g., Structured Query Language [SQL] queries) to be performed to retrieve allowed subsets of the protected data, such as in a database. Finally, in Section 10.5, we describe tools such as *homomorphic encryption* (HE), *verifiable computation* (VC), *secure multiparty computation* (MPC), and *functional encryption* (FE) for protecting computation over sensitive data.

For each of the presented technologies, we use the following outline. We begin with a high-level overview of what protections the cryptographic technology provides and how it can be used. Next, we give a more detailed definition of the security achieved and the critical properties of the technology. Finally, we give an in-depth summary of the history and state-of-the-art developments for the technology to illustrate the differences between individual schemes and their potential use cases.

10.2 Basics of Data Security

The chapter begins with an overview of some of the basic goals and tools of data security. First, we discuss the typical adversary models used to capture the threats to data that need to be addressed. Then, we briefly review common security goals and the tools used to achieve them.

10.2.1 Modeling the adversary

In this chapter, we restrict our attention to technologies with *provable* statements of security. These provable statements require that we formally model the threats addressed in the form of an adversary. This adversary formally captures the types of attacks as well as the limits on the attacks that we want the technology to withstand. Some properties typically addressed by an adversary model include the following:

- *Types of adversarial behavior*: Cryptographic protocols typically aim to provide security against one of two adversarial behaviors. An *honest-but-curious* or *semi-honest* adversary follows the protocol description but tries to learn unauthorized information from the messages he or she receives. A *malicious* adversary, on the other hand, may deviate arbitrarily from the protocol in order to learn private information or disrupt the protocol. These, roughly, model an eavesdropper who observes part of the protocol execution in an effort to learn private information and a malicious insider or hacker trying to actively interfere with the computation.

- *Amount of collusion*: In cryptographic protocols involving multiple parties, it is necessary to consider the number of parties that may behave adversarially and whether those parties will collude with each other during their attacks. Typically, cryptographers limit the total number of adversarial parties but assume that all such parties will collude. In some settings, such as searchable encryption, weaker models of collusion restrict parties playing different roles in the protocol from colluding with each other. Such restrictions are often justified by the organizational structure of the participating parties and possible adversaries.

- *Computational limitations*: Most provably secure cryptographic protocols rely on the mathematical difficulty of certain problems like factoring integers or computing discrete logarithms in finite groups. An important consideration for any cryptographic primitive is how strong an assumption is necessary to prove its security. Typically, schemes secure that are based on weaker and more well-studied mathematical assumptions are preferable, and much effort in cryptography is devoted to reducing the assumptions needed for desired cryptographic tasks.

10.2.2 Security goals

Having defined the adversary we want to protect against, we need to describe the security goals. The three most fundamental security goals are *confidentiality*, *integrity*, and *availability*, collectively known as the *CIA triad*.

10.2.2.1 Confidentiality

Confidentiality is the goal of keeping all sensitive data secret from an adversary. More formally, traditional definitions of confidentiality guarantee that an adversary should learn no information about the sensitive data, other than its length. Confidentiality is critical in big data applications to guarantee that sensitive data are not revealed to the wrong parties.

10.2.2.2 Integrity

Integrity is the goal that any unauthorized modification of data should be detectable. That is, a malicious adversary should not be able to modify such data without leaving a trace. This is very important to help guarantee the veracity of data collected in big data applications.

10.2.2.3 Availability

Availability is the goal of always being able to access one's data and computing resources. In particular, an adversary should not be able to disable access to critical data or resources. This is a very important security goal in big data processing, as the sheer volume and velocity of the data make guaranteeing constant access a difficult task. However, in today's big data systems, availability is typically guaranteed via non-cryptographic means such as replication, and we will not discuss it further in this chapter.

10.2.3 Basic cryptographic tools

We now give a very brief overview of the basic cryptographic tools used to ensure the aforementioned security goals in simple applications such as data in transit. Then, in the remainder of this chapter, we present techniques that build on these to guarantee CIA of data while enabling richer uses of the data. For a much more complete presentation of these basic tools, their formal definitions, and detailed constructions, we refer the interested reader to an excellent book by Katz and Lindell [97].

10.2.3.1 Encryption

The main tool for guaranteeing confidentiality of data is data encryption. Encryption takes a piece of data, commonly called the *plaintext*, together with a cryptographic key and produces a scrambled version of the data called the *ciphertext*. Using the key, it is possible to decrypt the data to recover the plaintext, but without the key, the ciphertext hides all information about the original data, other than its length. This security property, commonly known as *semantic security* [78], guarantees that, without the key, an adversary cannot learn any (potentially sensitive) property of the underlying data even if he or she has a lot of insight as to what the data may be. This is critical in applications where data may have some predefined structure, such as in financial transactions or if partial information about the underlying distribution of data is known, such as when the data are measuring some real-world phenomenon.

A little more formally, encryption consists of the following three protocols:

- KeyGen: A key generation algorithm that generates the necessary cryptographic keys

- $Enc(k, p) = c$: An encryption algorithm that uses a key k to scramble the plaintext p into cipher-text c

- $Dec(k, c) = p$: A decryption algorithm that uses the key k to recover the plaintext p from the ciphertext c

Encryption schemes come in two flavors: secret-key encryption, described above, and public-key encryption. In secret-key encryption, the same key is used for encrypting and decrypting data. In public-key encryption, KeyGen produces two keys: a public key and a secret key. The public key is used to encrypt the data, but cannot be used to decrypt the data and thus can be made public. The secret key, which must be kept private, is used to decrypt the data.

Secret-key encryption has been around for thousands of years with many protocols developed over the years [94]. Today, there are well-established standards for secret-key encryption such as schemes based on the Advanced Encryption Standard block cipher [128]. In the 1970s, public-key encryption was introduced by Diffie and Hellman [55] and Rivest et al. [148] and was formalized by Goldwasser and Micali [78].

10.2.3.2 Message authentication codes and digital signatures

The primary tools for guaranteeing integrity of data are message authentication codes (MACs) and digital signatures. Roughly, they take both a message and a key and generate an authentication value that can be used to verify the integrity of the data. The standard security property, *existential unforgeability*, of these primitives states that without the key it is impossible to forge a valid MAC or signature on any piece of data even after having seen MACs or signatures on other data items. Thus, an adversary cannot create or modify data without being detected.

A little more formally, MACs and signatures consist of the following three protocols:

■ KeyGen: A key generation algorithm

■ $\text{Sign}(k,m) = \sigma$: A signing algorithm that uses the key k and the message m to generate an authentication object σ

■ $\text{Verify}(k,m,\sigma) = b$: A verification algorithm that uses the key k, message m, and the authentication object σ and returns a Boolean value b depending on whether or not the message was signed using the corresponding key

As with encryption, there are two flavors of these schemes: secret key and public key. MACs are the secret-key variant, where the same key is used both to sign the data and to verify that the signature is valid. Digital signatures are the public-key variant, where a secret key is needed to sign the data, but a public verification key allows anyone to check the validity of the signature.

The most commonly used MAC, known as hash-based MAC (HMAC), was originally proposed and analyzed by Bellare et al. [12]. Public-key-based digital signatures were originally proposed by Diffie and Hellman [55] and Rivest et al. [148] and formalized by Goldwasser et al. [79].

10.2.3.3 Implementations and standards

All of the cryptographic primitives described above are already widely in use and there are multiple available implementations and standards for how they should be used. The National Institute for Standards and Technology maintains a list of standards for all of the above primitives as part of their cryptographic toolkit [127]. For US government applications, the National Security Agency maintains what is known as the *Suite B standard* [125] for which primitives must be used to protect sensitive US government data. Finally, the use of public-key primitives is standardized by the RSA Laboratories' PKCS #1 standards [108]. We strongly urge readers to follow recommendations and implementations described in these standards and avoid implementing their own solutions, as it is very difficult to implement bug-free cryptographic protocols and these standards are the results of decades of improvements and bug fixes.

10.3 Secure Block Storage and Access Control

We now turn to cryptographic techniques to secure data in storage. The primary goal of these techniques is to enforce access control to data stored in potentially untrusted repositories. That is, we want to give authorized parties access to the data they need while ensuring that unauthorized parties, either outsiders trying to gain access or malicious insiders in the organization managing the repository, cannot access sensitive data. In this section, we focus on systems where data are stored in blocks that are stored and retrieved by a unique identifier, such as in a file system. In such systems, we want authorized parties to be able to retrieve data by its identifier, but do not need to enable complex search queries to retrieve subsets of the data.

As may be expected, the first step in any such solution is to encrypt the data to hide it from unauthorized parties. However, the problem now becomes how to give authorized parties access to the data. We discuss two techniques for enabling such access control. The first of these, commonly called *group keying* or *broadcast encryption*, achieves this through careful key management to ensure authorized parties have the necessary keys to decrypt the data. The second of these, a technique known as *attribute-based encryption (ABE)*, instead relies on more powerful cryptographic techniques to automatically enforce the access permissions.

10.3.1 Key management for access control

10.3.1.1 Introduction

Key management includes generating and distributing cryptographic keys to system users in such a way that only authorized parties have the necessary keys to decrypt sensitive data. Most modern systems include some form of key management for controlling access to data in this way, and there are many commercially available, standardized solutions for generating and managing keys.* These typically use a *trusted* key management server to manage all keys in the system and to distribute the necessary keys to authorized parties. Here, we instead focus on a cryptographic technique called *broadcast encryption* or *group keying*, which allows a data owner to encrypt data to a designated set of recipients without having to rely on a trusted key manager. This is particularly important in big data applications where the storage may be handled by an untrusted repository on which a trusted key manager may not be available.

10.3.1.2 Definition

Broadcast encryption or group keying is a cryptographic technique for establishing cryptographic keys shared among a designated set of parties, thus giving these authorized parties access to encrypted data. Specifically, broadcast encryption gives a way to establish a cryptographic key such that all authorized parties receive the key and all unauthorized parties have no information about the key. In particular, even if some number of unauthorized parties collude, they should not be able to learn data that none of them is individually authorized to learn. An important goal of this primitive is to minimize the total size of encrypted data that must be generated and the amount of key material that must be stored by each of the participating parties.

10.3.1.3 Survey

Imagine that a data owner wishes to share data with some subset of the users of a system. One trivial solution is to have him share a different cryptographic key with each of the recipients

---

*See, for example, the various vendor implementations of the OASIS Key Management Interoperability Protocol (KMIP) standard for key management systems [129].

and separately encrypt the data to each of them. However, this requires the data owner to store a large number of keys, and also the size of the encrypted data grows linearly in the number of recipients. Broadcast encryption gives techniques to achieve this functionality without incurring these costs in key and data storage.

10.3.1.3.1 Single sender broadcast encryption

Broadcast encryption was first considered by Fiat and Naor [59] in a setting where there is one data owner who wants to share data with a set of authorized recipients. This construction was able to achieve much shorter keys and ciphertexts when the number of unauthorized users is small. This was further improved by Naor et al. [123], who showed a protocol able to handle an arbitrary number of unauthorized users while only incurring a logarithmic (in the number of parties) overhead both in key and ciphertext size when the number of adversaries is not too large. Roughly, both of these schemes work as follows. First, they generate cryptographic keys and then distribute these keys among all the possible users. Then to encrypt data to an authorized set of users, they encrypt the data under an appropriately chosen subset of the keys. The partitions of the keys and the subset used to encrypt are chosen to guarantee that all authorized user will know at least one key enabling it to decrypt the data, while all unauthorized users will not know any of these keys. See Figure 10.1 for an example of the mechanism given by Naor et al. One critical limitation of both of these schemes is that they only allow for one data owner who can share data, but in most big data scenarios there are multiple data providers.

10.3.1.3.2 Public-key broadcast encryption

In order to overcome this single sender limitation, the literature turned to public-key broadcast encryption. Such schemes allow anybody to share data, but rely on much stronger computational assumptions. The first scheme to do this was a scheme by Dodis and Fazio [56], who showed how to achieve parameters similar to the Naor et al. scheme in the public-key setting. An alternative construction by Boneh et al. [26] reduces the size of the secret keys needed by the data recipients down to a constant independent of the number of users, at the cost of (somewhat) increasing the size of the public key and encrypted data. One major drawback of these schemes is that they rely on a relatively novel, powerful, and nonstandardized cryptographic building

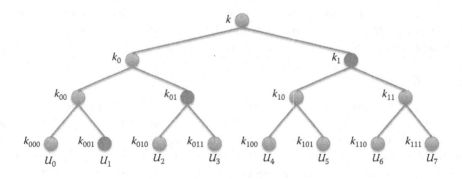

Figure 10.1: The Naor et al. tree for broadcast encryption for eight users. Each k_i represents an independent cryptographic key. Each user U_j receives all the keys on the path from the root to their leaf node. To encrypt the data, encrypt a copy of the data under the keys at the roots of subtrees containing only authorized users. For example, to encrypt to the set of all users except U_0, one would use the keys corresponding to the nodes with keys k_{001}, k_{01}, k_1.

block called *bilinear groups*. Further work [30,31] was able to significantly improve on the parameters of these schemes by relying on even stronger computational assumptions.

10.3.1.4   State of the art

There are several available implementations of these broadcast encryption protocols. First, the Lincoln Open Cryptographic Key Management Architecture (LOCKMA) [48,99] leverages the Naor et al. protocol to establish group keys for efficient communication over tactical networks. Also, an implementation of the public-key scheme by Boneh et al. has been demonstrated on top of the pairing-based cryptography library [115].

10.3.2   Attribute-based access control

10.3.2.1   Introduction

Key management-based solutions such as above have an inherent limitation. In order to share data with a set of users, it is necessary to know the identities (and keys) of all the authorized users. This is problematic in large systems or in systems with several organizational structures (as is very common in big data architectures where the data are collected, stored, and used in different environments) as the data owner is unlikely to know the identities of all the authorized users. An alternative approach to access control in such settings is a technique called *attribute-based access control (ABAC)*. In ABAC, data are encrypted together with a policy describing the attributes of users authorized to access the data. The users receive keys for the attributes they possess and are able to access the data if and only if those attributes are authorized. This allows for enforcing access to data without knowing the full set of users with the authorized attributes. For example, to encrypt data for analysis by NIH scientists, a data provider could encrypt the data with the policy (NIH and scientist) and only someone possessing both these attributes would be able to decrypt the data.

One approach to ABAC that has gained traction in government and commercial uses is to have a trusted server evaluate the access policy over a user's attributes and grant or restrict access to data accordingly [131,165]. However, this requires trusting the server to correctly administer these permissions and is problematic in scenarios where there is no such trusted entity, such as in outsourced storage. We instead focus on solutions for ABAC that do not require a trusted server to evaluate the access policies. Specifically, we present a powerful cryptographic technique known as *ABE* that can be used to solve this problem cryptographically.

10.3.2.2   Definition

(Ciphertext-policy) ABE, introduced by Sahai and Waters [151], is a form of encryption in which keys are associated with attributes and data are encrypted with a policy specifying which attributes are needed to decrypt the ciphertext. The security of this primitive guarantees that a key will succeed in decrypting a ciphertext if and only if the attributes in the decryption key satisfy the access policy specified in the ciphertext. More formally, the ciphertext contains a Boolean formula f over attributes and a key will successfully decrypt the ciphertext if f evaluates to 1 on the attributes contained in the key.

There are two properties that must be addressed in any ABE scheme. The first is the expressivity of the supported policy formulas. Schemes achieving more expressive policies can be used to enforce access control in more diverse settings. The second property is *collusion resistance*. That is, unauthorized users should not be able to combine their keys in order to decrypt data for which neither of their keys individually satisfies the access formula. This is typically the hardest feature of ABE to achieve and requires stronger cryptographic assumptions.

10.3.2.3 Survey

10.3.2.3.1 A first attempt

As a first attempt to constructing ABE, we can do the following. Let each cryptographic key correspond to an attribute, so k_A is given to all users with attribute A. Now, to encrypt a piece of data so that only those possessing attributes A and B can decrypt, we can encrypt the data first under k_A and then under k_B, resulting in $c = \text{Enc}_{k_B}(\text{Enc}_{k_A}(x))$. Clearly, both keys are needed to decrypt the data. To encrypt to users with either attribute A or B, encrypt the data separately under each key, resulting in $c = (\text{Enc}_{K_A}(x), \text{Enc}_{K_B}(x))$. Now either key will be able to decrypt. It is possible to repeat this procedure to enforce access policies represented by arbitrary Boolean formulas over the attributes.

However, this simple solution is not collusion resistant. If one user with attribute A (but not B) colludes with a user with attribute B (but not A), they can decrypt data encrypted to policy (A AND B) even though neither of them satisfies this policy. More complicated cryptography is needed to achieve the necessary collusion resistance.

10.3.2.3.2 Attribute-based encryption

The first ABE scheme to satisfy full collusion-resistance was given in 2005 by Sahai and Waters [151] supporting a limited set of policies known as *threshold policies*, where an authorized user's key has to have large overlap with the set of keys specified in the policy. An implementation of this scheme describing how this can be used for access control was given by Pirretti et al. [137]. The class of supported policies was extended to arbitrary Boolean formulas by Goyal et al. [82] and Bethencourt et al. [18].

All of these schemes take the following approach to prevent collusion attacks. As in the no-collusion scheme above, they generate cryptographic keys corresponding to the possible attributes. However, each generated key is also personalized to the specific user. So, the key for attribute A given to user 1 would be different from the key for attribute A given to user 2. This prevents keys from different users from decrypting the data since the personalized components will not match. However, this personalization approach requires the use of a nonstandardized cryptographic building block called *bilinear groups* and it is unlikely that (collusion resistant) ABE can be based on standard encryption [98].

Since these works, there have been a number of improvements optimizing performance, for example, [100], achieving richer classes of access policies [81], allowing for delegation of access permissions [171], and giving constructions relying on alternative security assumptions, for example, [64].

10.3.2.4 State of the art

There are now a number of available implementations of ABE. The first such implementation was given by Bethencourt et al. [18] on top of the pairing-based crypto library [115]. The latest implementation of ABE, due to Khoury et al. [100], achieves 3 ms for ABE encryption and 6 ms for decryption, making this primitive fairly practical for securing data storage. On the functionality side, we now have ABE schemes capable of enforcing very rich classes of policies such as policies specified by arbitrary (polynomial-size) Boolean circuits over the attributes [81]. We note that all of these schemes still rely on nonstandard security assumptions.

10.4 Secure Search

The previous section dealt with security and access control for block storage where data can only be retrieved via its unique identifier. However, in any big data system, users rarely want to retrieve all available data and instead usually only fetch a subset of the available data based on some specified search criteria. We now switch topics to discuss cryptographic techniques to enable *secure search* allowing for complex, database-style queries to be performed on stored data.

10.4.1 Introduction

Searchable encryption refers to a collection of cryptographic techniques that enable a server to perform searches directly over encrypted data. Most database technologies in use today require a server to be able to read all data contents in order to perform a fast search over it. By contrast, searchable encryption technology permits a server to perform basic database operations directly over encrypted data. Even though the server cannot read the data, it can still return encrypted versions of matching results back to a querier. In summary, searchable encryption technology provides stronger confidentiality and privacy guarantees than those in databases today by separating the roles of providing and accessing data.

There are a myriad of use cases for this technology in today's interconnected world; we provide just a few examples here. First, in a secure cloud outsourcing scenario, a client can upload files to the cloud and then make searches on the data later, without giving the cloud provider access to read the data. Second, in a secure publish–subscribe system, a broker can properly route messages from many providers to interested subscribers without being able to read the data itself.* Third, in a secure email scenario, many senders can send encrypted mail to one receiver, and then, the receiver can make searches to his/her email host to pull only the emails of interest to his/her without giving the host access to the contents of his/her emails. Fourth, in a secure database scenario, a data provider can allow multiple clients to search over her data without knowing the locations of data being requested; note here that preventing the data provider from learning the queriers' access patterns (e.g., frequency of requested records and overlaps between sets of records returned in response to different queries) is critical since the provider already knows the contents of her own data, so encryption alone does not suffice here.

Research into searchable encryption has been ongoing for the past 15 years. While there is still work to be done, the current state of this technology is quite usable. First, modern searchable encryption technologies are quite fast: their performance is only about 20%–200% slower than an unprotected search over plaintext data. Second, the searchable encryption works described below cover a large subset of the types of searches typically performed in databases, such as those permitted by the where clause in the SQL. For applications where one is willing to sacrifice a bit in security in exchange for performance, there are alternative searchable encryption schemes that work on top of existing database back-ends (e.g., MySQL and Accumulo) achieving much better performance in exchange for revealing some additional information about the stored data [62,139].

10.4.2 Definition

A formal definition of searchable encryption is both complex and ever-changing. In this chapter, we present an overview of the concepts behind searchable encryption here and then describe some of the complexities in the next section.

---

*Note that broadcast encryption, described in Section 10.3.1, achieves a similar goal without using a proxy, but instead requiring providers to know all interested subscribers.

10.4.2.1 Parties involved

The standard model of searchable encryption considers three types of parties: an *owner* who initially creates or possesses the data, a *querier* who wants to learn something about the data, and a *server* who handles the bulk of the storage and processing work. We note that in many scenarios, such as the publish–subscribe and email scenarios above, there may be multiple data owners or queriers.

We note that some scenarios envision only two parties, which we call the querier and server. For instance, the cloud outsourcing scenario above considers a client who acts as both the initial data owner and subsequent querier, and the secure database scenario considers a data owner who does her own server processing.*

We can fit these scenarios into our three-party framework by thinking of them as instances of *collusion*† between two of the three parties: in the cloud outsourcing scenario, the data owner colludes with the querier, and in the second scenario, the data owner colludes with the server.

10.4.2.2 Security guarantees and limitations

In this chapter, we restrict our attention to searchable encryption technologies with *provable* claims about their security. Here, we provide an overview of some of the provable security guarantees provided by searchable encryption technology. First, the querier only learns the contents of records that match her query; he or she learns nothing about the rest of the owner's dataset, except in some cases its size. Second, the data owner does not learn anything about queries being performed on her data. Third, the server performs searches in a relatively oblivious fashion, learning almost nothing about either the data or queries involved.

However, most searchable encryption schemes do reveal or *leak* some information about data and queries to the server. We describe two common leakages that most searchable encryption schemes reveal. First, *search pattern leakage* permits the server to determine if two different queries are identical, even if she cannot learn what the queries are. Second, *access pattern leakage* permits the server to determine which records in the dataset are returned in response to each query, even if he or she cannot read the records. As a consequence, the server can also determine which records are *popular* and being returned in response to many (potentially different) queries.

The precise cryptographic definitions that formalize these security notions have evolved over the years. We describe some of the nuances of these security definitions in Section 10.4.3 and refer interested readers to Bosch et al.'s survey [32] for the formal definitions of searchable encryption.

10.4.2.3 Trust requirements

In order to make provable statements of security like the ones above, searchable encryption schemes (and indeed, all of cryptography) require some preconditions to be met: they require users to place their trust in certain aspects of the behavior of other technologies or parties involved in the cryptographic protocol. Specifically, the searchable encryption schemes commonly rely on assumptions about the allowed adversary behavior, limits on collusion between the involved parties, and computational assumptions for security (see Section 10.2.1 for a detailed discussion).

---

*The two-party cloud outsourcing scenario in which the owner and querier are the same entity is also used in the model of fully homomorphic encryption (FHE), described in detail in Section 10.5.1.1.

†Collusion is typically used in cryptography to model a collection of malicious users who are working together, as described in Section 10.2.1. But, the same concept applies just as well to the benign setting in which two entities are *working together* because they are really the same person.

In addition to the above, there are several protocols for secure search that do not require trust in the veracity of computational assumptions but rather in the inability of an adversary to perturb a specialized piece of hardware that performs computations inside of a (hopefully) tamper-proof shell. Put bluntly, these schemes place their trust in an adversary's physical limitations rather than his or her mathematical ones. However, as this chapter is focused on software-based solutions, we will only briefly mention one example of such schemes and refer interested readers to the original papers [8,19,160].

10.4.3 Survey

In this section, we first provide a brief history of three important cryptographic primitives that preceded searchable encryption. Next, we describe some of the initial works on searchable encryption. Finally, we detail several innovations that improve the security of searchable encryption or add new dimensions to the functionality, performance, or security of this technology.

We stress that this section only provides a brief overview of the major thematic advances in searchable encryption technology over the years. We purposely do not delve into all technical details and only present the first few papers that tangibly influenced each theme presented below. For a more detailed technical review of the (more than 100) research works on searchable encryption, we refer readers to the comprehensive survey by Bösch et al. [32].

10.4.3.1 History: Precursors to searchable encryption

We begin by describing three cryptographic technologies initially developed in the 1990s that solve related problems. All of these works utilize a two-party setup with a querier and a server. Some of them are better suited to the cloud outsourcing scenario (in which the querier provides the data) and some are better suited to the database setting (in which the server provides the data).

First, *private information retrieval*, or PIR, considers the cloud outsourcing problem in which a data owner outsources his/her data to an untrusted server and later wants to query for a single record. While the database contents themselves may be clearly readable by the server, the querier wants to hide his/her interests from the server (i.e., which record is being returned in response to a query). An important goal in PIR works is to minimize the amount of communication required between the server and querier. Chor et al. [47] constructed the first PIR scheme in an information-theoretic model with multiple non-colluding servers. Shortly thereafter, several works provided PIR under the assumption that certain cryptographic problems are hard [37,106,107]. We refer interested readers to Gasarch's excellent survey [66] for more information about PIR.

Second, Chor et al. [46] introduce *private information retrieval* (PIR) *by keywords*. This work introduces the concept of searching (as we think of it today) to the cryptographic literature. Specifically, it uses multiple rounds of PIR to search by keyword rather than by the index (i.e., pointer or unique identifier) of the record that the client wishes to retrieve. Nevertheless, data are still stored unencrypted on the server.

Third, *symmetric private information retrieval*, or SPIR, extends PIR's security guarantees to protect the server as well. Specifically, while PIR permits the client to view the entire database, SPIR limits the number of records that a querier may retrieve in response to any query. Hence, SPIR is useful both in the cloud outsourcing scenario and the database scenario, and it can also support multiple queriers securely by using its access controls to restrict each querier to reading only her own data on the server. SPIR was introduced by Gertner et al. [72], though it has its roots in another cryptographic primitive (called *oblivious transfer*) developed by Rabin in the 1970s [142].

Finally, *oblivious random-access memory* (ORAM), introduces the concept of a server performing keyword searches directly over encrypted data in an *oblivious* fashion, that is, without learning either the contents of any records stored at the server or the access patterns of data returned to the client (such as the simple question "is this identical to data that the server returned to the client in the past?"). Goldreich and Ostrovsky's seminal work [75] introduces multiple ORAM techniques, including an asymptotically optimal logarithmic-time search and a square-root-time search that turns out to be faster in practice [91]. ORAM has been extensively studied by the cryptographic community since the 1990s, and recent works have substantially improved the server's computational burden [53,54,120,121,146,163,164].

One performance impediment of ORAM schemes is the large number of rounds of communication that they require between the querier and server. By contrast, most of the searchable encryption works discussed below only require one round of communication. In order to achieve this performance improvement, however, most searchable encryption works reveal some search pattern information to the server.

10.4.3.2 Searchable encryption

In 2000, Song et al. published the seminal work on searchable encryption [161]. They presented a cryptographic protocol in a two-party setting (with a querier and server) to search over encrypted data that provided the following four properties:

1. *Provable security for encryption*: The untrusted server stores ciphertext data that have been encrypted using a semantically secure encryption scheme (as described in Section 10.2.3), so the server cannot learn any information about the corresponding plaintexts.

2. *Controlled searching*: The untrusted server can only perform searches that have been authorized by the querier. The server cannot make searches on her own.

3. *Hidden queries*: Song et al. support keyword searches. During a query, the server does not learn the keyword. (This concept is later formalized by Goh [73] as *chosen keyword security* and defined in a similar style to semantic security.)

4. *Query isolation*: While the server does learn which records are returned to the querier, the server learns nothing more about plaintexts than this information.

Additionally, the Song et al. protocol provides good performance by adhering to a few simple guidelines that subsequent works also follow:

1. The search protocols are simple and reminiscent of their insecure counterparts.

2. The scheme relies on faster symmetric-key cryptography rather than slower public-key cryptography. Song et al., in fact, use symmetric-key cryptography exclusively: schemes that follow their lead are collectively referred to as *searchable symmetric encryption*, or SSE. While some of the subsequent works described below do utilize public-key cryptography to provide additional functionality, they often use public-key cryptography as sparingly as possible.

3. Like most unprotected database search technologies, Song et al. bolster the performance of queries by precomputing an *index* mapping keywords to records that match the query. While the bulk of Song et al.'s work is on non-indexed search, this observation foreshadows many of the subsequent works on SSE.

After the initial Song et al. result, a myriad of works improved the functionality, performance, and security of searchable encryption technology. We now present the various dimensions along which these works improved searchable encryption. Understanding these dimensions will be important for understanding the state-of-the-art in searchable encryption schemes.

10.4.3.2.1 Secure indices

Goh [73] formally defines the concept of index-based SSE. We note that indexing induces a performance tradeoff: to achieve faster query throughput, SSE schemes may require a long time to compute the index whenever the database is created or modified. In fact, Song et al. and Goh do not support efficient database modifications (insertions, updates, or deletions) into their indices due to the difficulty of doing so in a confidential manner (i.e., without revealing the desired modification to the server).

Chang and Mitzenmacher [41] make the valuable observation that indexing provides a clean conceptual separation between two acts that the server must perform: determining the indices of records that match a query and fetching the contents of records at those indices. The first action might require complicated data structures to support sophisticated queries, but the contents of those data structures (i.e., the indices themselves) are small. The second action requires a large data structure to hold record contents but only requires that this data structure supports the simplest of queries: a simple pointer lookup.

With this observation, Chang and Mitzenmacher are able to use standard compression and encryption schemes on the record contents. Their observation is utilized in many of the SSE schemes that follow, including all of the works presented in Section 10.4.4.

10.4.3.2.2 Error rates

Goh [73] constructs a secure index mechanism whose search time is independent of the number of keywords in the database. To do so, Goh utilizes *Bloom filters*, a simple fixed-size data structure that can check for membership in a variable-length set. Goh's indexing mechanism is simple: it builds one Bloom filter per record that includes all of its keywords. Because Bloom filters have a one-sided error, it is possible that the server could return records to the querier that do not actually match the keyword being searched. Subsequent SSE works that utilize Bloom filter technology, such as the Blind Seer [60,133] scheme discussed in Section 10.4.4 below, also have nonzero false-positive rates.

10.4.3.2.3 Modifications

Because of their separation of searching and file storage, Chang and Mitzenmacher [41] provide keyword search data structures that are much simpler than those of Song et al. or Goh. As a result, their index structure supports modifications: insertions, updates, and deletions. While Chang and Mitzenmacher can only support modifications if they have an *a priori* dictionary or universe of all possible keywords that could be inserted later, subsequent works [39,95,96, 133,139,167] overcome this restriction and support faster modifications directly over the data structures they use for indexing.

10.4.3.2.4 Multiple queriers

Curtmola et al. [50] formally define SSE with multiple queriers. They do so in two settings: a *nonadaptive* setting in which security is only proved when the adversary must commit to all queries before seeing any responses, and a stronger *adaptive* setting in which the adversary is permitted to choose queries based on the responses to prior ones. While adaptively secure cryptography is preferable for its relevance to real-world usage of SSE, nonadaptive protocols are faster with no known attacks in practice.

To aid in the multiple-querier setting, Chase and Kamara [42] provide an access control mechanism called *controlled disclosure*. This mechanism gives a data owner the ability to grant someone else access to a portion of a large dataset.

10.4.3.2.5 Public-key searchable encryption and outsourced SPIR

Public-key cryptography enables three-party scenarios (such as the secure publish-subscribe and email scenarios above) in which the server that stores data never has the ability to read or perform searches over it. Boneh et al. [25] provide the first construction of public-key searchable encryption in which the querier can produce a *trapdoor* token that permits the server (non-interactively) to locate all records containing a given keyword without learning anything else. We refer interested readers to Boneh and Waters [29] for a formal, generic definition of public-key searchable encryption. Many of the works in this space use a relatively novel, powerful, nonstandardized cryptographic building block called *bilinear groups.*

More recently, some of the works described in Section 10.4.4 attempt to satisfy the security constraints of the three-party scenario (namely, having a server who is not trusted to learn either the data or queries, as described in Section 10.4.2.2) while minimizing or removing the use of public-key cryptography in order to improve performance [39,40,60,93,133]. Jarecki et al. [93] refer to this concept as *outsourced SPIR*.

10.4.3.2.6 Searching over different data structures

Inspired by traditional databases, SSE focuses on the problem of record retrieval via text-based searches. But, just as with the rest of the big data community, encrypted search has progressed beyond this simple setup. Chase and Kamara [42] define *structured encryption*, which enables searches over arbitrary structured data such as matrices and graphs. Chase and Shen [43] generalize this notion even further to *queryable encryption*, which permits queries over potentially unstructured data. Chase and Shen provide one form of queryable encryption: the ability to do a substring search over a large file.

10.4.3.2.7 Query expressivity

Researchers have steadily increased the expressivity of searchable encryption schemes. Collectively, modern schemes support a large percentage of the selections possible in, for instance, the where clause of the SQL.

From the beginning, searchable encryption schemes (and their predecessors in Section 10.4.3.1) supported equality keyword searches such as select * where name = Alice. Researchers subsequently added support for other types of queries, often by finding a new data structure to traverse in a secure fashion. We describe a few of the innovations in data structures below and explain the new query types that they enable.

- *Boolean queries*: Goh's usage of Bloom filters [73] facilitates secure searching for Boolean combinations (i.e., ANDs and ORs) of keywords. The same technique also permits searches for the number of occurrences of a keyword inside a record. Finally, by building a tree of Bloom filters, Goh's technique also facilitates searches for words that occur infrequently in a large dataset, which can be useful for statistical analyses of large datasets. Pappas et al. [133] use a similar Bloom filter technique, and Cash et al. [40] support Boolean queries in a different fashion.

- *Inequality and range comparisons*: A technique called *order-preserving encryption*, in which the order of ciphertexts matches the order of underlying plaintexts, enables a weak but incredibly fast form of range searches. Research into this technology began in the database community with the work of Agrawal et al. [3], and it has continued in the cryptographic community with several works that improve the performance and characterize the security limitations of this technology [22,23,118,138]. On the other end of the spectrum, Boneh and Waters [29] provide a stronger but slower form of range comparison, subset, and conjunction searches.

- *Substrings*: Chase and Shen [43] use a suffix tree data structure to support searches for substrings in a single, large file. Their work operates in the cloud outsourcing two-party scenario.

- *Similarity searches*: A few works provide for similarity or fuzzy searches. Park et al.'s scheme can be used for approximate string matching [134], and Adjedj et al.'s scheme can be used for biometric identification [2].

- *Inner products*: A few works provide for searches that involve basic arithmetic operations such as inner product computations [33,155].

With support for a broad array of query types, searchable encryption technology is approaching the point of commercial interest [24].

10.4.3.2.8 Performance and scale improvements

Finally, we examine two methods utilized to improve the performance and scalability of searchable encryption: preprocessing and parallelization.

First, Freedman et al. [61] developed a technique to save online query processing time by performing some preprocessing work between the querier and server (a technique that is also used in ORAM schemes). Second, Kamara and Papamanthou [95], Cash et al. [39], and Pappas et al. [133] discover opportunities for parallelization, both within a single query's execution and between several queries that are executed concurrently. With these performance improvements, these works are able to handle significantly larger databases, with the latter two works scaling to achieve good performance on terabyte-sized datasets. We discuss these two works, along with other modern, usable searchable encryption techniques, next.

10.4.4 State of the art

The state-of-the-art secure search schemes all make different choices in the dimensions described above resulting in different trade-offs between security, functionality, and performance. We now briefly describe several of the latest searchable encryption schemes. We focus on schemes supporting rich sets of query functionality rather than those focused on only one query type. Such schemes largely fall into one of two camps: Schemes focused on achieving very strong security, with minimal leakage, potentially at the cost of performance, and schemes focused on achieving performance close to unencrypted databases at the cost of allowing more leakage to an adversary. We now describe several schemes following each of these approaches.

10.4.4.1 Security-focused schemes

In the past few years, there have been several efforts to build searchable encryption schemes providing strong security guarantees while providing rich query expressiveness and being able to scale to terabyte size databases. This work has largely been driven by the Intelligence Advanced Research Projects Activity (IARPA)—Secure and Privacy Assurance Research (SPAR) program [88], which aimed to produce prototype implementations of searchable encryption that could perform within an order of magnitude of standard SQL databases while minimizing the amount of leakage to the parties involved.

The two protocols to come out of this project are Blind Seer [133] and Cash et al.'s scheme [40]. Both of these work in the three-party model described earlier and assume no collusion between the three parties: server, querier, and data owner. Both of these protocols achieved strong security against all three parties. Specifically, the data owner cannot learn the querier's queries and the querier can learn only information returned by authorized queries. The server can only learn limited information about the data it is storing and the queries it is serving, such as query search patterns. The exact details of the leakage provided to the server vary between the two protocols resulting in slightly different security guarantees.

Both of these protocols work by constructing secure search indices to allow for sublinear searching. This is done by combining cryptographic techniques with indexing structures from the database community. While the two schemes share this basic premise, the underlying technologies are drastically different and use different cryptographic primitives. We now briefly sketch these two protocols.

10.4.4.1.1 Blind Seer

The Blind Seer protocol constructs an index consisting of an encrypted search tree. The leaves of this tree correspond to the individual records and contain all the searchable keywords from that record. Internal nodes contain a Bloom filter* storing the set of all keywords stored in their descendants. Now, a Boolean query (e.g., A AND B) can be executed as follows:

- Beginning at the root, check to see if the Bloom filter of the current node contains the keywords being searched for in the query. If not, terminate the search.

- If so, visit all the children of the node recursively for the same check until the leaf nodes are reached.

- Return all leaf nodes containing the searched for keywords.

To additionally support range queries and queries including negations, the set of keywords in the leaf nodes is extended to include negated terms and also special range keywords.

However, just performing the queries as described above would leak both the query and the data to the server performing the search. Instead, the Bloom filters are encrypted to protect their content, and the Bloom filter check is performed by a secure computation between the client and server, which allows one to check membership in a Bloom filter while hiding the content of the Bloom filter and the value being looked for. Now the client and server can together traverse the search tree as before with the server only learning the search pattern of which tree nodes are visited by the search, but not the keywords contained in those nodes or the search terms.

Blind Seer also supports modifications (which includes inserts, deletions, and updates) within their secure index. Inserts are handled by creating a separate Bloom filter which is included as part of the search. Deletions are handled by marking the corresponding nodes, both in the original and inserted Bloom filters, as deleted. Updates are implemented as simply an insertion and deletion. From time to time, the entire tree of Bloom filters is recreated to include these inserted records.

---

*A data structure allowing for efficient set membership queries.

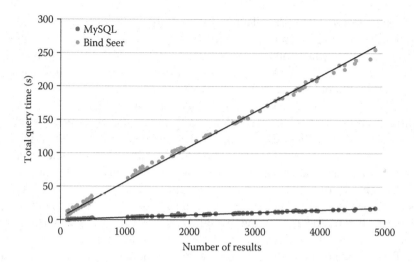

Figure 10.2: Performance of Blind Seer compared to a MySQL database. Performance cost is within 15x—mostly due to network communication costs. (Reproduced with permission from Pappas, V. et al., Blind Seer: A scalable private DBMS, in *IEEE Symposium on Security and Privacy*, Berkeley, CA, May 18–21, 2014, IEEE Computer Society, pp. 359–374.)

Blind Seer's performance was measured as part of the SPAR program by MIT Lincoln Laboratory [168] and can be seen in Figure 10.2. For the most part, they function within an order of magnitude of a MySQL baseline system.

10.4.4.1.2 Cash et al. protocol

The Cash et al. protocol constructs an encrypted index based on the idea of an expanded inverted index. In their basic scheme, without factoring in security, they store two different indices. The first is an inverted index that maps keywords to record identifiers, and the second is an index that maps record identifiers to keywords contained within. In order to do a conjunctive search, the server looks in the first index for the first keyword in the conjunction. For each of the records that matches the keyword, the server checks to see if the rest of the terms of the query are matched by checking the second index. If they do, that particular record is returned as part of the query.

To add security to the above search protocol, first the records in the database are permuted and encrypted. Then, the indices are built for the permuted encrypted database. Furthermore, during a search, careful cryptographic techniques guarantee that the server can check the second index for keywords included in the query without learning what the keywords are or learning anything about the remaining keywords contained in the record.

Like the Blind Seer protocol, the Cash et al. protocol is able to achieve reasonable performance compared to standard search technologies. Specifically, the query time for queries returning small result sets is essentially independent of the size of the database for databases up to one billion records. For such queries, this protocol is almost comparable to the runtime of an unencrypted MySQL backend.

10.4.4.1.3 Bajaj and Sion approach

An alternative approach for security-focused secure search was given by Bajaj and Sion [8], who use trusted hardware such as IBM 4758 and 4764 PCI cards in place of cryptographic assumptions. Here, security stems from requiring the server to perform the sensitive portion of their searches properly on a trusted hardware device thus guaranteeing security of the data and correctness of the searches. Specifically, security is guaranteed as long as adversaries cannot tamper with the internal state of the trusted hardware.

Bajaj and Sion use this approach to build a full relational database whose data structures and query processing methods are split into secure pieces run in a trusted enclave and insecure components that run on commodity hardware. To analyze the efficiency of this approach, Bajaj and Sion estimate the cost of computations, storage, and networking in both small-scale and large-scale (*the cloud*) computing environments. They conclude that large-scale computations inside secure processors is only 2–4× more costly than computation in small-scale insecure enterprises and thus orders of magnitude faster than any cryptographic operation. Thus, this approach allows them to build trusted hardware-based secure search schemes achieving strong security while also getting all the desired query functionality and scalability.

10.4.4.2 Performance-focused schemes

On the other end of the spectrum, there have been several efforts trying to add a minimal level of security to database search while achieving minimal performance overhead. Moreover, these techniques allow the use of an unmodified traditional database as the server in the secure search protocol, thus allowing backward compatibility with existing infrastructure.

10.4.4.2.1 CryptDB

CryptDB, built by Popa et al. [139], provides a client side library that supports executing SQL queries over encrypted data. The key insight that allows them to do this is that various forms of encryption allow for different queries to be performed directly on encrypted data. Thus, by encrypting data accordingly, it is possible to allow searching over encrypted data. Specifically, Popa et al. use the following encryption schemes:

- *Semantically secure encryption* provides the strongest security guarantees; the encryption process is randomized and encryptions of different messages are indistinguishable. However, this type of encryption, while the most secure, does not allow for any specific queries to be made across it.

- *Deterministic encryption (DET)* [11] provides a weaker security guarantee; two equal values always encrypt to the same ciphertext, which allows the server to check equality of the underlying plaintexts to perform frequency analysis. However, this allows a client to make targeted keyword search across deterministically encrypted columns as these only require checking equality.

- *Order preserving encryption (OPE)* [138], in addition to being deterministic, also preserves the relative order of the underlying plaintexts. So, if $a < b$, then $\mathsf{Enc}(a) < \mathsf{Enc}(b)$. This means it provides an even weaker security guarantee than deterministic encryption, allowing the server to learn the order of the underlying values. It does, however, allow a user to make range queries.

- *Additively homomorphic encryption* has similar security to semantically secure encryption but has the added property of allowing limited computations, in this case addition, across the encrypted data in the columns. This is further discussed in Section 10.5.1.1.

CryptDB uses these different encryption types by encrypting the different fields in an SQL database using different types of encryption. Fields over which ranges may be desired are encrypted using OPE, fields that need to support keyword queries are encrypted using DET, and raw data fields that cannot be searched over can be encrypted using semantically secure encryption. Additionally, CryptDB allows for joins over fields encrypted under OPE and DET as this just requires equality checks. In order to reduce the amount of leakage from such encryptions, all fields encrypted using OPE and DET are *onion encrypted* with a semantically secure encryption scheme. That is, the these ciphertexts are again encrypted under a semantically secure encryption, thus removing all leakage. They remain encrypted in this way until a query is made that requires equality or range search over this field at which point the semantically secure layer is decrypted off to allow the query to be performed. The benefit of this is that no information is revealed about values in fields that are not searched over.

Even with this measure, CryptDB offers a much more limited form of security than that found in the Blind Seer's and Cash et al.'s solutions since once the outer layer of encryption is removed on some field, the server learns some information about the values in that field for all records, even the ones not returned by the query. However, the performance of CryptDB is much closer to traditional search technologies, performing within $2\times$ for most query types as can be seen in Figure 10.3.

10.4.4.2.2 Monomi

Monomi [166] takes CryptDB's advances and extends its performance results by introducing a query and schema planner. This planner, based on the variables of client storage and allowed communication complexity (number of rounds of communication and the size of the portion of the database sent over for each query), splits the query execution between server-supported portions (such as range queries over an order preserving column) and client-side computation. The schema and additional encrypted columns are also automatically generated based on a sample set of queries that the client inputs at generation time. This effort to apply traditional

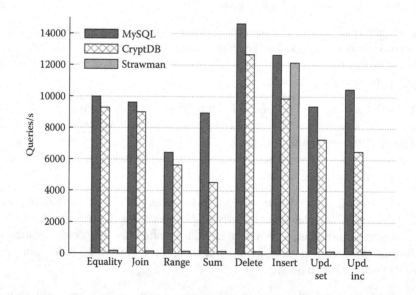

Figure 10.3: Performance of CryptDB for a variety of query types. Strawman is a naive implementation of the algorithm without optimizations to speed up the construction of the encrypted indexes. (Data from Popa, R.A. et al., CryptDB: Protecting confidentiality with encrypted query processing, in: *ACM Symposium on Operating Systems Principles*, 2011.)

query planning to encrypted search technology allows for more expressive queries and better performance than CryptDB demonstrates.

10.4.4.2.3   Computing on masked data

Computing on masked data (CMD) [62] extends the ideas of CryptDB to NoSQL databases to support big data applications. Specifically, CMD applies these protections to the Accumulo database [165], a distributed database noted for its high ingest rates and native cell-level security. CMD supports a similar set of encryption types to CryptDB, including semantically secure, deterministic, and order preserving encryption. CMD targets, and achieves, a performance goal of computing on the encrypted data within a factor of $2\times$ of the unencrypted system. They note that compared to the performance cost of performing the query, the performance cost of encrypting and decrypting the data is relatively small.

10.5   Secure Data Processing

In the previous section, we discussed performing secure searching on encrypted data. However, in many big data applications it is not enough to be able to simply retrieve stored data. Instead, it is desirable to perform analytic computations over the data and return only the result of these computations rather than the original data. We stress that it is very important that the data and the computations be protected even while the processing is being performed. In this section, we describe four cryptographic techniques that enable generic secure computation: HE, VC, MPC, and FE.

All of these can be used to securely outsource the processing of data but in different scenarios. HE allows computing on encrypted data while maintaining confidentiality; it can be used to outsource processing of sensitive data to another entity that is trusted to perform the computation correctly but should not learn the data. VC allows computing on data and allowing the integrity of the computation to be checked; it can be used to outsource processing of data to another entity that is allowed to learn the data but not trusted to perform the computation correctly. It is possible to combine HE and VC to achieve both confidentiality of the input and output and integrity of the computation. MPC allows performing a distributed computation on sensitive inputs held by multiple parties while maintaining confidentiality of each party's inputs from every other party and ensuring that the computation was performed correctly. MPC guarantees these properties as long as no more than some threshold of computing parties are adversarial. MPC can be used in settings where different sensitive inputs are held by different parties or in settings where a single client wants to outsource computation on its sensitive input by distributing the computation over multiple compute nodes. Finally, FE allows a data owner to let others learn specific functions of her sensitive data and nothing else.

10.5.1   *Homomorphic encryption*

10.5.1.1   *Introduction*

Imagine that we own a computationally weak computer and wish to outsource some work to a server that we do not trust. The data on which the work is done is sensitive, and we want to make sure that the server cannot learn anything about it. Although the idea of delegating a computation without revealing anything about the inputs or outputs may seem very unintuitive, cryptography has a tool enabling just that: this tool is *homomorphic encryption (HE).*

10.5.1.2 Definition

HE is a type of public-key encryption that allows anyone to compute functions of data while it is encrypted without decrypting it or learning anything about it. Specifically, in addition to the algorithms KeyGen, Enc, and Dec associated with any encryption scheme, HE also has a fourth algorithm Eval. This algorithm takes in the public key pk, a function f, and encrypted data c_{in} and returns an encrypted output $c_{out} = \text{Eval}(pk, f, c_{in})$ such that c_{out} is an encryption of the result of applying f to the originally encrypted data, that is, $f(\text{Dec}(sk, c_{in})) = \text{Dec}(sk, c_{out})$. Anyone can evaluate Eval, because it does not require the secret key sk. Note that the result is an encryption of the result; retrieving the result itself requires the secret key. Moreover, the evaluation does not reveal anything about the data c_{in} or the computation output c_{out}, since that would violate semantic security.

10.5.1.3 Survey

Many public-key encryption schemes are naturally homomorphic with respect to a limited set of functions f, such as only addition or only multiplication [58,132,141,148]. An intriguing question, initially posed in 1978 by Rivest, Adleman, and Dertouzos [147], is whether there exists an encryption scheme that simultaneously permits the evaluation of (an unbounded number of) *both* addition and multiplication operations on the underlying plaintexts. Since these two operations constitute a logically complete set of operations, such a scheme would be a FHE scheme, meaning that it is homomorphic with respect to all functions f. Boneh, Goh, and Nissim [27] give an encryption scheme that is homomorphic with respect to quadratic functions f, that is, their encryption scheme supports the evaluation of arbitrarily many additions, but only one multiplication, thus falling short of this goal.

In 2009, the seminal work of Gentry [67] affirmatively answered this long-standing question by demonstrating an encryption scheme that provides Eval operations for an unbounded number of both additions and multiplications. Gentry splits the FHE problem into two components: the design of a *somewhat homomorphic encryption* (SWHE) scheme that permits a limited number of Eval operations, and the insight of a *bootstrapping* algorithm that achieves FHE through the use of multiple applications of SWHE. A slight tweak of this initial scheme was implemented by Gentry and Halevi [68].

10.5.1.3.1 Somewhat homomorphic encryption

There are many versions of Gentry's SWHE scheme. One of the most straightforward ones uses integers modulo a large prime p, which is the secret key. Ciphertexts take the form of large integers, which are encryptions of 0 if their remainder modulo p is even, and encryptions of 1 if their remainder modulo p is odd. It is straightforward to see that adding (or multiplying) two ciphertexts gives an encryption of the sum (or product) of the plaintexts, as long as the remainders modulo p (i.e., the *noise*) of the two original ciphertexts are small enough. Informally, this scheme is secure because of the *learning with errors* assumption [145], which says that if p is unknown, then multiples of p with some small additive noise look completely random, and an observer cannot learn p.

10.5.1.3.2 Bootstrapping

In the scheme described above, the error keeps growing until it is impossible to perform any more homomorphic evaluations. However, Gentry made the key observation that if a scheme is homomorphic enough to evaluate its own decryption circuit in addition to at least one other operation, then it can be made fully homomorphic by means of *bootstrapping*. Along with the public key, an encryption under that public key of the corresponding secret key is published. When the ciphertext noise grows so large that no more homomorphic operations can be performed, the ciphertext is encrypted again. This results in two layers of encryption—the original one and the new one. Under the new layer of encryption, the old ciphertext is homomorphically decrypted (using the encryption of the secret key), resulting in a ciphertext with just one layer of encryption—the new layer. The noise of this new ciphertext is only whatever is accrued over the course of homomorphically evaluating a single decryption circuit. So, as long as the decryption circuit isn't deep enough to generate too much error, one or more additional homomorphic operations can be performed before the error grows too large again.

10.5.1.4 State of the art

After Gentry's initial work, many cryptographers have designed new FHE algorithms that make substantial improvements along several dimensions:

■ Decreasing the ciphertext and key sizes [158].

■ Optimizing the FHE for specific common operations such as single instruction, multiple data (SIMD) operations [69,159]. SIMD operations involve performing the same operations on multiple bits in parallel.

■ Using hardware (e.g., field-programmable gate arrays [FPGAs] [140]) in order to make FHE faster.

■ Increasing the number of Eval operations possible in a SWHE scheme before bootstrapping, eliminating bootstrapping altogether, or making bootstrapping more efficient [34,71]. Bootstrapping is the most expensive part of homomorphic evaluation, and reducing the frequency with which it needs to be done greatly increases evaluation efficiency. One of the approaches to this is reducing the noise a limited number of times by changing the ciphertext in a different way (*modulus reduction*) between every two bootstrapping operations.

■ Basing the cryptography upon weaker, more accepted assumptions [35].

Some of these improved algorithms have been implemented in software [57,70,83,84]. Table 10.1 describes the homomorphic evaluation time of some operations in the HElib implementation [83].

Table 10.1 Evaluation time (in seconds) of some operations in HElib

Gate Type	Count	Mean	Std Dev	Min	Max
Add	101	$2.04 \cdot 10^{-4}$	$1.99 \cdot 10^{-4}$	$1.10 \cdot 10^{-5}$	$6.18 \cdot 10^{-4}$
Add constant	101	$1.85 \cdot 10^{-4}$	$1.75 \cdot 10^{-3}$	$6.00 \cdot 10^{-5}$	$4.43 \cdot 10^{-3}$
Multiply	101	$1.45 \cdot 10^{-2}$	$1.39 \cdot 10^{-2}$	$4.76 \cdot 10^{-4}$	$3.00 \cdot 10^{-2}$
Multiply by constant	101	$1.92 \cdot 10^{-3}$	$1.82 \cdot 10^{-3}$	$7.60 \cdot 10^{-5}$	$5.19 \cdot 10^{-3}$

The rows describe the operations on ciphertexts that correspond to the following operations: addition of two plaintexts, addition of a public constant to a plaintext, multiplication of two plaintexts, and multiplication of a plaintext by a public constant. The columns describe statistics of the time it takes to evaluate a single gate of each type, that is, the number of evaluations performed (count), the mean evaluation time, the standard deviation, the minimum, and the maximum. The numbers are taken from Varia et al. [119,169].

10.5.2 Verifiable computation

10.5.2.1 Introduction

Imagine that, as in the previous section, we own a computationally weak computer and wish to outsource some work to a server that we do not trust. Unlike the FHE scenario though, suppose we now do not care about the confidentiality of the data (perhaps these data are not sensitive), but we do care that the work is done correctly.

There are a number of ways that one might approach this without using cryptographic tools. The most straightforward approach is *replication*—outsourcing the computation to several different servers and then taking the most common answer as correct. However, this only works as long as the failures of the servers are not correlated (as in Byzantine fault tolerance [109], which is not always the case). It could be that most of the servers in the cluster are controlled by the same adversarial entity or that they are running a single operating system that is faulty in a specific way. Another way is *auditing*—after outsourcing the work, also performing it oneself with some probability, and if the answer does not agree with the server's answer, ceasing to trust the servers to do any of the work. This only works as long as one believes the servers will fail on a noticeable fraction of the computations and not on just a few of them (which may be vitally important to get correct). If we use cryptographic tools, the landscape is much more promising. VC is a cryptographic tool that guarantees the integrity of outsourced computations without making any assumptions about the server failure rate or correlation of failures.

10.5.2.2 Definition

Typically, VC is defined in the context of two parties: a computationally weak *verifier* (or *client*), and a computationally strong but untrusted *prover* (or *server*) to whom the verifier delegates some work. Given an input x and a function f to evaluate on x, the prover is expected to produce an output y, as well as a proof π that $y = f(x)$ that the verifier can use to confirm the correctness of the computation. In order for the verifier to want to outsource the computation to the prover in the first place, it should be much more efficient for the verifier to check the proof of correctness π of the output y than for the prover to perform the computation itself. VC schemes should make it infeasible for a prover to come up with an incorrect output $y^* \neq f(x)$ and a proof π^* such that π^* convinces the verifier that $y^* = f(x)$.

10.5.2.3 Survey

Most VC constructions are based on *probabilistically checkable proofs* or *PCPs* [7]. A PCP is a (possibly long) string produced by the prover to prove the validity of some statement. The PCP itself can be viewed as the proof of the statement's validity; however, having to read the entire PCP may be too big a burden on the computationally weak verifier. Instead, the special property of a PCP is that the verifier can check the validity of a PCP by looking only at a constant number of random locations in the PCP. This works because any invalid PCP is necessarily inconsistent in a large number of locations, so it is detectable as invalid by the verifier with high probability. This powerful machinery has found ample uses in cryptography and theoretical computer science, and we refer readers to the book by Arora and Barak [6] for an excellent overview.

However, PCPs alone do not give us VC; there has to be some way for the prover to produce and *fix* a PCP without sending the whole thing to the verifier. Having the prover store the PCP and answer queries from the verifier on the fly will not work, because the prover could then cheat by changing parts of the PCP in reaction to the verifier's queries. In the next few paragraphs, we describe several ways to build VC on top of PCPs, as presented in [170].

10.5.2.3.1 Verifiable computation using interaction

Goldwasser et al. [77] and later Cormode et al. [49] describe using interactive proofs for verifying computation. Allowing the prover and verifier to interact (instead of requiring the prover to send a fixed string π to the verifier) makes it hard for the prover to lie in a way that the verifier will not eventually catch him/her. Instead of *naively* asking questions about values at specific locations of the PCP, the verifier asks queries in an adaptive fashion such that without a valid PCP, the prover will eventually be forced to contradict himself/herself. Roughly, the prover's early answers commit him/her to a lie and the verifier's later queries make it impossible for the prover to stick to that lie.

10.5.2.3.2 Verifiable computation using commitments

Another line of work (e.g., Killian [101]) uses *cryptographic commitments*. A cryptographic commitment is a digital object that binds the prover to a certain statement without revealing the statement. The commitment can be much smaller than the statement itself. Later, when the prover produces the statement itself (or, for some commitment schemes, even part of the statement), the verifier can use the commitment to check that the statement is in fact the one the prover committed to earlier. If the prover computes a commitment c to the entire PCP and sends c to the verifier, the verifier can then query the prover on the parts of the PCP he/she wishes to view, and the prover will have to answer honestly, because if he/she changes parts of the PCP the verifier will be able to tell that they no longer match the commitment c.

10.5.2.3.3 Verifiable computation using homomorphic encryption

Additively or multiplicatively HE (described in Section 10.5.1.1) can be used to conceal the verifier's queries about the PCP, but still allow the prover to answer them. Since the prover only sees the queries in encrypted form, he/she has no way to know how to adapt his/her PCP in response to them. A few constructions based on this paradigm include Ishai, Kushilevitz, and Ostrovsky [92], Ben-Sasson et al. [16], and Bitansky et al. [20]. One advantage of this approach is that it allows the verifier's queries to be reused, reducing the amount of interaction required.

10.5.2.4 State of the art

A great survey of practical VC implementations is given in [170]. Implementations include Pinocchio [135], SNARKs for C [16], and many others. Table 10.2 describes some aspects of their relative performance.

10.5.3 Multiparty computation

10.5.3.1 Introduction

In many situations, several parties with sensitive data wish to jointly compute some function of the collection of their data. For example, hospitals may want to perform medical research on their combined patient data, companies may want to predict cyber threats by analyzing

Table 10.2 **This table describes some aspects of the relative performance of Pinocchio [135] and SNARKs for C [16]**

	Evaluation and Proof Generation Time	*Verification Time*
Pinocchio [135]	144 s for 277,745 multiplications	10 ms
SNARKs for C [16]	31 s for 262,144 operations	5 s

threat information from companies in the same sector, or individuals may wish to participate in an auction with sealed bids. In all of these situations, each party wants to learn the result of the computation on the data held by all the parties but without sharing their own sensitive information. Furthermore, there may not be a trusted party to whom everyone is willing to give their information and that everyone trusts to perform the computation securely and correctly.

MPC is an area of cryptography that addresses this problem. MPC protocols allow people to perform distributed computation on their private data without revealing their data and without the use of a trusted party. As we will describe below, MPC protocols exist for any computable function and for any number of parties. Thus, MPC can not only provide privacy for existing applications in which parties currently share their data but would prefer not to; MPC can also enable new applications that are currently impossible because parties are unwilling or unable to share their data.

10.5.3.2 Definition

MPC allows two or more parties to compute a joint function of their inputs in such a way that everyone learns the correct output of the function but no one learns any other information, even when some of the parties may be adversarial.

In an ideal world, there would be a party that everyone trusts with their private inputs and trusts to perform the computation correctly. Then, the MPC problem would be solved trivially: everyone would send their inputs to the trusted party, and the trusted party would perform the computation and distribute the outputs. However, in the real world, we cannot assume there is a trusted party. The security of an MPC protocol guarantees essentially that it emulates the ideal-world behavior.

MPC generally considers two types of adversaries: semi-honest and malicious (as described in Section 10.2.1). We assume collusion among adversarial parties (as described in 10.2.1). In fact, we can think of there being a single adversary who *corrupts* some subset of the parties, observing their inputs and outputs and (in the case of a malicious adversary) controlling their behavior. We will also distinguish between computational security, which is based on the assumed hardness of some computational problem, and information-theoretic security, which is unconditional security based on bounds on the amount of information an adversary can learn.

10.5.3.3 Survey

The fundamental results of MPC have demonstrated that protocols exist to compute *any* function securely. In the rest of this section, we will consider *generic* MPC protocols, that is, protocols that describe how to compute any function securely, as opposed to ones that are specific to restricted classes of functions. Generic MPC protocols are typically designed for *Boolean* or *arithmetic circuits*. A circuit is a directed acyclic graph consisting of gates, where each gate takes input values and produces an output value that may feed into other gates as input. One can think of a Boolean circuit as a circuit operating on bits with XOR and AND gates. An arithmetic circuit operates on elements of a finite field with addition and multiplication gates.

Early results showed that generic MPC is possible. Yao [172] gave the first two-party protocol for computing functions represented as Boolean circuits using a technique called *garbled circuits*, providing computational security against a semi-honest adversary. Yao's protocol requires a constant number of rounds of communication. Goldreich, Micali, and Wigderson [74] made two contributions: (1) the first multiparty protocol for two or more parties, also for Boolean circuits, with computational security against a semi-honest adversary and (2) a general compiler for transforming any protocol with semi-honest security to one with malicious security. These protocols require communication for every multiplication gate, so the number of rounds of communication is proportional to the depth of the circuit. Next, Ben-Or et al. [15] and Chaum, Crépeau, and Damgård [44] simultaneously presented the first multiparty protocols with information-theoretic security; these protocols work on functions represented as arithmetic circuits. We now describe the main ideas behind some of the foundational protocols and subsequent improvements for Boolean and arithmetic circuits.

10.5.3.3.1 Boolean circuits

Most MPC protocols for securely evaluating Boolean circuits are based on Yao's garbled circuits. Yao's garbled circuit approach uses a primitive called *oblivious transfer*. In its simplest form, oblivious transfer involves a sender with two private values and a receiver with a private bit; the result of the oblivious transfer is that the receiver learns the value specified by its bit, while the sender learns nothing about which of its two values the receiver obtained.

Yao's protocol for securely evaluating Boolean circuits works as follows. One party will be the circuit generator (call her Alice) and the other will be the circuit evaluator (call him Bob). For each wire in the circuit, Alice randomly chooses two secret keys, one corresponding to each of the input bit choices 0 and 1. For each gate, Alice creates a truth table in which the output wire key is encrypted under the pair of the corresponding input wire keys. Alice permutes the truth table to hide the order of the rows and sends it to Bob along with the keys corresponding to her inputs. These keys are random, so they do not reveal to Bob anything about Alice's inputs. Then, using oblivious transfer, Bob obtains the keys from Alice corresponding to each of his inputs without Alice learning anything about Bob's inputs. For each gate, Bob decrypts the appropriate entry in the truth table to get the output wire key. After evaluating the entire circuit in this way, Bob tells Alice the final output key and Alice determines whether the output is 0 or 1. Throughout the computation, neither party learns any information about each other's inputs except for the output of the function at the end.

Subsequent works have improved on Yao's protocol in several dimensions, including reducing the ciphertext size, reducing the computational cost, and adding malicious security.

Ciphertext size: While Yao's original protocol requires transmitting 4 ciphertexts for the truth table of each gate, subsequent works have reduced the required number of ciphertexts. Naor et al. [124] showed how to use only 3 ciphertexts per truth table, and Pinkas et al. [136] further improved this to only 2 ciphertexts per truth table. Kolesnikov and Schneider [103] showed how to garble XOR gates *for free* without any ciphertexts; however, in this method AND gates still require 3 ciphertexts. Kolesnikov et al. [102] gave a method that garbles XOR gates using between 0 and 2 ciphertexts while AND gates require only 2 ciphertexts. Finally, Zahur et al. [173] show how to garble XOR gates with no ciphertexts and AND gates with only 2 ciphertexts.

Computational cost: There has been progress reducing the computational cost of garbling by using different techniques for encrypting a wire label under two keys. Naor et al. [124] showed how to do this using two hash evaluations. Lindell et al. [112] reduced this to one hash evaluation. Shelat and Shen [154] showed how to use a single block cipher evaluation, and Bellare et al. [13] further improved the efficiency by eliminating the key schedule, an expensive part of the block cipher evaluation. Another line of work reducing the computational cost of garbling was initiated by Huang et al. [87], who observed that it is possible to pipeline the circuit generation and circuit evaluation, so that the entire circuit does not have to be generated and stored all at one time before the computation.

Malicious security: The original protocol of Yao is not secure against malicious adversaries because the circuit generator could cheat by sending a garbled circuit for a different function. One technique for providing malicious security for Yao's garbled circuits, introduced by Lindell and Pinkas [111], is called *cut and choose*. In this technique, the circuit generator generates many copies of the circuit and sends them to the circuit evaluator. The circuit evaluator randomly chooses a subset of these to be opened to check that they indeed evaluate the correct function and evaluates the other copies of the circuit. If the circuit generator cheats, the circuit evaluator will detect it with high probability.

10.5.3.3.2 Arithmetic circuits

Many MPC protocols for evaluating arithmetic circuits are based on a technique called *secret sharing*. Here, we describe secret sharing and then present the high-level idea of one secret-sharing-based MPC protocol.

A t-out-of-n secret sharing of a secret value x creates n values (shares) such that the combination of any t or more shares can be used to reconstruct x, while any fewer shares reveal no information about x. Secret sharing can be performed using polynomials [153], as shown in Figure 10.4. To do this, choose a random degree $(t-1)$ polynomial whose constant term is the secret value and give each party a distinct point on the polynomial. Then, any t parties can perform polynomial interpolation and recover the polynomial and the secret value, while any fewer parties learn no information about the polynomial or the secret value.

The Ben-Or, Goldwasser, Wigderson (BGW) [15] protocol is based on secret sharing. First, each party secret-shares his or her input with the other parties. Then, the parties perform arithmetic operations on the secret-shared values by manipulating the shares. Specifically, addition of two secret-shared values can be performed by each party adding their shares locally and then jointly reconstructing using the resulting shares. Similarly, multiplication of two values can be performed by multiplying the corresponding shares; however, the resulting shares then lie on a higher-degree polynomial, so an interactive degree reduction procedure is required. Thus, the protocol for any function represented in terms of addition and multiplication operations can be evaluated securely.

Assuming secure pairwise communication channels between the parties, the BGW protocol provides information-theoretic, semi-honest security against up to $1/2$ of the parties being corrupted. Using additional techniques, BGW provides malicious security against up to $1/3$ of the parties being corrupted. Improvements by Rabin and Ben-Or [143] and by Beaver [10] provide malicious security against up to $1/2$ of the parties being corrupted, assuming that the parties have a broadcast channel. A more recent line of work has addressed malicious security against

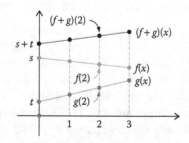

Figure 10.4: Graphical illustration of the addition of shares in a secret sharing scheme.

more parties being corrupted. Bendlin et al. [17] and Damgård et al. [52] achieve malicious security against up to all but one of the parties being corrupted, using a preprocessing *offline* phase that contains the expensive public-key operations and is computationally secure, followed by an efficient *online* phase for the actual computation that is information-theoretically secure.

10.5.3.4 State of the art

Several MPC compilers and programming frameworks have been implemented and made publicly available.

10.5.3.4.1 Two-party computation

Fairplay [116] was the first compiler for secure two-party computation, using garbled circuits. TASTY [85] combines garbled circuits and HE. Since the development of these early compilers, there have been many implementations of garbled circuits with various optimizations. These include FastGC [87], Kreuter et al. [105], CBMC-GC [86], PCF [104], JustGarble [13], and TinyGarble [162]. A couple compilers such as SCVM [113] and ObliVM [114] have support for oblivious RAM (described in Section 10.4), evaluating secure dynamic memory accesses more efficiently than performing a linear scan of memory as in circuit representations. Compilers such as PCF, SCVM, ObliVM, and Wysteria [144] use an intermediate representation of circuits that allows generating the circuit on the fly instead of all at once before computation, reducing the compilation time.

10.5.3.4.2 Multiparty computation

FairplayMP [14] extended Fairplay to more than two parties. VIFF [51], SEPIA [36], Choi et al. [45], Sharemind [21], PICCO [174], and Wysteria [144] compile code to secret-sharing-based schemes for more than two parties. SEPIA, PICCO, and Sharemind distinguish between input parties, who secret-share their inputs among the parties but do not necessarily participate in the computation, and computational parties, who perform the actual computation. Sharemind works specifically in the setting of three computational parties. Wysteria offers formal security guarantees through a special type system.

10.5.4 Functional encryption

10.5.4.1 Introduction

We now consider a situation in which a data owner wishes to allow certain computations to be performed on their sensitive data, but wishes to retain strict control over these data and computations. For example, consider a cloud-based email service where users may want the cloud to be able to perform spam filtering on their encrypted emails but learn nothing more about the contents of these emails. Here the user wants to allow the cloud to compute and learn the output of a specific function on their data without giving the cloud any more access to their emails. A cryptographic primitive that enables this capability is FE, which achieves both the access control properties of ABE (described in Section 10.3.2) and the computing on encrypted data of HE (described in Section 10.5.1.1), allowing a data owner to control exactly who can perform computation on their data and which functions they can compute.

10.5.4.2 Definition

FE is a public-key encryption scheme where, in addition to regular secret keys used to decrypt the data, there are functional secret keys. These are keys that, instead of decrypting the data, give access to the result of the corresponding function evaluated on the data. More formally, the KeyGen procedure now takes a function f and returns a key sk_f. Then, $\text{Dec}_{sk_f}(c)$ returns $f(x)$ where $c = \text{Enc}(x)$. The security guaranteed by FE is that someone holding the key for function f learns no more information about the data x than is revealed by $f(x)$. As in the case of ABE, the critical property here is collusion resistance. Specifically, FE guarantees that even someone holding keys for multiple functions cannot learn more than the outputs of the corresponding functions.

10.5.4.3 Survey

The concept of FE was first proposed in 2005 by Sahai and Waters [151] as a generalization of ABE. The first formal treatment of this powerful primitive was given by Boneh, Sahai, and Waters [28] and O'Neill [130]. However, it was quickly pointed out that there are some very strong limitations to the types of FE schemes that can be constructed in terms of efficiency, security, and generality [4,38].

It turns out that finding the *correct* definition for security of FE is quite nontrivial. The aforementioned work demonstrated that some natural definitions known as *indistinguishability-based* definitions are in some ways too weak to properly capture the desired security, while other definitions known as *simulation-based* definitions are too strong in that they cannot be satisfied by any scheme that works for all functions, as most existing schemes do. Thus, a lot of work has gone into understanding the relative power and security of these definitions [4,38].

Despite these limitations, a number of constructions of FE for arbitrary functions have appeared in the academic literature. Most of these focus on a restricted class of schemes, where only a single functional key can be given out over the course of the scheme [76,150]. This limitation was improved slightly by Gorbunov et al. [80] to allow the scheme to issue a bounded number of keys and thus tolerate a bounded number of collusions. However, the parameters of this scheme grow with the collusion bound, making it unlikely to be practical unless one is willing to assume a very small bound on the number of colluding parties. A somewhat different approach is taken by Naveed et al. [126] who instead require interaction between the data owner, who generates the keys for functions over his or her data, and any party wishing to evaluate a function on the encrypted data. This solution also seems unlikely to be practical over big data sets unless only a small subset of the data will ever be computed on. Several very recent works have shown ways to achieve security for FE against unbounded collusion. However, the security of all such schemes relies on very strong, though not fully accepted assumptions, such as the existence of indistinguishability obfuscation [63] or the hardness of certain problems over multilinear maps [65].

10.5.4.4 State of the art

FE offers great promise to enable very tightly controlled access to computation over sensitive data. However, it is currently still mostly in the realm of theoretical research. Existing schemes are most likely too inefficient to be used in practice and rely on very strong security assumptions. Additionally, very few reference implementations of FE schemes exist, making it very difficult to evaluate their performance and applicability for real-world use cases. In fact, at the time of writing, we are only aware of one prototype implementation of FE in existence today [126].

Thus, FE is not currently in a state ready to be used in today's big data applications. However, we believe that this primitive will play an important role in securing computation on sensitive data in the future. By including FE in our survey, we hope to raise awareness about this powerful primitive and draw interest to push researchers to identify appropriate applications for it and to develop schemes achieving appropriate performance for these applications in real-world deployments.

10.6   Conclusion

In this chapter, we have presented a wide variety of cryptographic techniques for securing data in transit, in storage, and in use. We believe that as security becomes a critical requirement for sensitive big data processing, these techniques will become an integral part of the big data ecosystem. We hope that the exposition in this chapter will raise awareness of the latest types of tools and protections available for securing big data. We believe better understanding and closer collaboration between the data science and cryptography communities will be critical to enabling the future of big data processing.

Acknowledgments

Ariel Hamlin, Nabil Schear, Emily Shen, Sophia Yakoubov, and Arkady Yerukhimovich's work is sponsored by the Assistant Secretary of Defense for Research and Engineering under Air Force Contract #FA8721-05-C-0002. Opinions, interpretations, recommendations, and conclusions are those of the authors and are not necessarily endorsed by the U.S. Government.

Mayank Varia's work is supported by the MACS project under NSF Frontier grant CNS-1414119.

References

1. Wenke Lee, Michael Backes, and Adrian Perrig. In *IEEE Symposium on Security and Privacy*, Berkeley, CA, May 19–22, 2013. IEEE Computer Society.

2. Michael Adjedj, Julien Bringer, Hervé Chabanne, and Bruno Kindarji. Biometric identification over encrypted data made feasible. In Atul Prakash and Indranil Gupta, editors, *Information Systems Security, 5th International Conference, Proceedings*, Kolkata, India, December 14–18, 2009, volume 5905 of *Lecture Notes in Computer Science*, pp. 86–100. Springer.

3. Rakesh Agrawal, Jerry Kiernan, Ramakrishnan Srikant, and Yirong Xu. Order-preserving encryption for numeric data. In Gerhard Weikum, Arnd Christian König, and Stefan Deßloch, editors, *Proceedings of the ACM SIGMOD International Conference on Management of Data*, Paris, France, June 13–18, 2004, pp. 563–574. ACM.

4. Shweta Agrawal, Sergey Gorbunov, Vinod Vaikuntanathan, and Hoeteck Wee. Functional encryption: New perspectives and lower bounds. In *Advances in Cryptology—CRYPTO 2013—Proceedings of the 33rd Annual Cryptology Conference, Part II*, Santa Barbara, CA, August 18–22, 2013, pp. 500–518.

5. Ross J Anderson. *Security Engineering: A Guide to Building Dependable Distributed Systems*, 2nd ed. Wiley, New York, 2008.

6. Sanjeev Arora and Boaz Barak. *Computational Complexity: A Modern Approach*. Cambridge University Press, Cambridge, 2009.

7. Sanjeev Arora and Shmuel Safra. Probabilistic checking of proofs: A new characterization of NP. *J. ACM*, 45(1):70–122, January 1998.

8. Sumeet Bajaj and Radu Sion. TrustedDB: A trusted hardware-based database with privacy and data confidentiality. *IEEE Trans. Knowl. Data Eng.*, 26(3):752–765, 2014.

9. Andrew Baumann, Marcus Peinado, and Galen Hunt. Shielding applications from an untrusted cloud with haven. In *11th USENIX Symposium on Operating Systems Design and Implementation*, Broomfield, CO, October 2014, pp. 267–283. USENIX Association.

10. Donald Beaver. Secure multiparty protocols and zero-knowledge proof systems tolerating a faulty minority. *J. Cryptology*, 4(2):75–122, 1991.

11. Mihir Bellare, Alexandra Boldyreva, and Adam O'Neill. Deterministic and efficiently searchable encryption. Cryptology ePrint Archive, Report 2006/186, 2006. http://eprint.iacr.org/.

12. Mihir Bellare, Ran Canetti, and Hugo Krawczyk. Keying hash functions for message authentication. In *Advances in Cryptology—CRYPTO '96, Proceedings of the 16th Annual International Cryptology Conference*, Santa Barbara, CA, August 18–22, 1996, pp. 1–15.

13. Mihir Bellare, Viet T Hoang, Sriram Keelveedhi, and Phillip Rogaway. Efficient garbling from a fixed-key blockcipher. In *IEEE Symposium on Security and Privacy*, Berkeley, CA, May 19–22, 2013, pp. 478–492.

14. Assaf Ben-David, Noam Nisan, and Benny Pinkas. FairplayMP: A system for secure multi-party computation. In *ACM Conference on Computer and Communications Security*, Alexandria, VA, October 27–31, 2008, pp. 257–266.

15. Michael Ben-Or, Shafi Goldwasser, and Avi Wigderson. Completeness theorems for non-cryptographic fault-tolerant distributed computation (extended abstract). In *Proceedings of the 20th Annual ACM Symposium on Theory of Computing*, Chicago, IL, May 2–4, 1988, pp. 1–10.

16. Eli Ben-Sasson, Alessandro Chiesa, Daniel Genkin, Eran Tromer, and Madars Virza. SNARKs for C: Verifying program executions succinctly and in zero knowledge. IACR Cryptology ePrint Archive, Report 2013/507, 2013.

17. Rikke Bendlin, Ivan Damgård, Claudio Orlandi, and Sarah Zakarias. Semi-homomorphic encryption and multiparty computation. In *Advances in Cryptology – EUROCRYPT '11, 30th Annual International Conference on the Theory and Applications of Cryptographic Techniques*, Tallinn, Estonia, May 15–19, 2011, pp. 169–188.

18. John Bethencourt, Amit Sahai, and Brent Waters. Ciphertext-policy attribute-based encryption. In *2007 IEEE Symposium on Security and Privacy (S&P 2007)*, Oakland, CA, May 20–23, 2007, pp. 321–334.

19. Bishwaranjan Bhattacharjee, Naoki Abe, Kenneth Goldman, Bianca Zadrozny, Vamsavardhana R Chillakuru, Marysabel del Carpio, and Chidanand Apté. Using secure coprocessors for privacy preserving collaborative data mining and analysis. In Anastassia Ailamaki, Peter A Boncz, and Stefan Manegold, editors, *Workshop on Data Management on New Hardware, DaMoN 2006*, Chicago, IL, June 25, 2006, p. 1. ACM.

20. Nir Bitansky, Alessandro Chiesa, Yuval Ishai, Omer Paneth, and Rafail Ostrovsky. Succinct non-interactive arguments via linear interactive proofs. In *Proceedings of the 10th Theory of Cryptography Conference on Theory of Cryptography, TCC'13*, pp. 315–333. Springer-Verlag, Berlin, Germany, 2013.

21. Dan Bogdanov, Sven Laur, and Jan Willemson. Sharemind: A framework for fast privacy-preserving computations. In *Computer Security - ESORICS 2008, 13th European Symposium on Research in Computer Security*, Malaga, Spain, October 6–8, 2008, pp. 192–206.

22. Alexandra Boldyreva, Nathan Chenette, Younho Lee, and Adam O'Neill. Order-preserving symmetric encryption. In Antoine Joux, editor, *Advances in Cryptology—EUROCRYPT 2009, Proceedings of the 28th Annual International Conference on the Theory and Applications of Cryptographic Techniques*, Cologne, Germany, April 26–30, 2009, volume 5479 of *Lecture Notes in Computer Science*, pp. 224–241. Springer.

23. Alexandra Boldyreva, Nathan Chenette, and Adam O'Neill. Order-preserving encryption revisited: Improved security analysis and alternative solutions. In Phillip Rogaway, editor, *Advances in Cryptology—CRYPTO 2011—Proceedings of the 31st Annual Cryptology Conference*, Santa Barbara, CA, August 14–18, 2011, volume 6841 of *Lecture Notes in Computer Science*, pp. 578–595. Springer.

24. Alexandra Boldyreva and Paul Grubbs. The cloud encryption handbook: Encryption schemes and their relative strengths and weaknesses, 2014. http://info.skyhighnetworks. com/WP-Cloud-Encryption-Handbook_Download_White.html.

25. Dan Boneh, Giovanni Di Crescenzo, Rafail Ostrovsky, and Giuseppe Persiano. Public key encryption with keyword search. In Christian Cachin and Jan Camenisch, editors, *Advances in Cryptology—EUROCRYPT 2004, Proceedings of the International Conference on the Theory and Applications of Cryptographic Techniques*, Interlaken, Switzerland, May 2–6, 2004, volume 3027 of *Lecture Notes in Computer Science*, pp. 506–522. Springer.

26. Dan Boneh, Craig Gentry, and Brent Waters. Collusion resistant broadcast encryption with short ciphertexts and private keys. In *Advances in Cryptology—CRYPTO 2005: Proceedings of the 25th Annual International Cryptology Conference*, Santa Barbara, CA, August 14–18, 2005, pp. 258–275.

27. Dan Boneh, Eu-Jin Goh, and Kobbi Nissim. Evaluating 2-DNF formulas on ciphertexts. In Proceedings of the Second International Conference on Theory of Cryptography, Springer LNCS 3378, Berlin, Heidelberg, 2005, volume 3378 of Lecture Notes in Computer Science , pp. 325–341.

28. Dan Boneh, Amit Sahai, and Brent Waters. Functional encryption: Definitions and challenges. In *Theory of Cryptography—Proceedings of the 8th Theory of Cryptography Conference, TCC 2011*, Providence, RI, March 28–30, 2011, pp. 253–273.

29. Dan Boneh and Brent Waters. Conjunctive, subset, and range queries on encrypted data. In Salil P Vadhan, editor, *Theory of Cryptography, Proceedings of the 4th Theory of Cryptography Conference, TCC 2007*, Amsterdam, the Netherlands, February 21–24, 2007, volume 4392 of *Lecture Notes in Computer Science*, pp. 535–554. Springer.

30. Dan Boneh and Brent Waters. Constrained pseudorandom functions and their applications. In *Advances in Cryptology—ASIACRYPT 2013—Proceedings of the 19th International Conference on the Theory and Application of Cryptology and Information Security, Part II*, Bengaluru, India, December 1–5, 2013, pp. 280–300.

31. Dan Boneh, Brent Waters, and Mark Zhandry. Low overhead broadcast encryption from multilinear maps. In *Advances in Cryptology—CRYPTO 2014—Proceedings of the 34th Annual Cryptology Conference, Part I*, Santa Barbara, CA, August 17–21, 2014, pp. 206–223.

32. Christoph Bösch, Pieter Hartel, Willem Jonker, and Andreas Peter. A survey of provably secure searchable encryption. *ACM Comput. Surv.*, 47(2):18:1–18:51, August 2014.

33. Christoph Bösch, Qiang Tang, Pieter H Hartel, and Willem Jonker. Selective document retrieval from encrypted database. In Dieter Gollmann and Felix C Freiling, editors, *Information Security—Proceedings of the 15th International Conference, ISC 2012*, Passau, Germany, September 19–21, 2012, volume 7483 of *Lecture Notes in Computer Science*, pp. 224–241. Springer.

34. Zvika Brakerski, Craig Gentry, and Vinod Vaikuntanathan. (Leveled) fully homomorphic encryption without bootstrapping. In *Proceedings of the 3rd Innovations in Theoretical Computer Science Conference, ITCS '12*, New York, 2012, pp. 309–325. ACM.

35. Zvika Brakerski and Vinod Vaikuntanathan. Efficient fully homomorphic encryption from (standard) LWE. In *IEEE 52nd Annual Symposium on Foundations of Computer Science, FOCS 2011*, Palm Springs, CA, October 22–25, 2011, pp. 97–106.

36. Martin Burkhart, Mario Strasser, Dilip Many, and Xenofontas Dimitropoulos. SEPIA: Privacy-preserving aggregation of multi-domain network events and statistics. In *19th USENIX Security Symposium*, Washington, DC, August 11–13, 2010, pp. 223–240.

37. Christian Cachin, Silvio Micali, and Markus Stadler. Computationally private information retrieval with polylogarithmic communication. In Jacques Stern, editor, *Advances in Cryptology—EUROCRYPT '99, Proceeding of the International Conference on the Theory and Application of Cryptographic Techniques*, Prague, Czech Republic, May 2–6, 1999, volume 1592 of *Lecture Notes in Computer Science*, pp. 402–414. Springer.

38. Angelo De Caro, Vincenzo Iovino, Abhishek Jain, Adam O'Neill, Omer Paneth, and Giuseppe Persiano. On the achievability of simulation-based security for functional encryption. In *Advances in Cryptology—CRYPTO 2013—Proceedings of the 33rd Annual Cryptology Conference, Part II*, Santa Barbara, CA, August 18–22, 2013, pp. 519–535.

39. David Cash, Joseph Jaeger, Stanislaw Jarecki, Charanjit S Jutla, Hugo Krawczyk, Marcel-Catalin Rosu, and Michael Steiner. Dynamic searchable encryption in very-large databases: Data structures and implementation. In *21st Annual Network and Distributed System Security Symposium, NDSS 2014*, San Diego, California, February 23–26, 2014. The Internet Society.

40. David Cash, Stanislaw Jarecki, Charanjit S Jutla, Hugo Krawczyk, Marcel-Catalin Rosu, and Michael Steiner. Highly-scalable searchable symmetric encryption with support for boolean queries. In Ran Canetti and Juan A Garay, editors, *Advances in Cryptology—CRYPTO 2013—Proceedings of the 33rd Annual Cryptology Conference, Part I*, Santa Barbara, CA, August 18–22, 2013, volume 8042 of *Lecture Notes in Computer Science*, pp. 353–373. Springer.

41. Yan-Cheng Chang and Michael Mitzenmacher. Privacy preserving keyword searches on remote encrypted data. In John Ioannidis, Angelos D Keromytis, and Moti Yung, editors, *Applied Cryptography and Network Security, Proceedings of the Third International Conference, ACNS 2005*, New York, June 7–10, 2005, volume 3531 of *Lecture Notes in Computer Science*, pp. 442–455.

42. Melissa Chase and Seny Kamara. Structured encryption and controlled disclosure. In Masayuki Abe, editor, *Advances in Cryptology—ASIACRYPT 2010—Proceedings of the 16th International Conference on the Theory and Application of Cryptology and Information Security*, Singapore, December 5–9, 2010, volume 6477 of *Lecture Notes in Computer Science*, pp. 577–594. Springer.

43. Melissa Chase and Emily Shen. Substring-searchable symmetric encryption. In *Privacy Enhancing Technologies Symposium (PETS)*, Philadelphia, PA, 2015, pp. 263–281.

44. David Chaum, Claude Crépeau, and Ivan Damgård. Multiparty unconditionally secure protocols (extended abstract). In *Proceedings of the 20th Annual ACM Symposium on Theory of Computing*, Chicago, IL, May 2–4, 1988, pp. 11–19.

45. Seung G Choi, Kyung-Wook Hwang, Jonathan Katz, Tal Malkin, and Dan Rubenstein. Secure multi-party computation of boolean circuits with applications to privacy in on-line marketplaces. In *Topics in Cryptology - CT-RSA 2012, The Cryptographers' Track at the RSA Conference*, San Francisco, CA, February 27–March 2, 2012, pp. 416–432.

46. Benny Chor, Niv Gilboa, and Moni Naor. Private information retrieval by keywords. IACR Cryptology ePrint Archive, Report 1998/3, 1998.

47. Benny Chor, Eyal Kushilevitz, Oded Goldreich, and Madhu Sudan. Private information retrieval. *J. ACM*, 45(6):965–981, 1998.

48. Joseph A Cooley, Roger I Khazan, Benjamin W Fuller, and Galen E Pickard. GROK: A practical system for securing group communications. In *Proceedings of The Ninth IEEE International Symposium on Networking Computing and Applications, NCA 2010*, Cambridge, MA, July 15–17, 2010, pp. 100–107.

49. Graham Cormode, Michael Mitzenmacher, and Justin Thaler. Practical verified computation with streaming interactive proofs. In *Proceedings of the 3rd Innovations in Theoretical Computer Science Conference, ITCS '12*, New York, pp. 90–112, 2012. ACM.

50. Reza Curtmola, Juan A Garay, Seny Kamara, and Rafail Ostrovsky. Searchable symmetric encryption: Improved definitions and efficient constructions. *J. Comput. Sec.*, 19(5):895–934, 2011.

51. Ivan Damgård, Martin Geisler, Mikkel Krøigaard, and Jesper B Nielsen. Asynchronous multiparty computation: Theory and implementation. In *Public Key Cryptography - PKC 2009, 12th International Conference on Practice and Theory in Public Key Cryptography*, Irvine, CA, March 18–20, 2009, pp. 160–179.

52. Ivan Damgård, Valerio Pastro, Nigel P Smart, and Sarah Zakarias. Multiparty computation from somewhat homomorphic encryption. In *Advances in Cryptology - CRYPTO 2012, 32nd Annual Cryptology Conference*, Santa Barbara, CA, August 19–23, 2012, pp. 643–662.

53. Jonathan Dautrich, Emil Stefanov, and Elaine Shi. Burst ORAM: Minimizing ORAM response times for bursty access patterns. In Kevin Fu and Jaeyeon Jung, editors, *Proceedings of the 23rd USENIX Security Symposium*, San Diego, CA, August 20–22, 2014, pp. 749–764. USENIX Association.

54. Srinivas Devadas, Marten van Dijk, Christopher W Fletcher, and Ling Ren. Onion ORAM: A constant bandwidth and constant client storage ORAM (without FHE or SWHE). IACR Cryptology ePrint Archive, Report 2015/5, 2015.

55. Whitfield Diffie and Martin E Hellman. New directions in cryptography. *IEEE Trans. Inform. Theory*, 22(6):644–654, 1976.

56. Yevgeniy Dodis and Nelly Fazio. Public key broadcast encryption for stateless receivers. In *Security and Privacy in Digital Rights Management, ACM CCS-9 Workshop, DRM 2002, Revised Papers*, Washington, DC, November 18, 2002, pp. 61–80.

57. Yarkin Doröz, Yin Hu, and Berk Sunar. Homomorphic AES evaluation using NTRU. IACR Cryptology ePrint Archive, Report 2014/39, 2014.

58. Taher El Gamal. A public key cryptosystem and a signature scheme based on discrete logarithms. *IEEE Trans. Inform. Theory*, 31(4):469–472, 1985.

59. Amos Fiat and Moni Naor. Broadcast encryption. In *Advances in Cryptology—CRYPTO '93, Proceedings of the 13th Annual International Cryptology Conference*, Santa Barbara, CA, August 22–26, 1993, pp. 480–491.

60. Ben A Fisch, Binh Vo, Fernando Krell, Abishek Kumarasubramanian, Vladimir Kolesnikov, Tal Malkin, and Steven M Bellovin. Malicious-client security in Blind Seer: A scalable private DBMS. In *IEEE Symposium on Security and Privacy*, San Jose, CA, May 17–21, 2015, pp. 395–410.

61. Michael J Freedman, Yuval Ishai, Benny Pinkas, and Omer Reingold. Keyword search and oblivious pseudorandom functions. In Joe Kilian, editor, *Theory of Cryptography, Proceedings of the Second Theory of Cryptography Conference, TCC 2005*, Cambridge, MA, February 10–12, 2005. volume 3378 of *Lecture Notes in Computer Science*, pp. 303–324. Springer.

62. Vijay Gadepally, Braden Hancock, Benjamin Kaiser, Jeremy Kepner, Peter Michaleas, Mayank Varia, and Arkady Yerukhimovich. Computing on masked data to improve the security of big data. *CoRR*, abs/1504.01287, 2015.

63. Sanjam Garg, Craig Gentry, Shai Halevi, Mariana Raykova, Amit Sahai, and Brent Waters. Candidate indistinguishability obfuscation and functional encryption for all circuits. In *54th Annual IEEE Symposium on Foundations of Computer Science, FOCS 2013*, Berkeley, CA, October 26–29, 2013, pp. 40–49.

64. Sanjam Garg, Craig Gentry, Shai Halevi, Amit Sahai, and Brent Waters. Attribute-based encryption for circuits from multilinear maps. In *Advances in Cryptology—CRYPTO 2013—Proceedings of the 33rd Annual Cryptology Conference, Part II*, Santa Barbara, CA, August 18–22, 2013, pp. 479–499.

65. Sanjam Garg, Craig Gentry, Shai Halevi, and Mark Zhandry. Fully secure functional encryption without obfuscation. IACR Cryptology ePrint Archive, Report 2014/666, 2014.

66. William I Gasarch. A survey on private information retrieval (column: Computational complexity). *Bull. EATCS*, 82:72–107, 2004.

67. Craig Gentry. A fully homomorphic encryption scheme. PhD thesis, Stanford University, Stanford, CA, 2009. crypto.stanford.edu/craig.

68. Craig Gentry and Shai Halevi. Implementing Gentry's fully-homomorphic encryption scheme. In *Advances in Cryptology - EUROCRYPT 2011, 30th Annual International Conference on the Theory and Applications of Cryptographic Techniques*, Tallinn, Estonia, May 15–19, volume 6632, 2011, pp. 129–148.

69. Craig Gentry, Shai Halevi, and Nigel P Smart. Fully homomorphic encryption with polylog overhead. In *Advances in Cryptology - EUROCRYPT 2012, 31st Annual International Conference on the Theory and Applications of Cryptographic Techniques*, Cambridge, UK, April 15–19, 2012, pp. 465–482.

70. Craig Gentry, Shai Halevi, and Nigel Smart. Homomorphic evaluation of the AES circuit. In *CRYPTO 2012*, volume 7417 of *LNCS*, 2012, pp. 850–867. Springer. Full version at http://eprint.iacr.org/2012/099.

71. Craig Gentry, Shai Halevi, and Nigel P Smart. Better bootstrapping in fully homomorphic encryption. In *Public Key Cryptography—PKC 2012*, volume 7293 of *LNCS*, 2012, pp. 1–16. Springer.

72. Yael Gertner, Yuval Ishai, Eyal Kushilevitz, and Tal Malkin. Protecting data privacy in private information retrieval schemes. *J. Comput. Syst. Sci.*, 60(3):592–629, 2000.

73. Eu-Jin Goh. Secure indexes. IACR Cryptology ePrint Archive, Report 2003/216, 2003.

74. Oded Goldreich, Silvio Micali, and Avi Wigderson. How to play any mental game or A completeness theorem for protocols with honest majority. In *Proceedings of the 19th Annual ACM Symposium on Theory of Computing*, New York, 1987, pp. 218–229.

75. Oded Goldreich and Rafail Ostrovsky. Software protection and simulation on oblivious rams. *J. ACM*, 43(3):431–473, 1996.

76. Shafi Goldwasser, Yael Tauman Kalai, Raluca A Popa, Vinod Vaikuntanathan, and Nickolai Zeldovich. Reusable garbled circuits and succinct functional encryption. In *Symposium on Theory of Computing Conference, STOC'13*, Palo Alto, CA, June 1–4, 2013, pp. 555–564.

77. Shafi Goldwasser, Yael T Kalai, and Guy N Rothblum. Delegating computation: Interactive proofs for muggles. In *Proceedings of the Fortieth Annual ACM Symposium on Theory of Computing, STOC '08*, New York, 2008, pp. 113–122. ACM.

78. Shafi Goldwasser and Silvio Micali. Probabilistic encryption & how to play mental poker keeping secret all partial information. In *Proceedings of the Fourteenth Annual ACM Symposium on Theory of Computing, STOC '82*, New York, 1982, pp. 365–377. ACM.

79. Shafi Goldwasser, Silvio Micali, and Ronald L Rivest. A digital signature scheme secure against adaptive chosen-message attacks. *SIAM J. Comput.*, 17(2):281–308, 1988.

80. Sergey Gorbunov, Vinod Vaikuntanathan, and Hoeteck Wee. Functional encryption with bounded collusions via multi-party computation. In *Advances in Cryptology—CRYPTO 2012—Proceedings of the 32nd Annual Cryptology Conference*, Santa Barbara, CA, August 19–23, 2012, pp. 162–179.

81. Sergey Gorbunov, Vinod Vaikuntanathan, and Hoeteck Wee. Attribute-based encryption for circuits. In *Symposium on Theory of Computing Conference, STOC'13*, Palo Alto, CA, June 1–4, 2013, pp. 545–554.

82. Vipul Goyal, Omkant Pandey, Amit Sahai, and Brent Waters. Attribute-based encryption for fine-grained access control of encrypted data. In *Proceedings of the 13th ACM Conference on Computer and Communications Security, CCS 2006*, Alexandria, VA, October 30–November 3, 2006, pp. 89–98.

83. Shai Halevi and Victor Shoup. HElib. https://github.com/shaih/HElib. Accessed: September 23, 2014.

84. Shai Halevi and Victor Shoup. Algorithms in HElib. In *Advances in Cryptology - CRYPTO 2014, 34th Annual Cryptology Conference*, Santa Barbara, CA, August 17–21, volume 8616, 2014, Par I, pp. 554–571.

85. Wilko Henecka, Stefan Køgl, Ahmad-Reza Sadeghi, Thomas Schneider, and Immo Wehrenberg. TASTY: Tool for automating secure two-party computations. In *Proceedings of the 17th ACM Conference on Computer and Communications Security*, Chicago, IL, October 4–8, 2010, pp. 451–462.

86. Andreas Holzer, Martin Franz, Stefan Katzenbeisser, and Helmut Veith. Secure two-party computations in ANSI C. In *ACM Conference on Computer and Communications Security*, Raleigh, NC, October 16–18, 2012, pp. 772–783.

87. Yan Huang, David Evans, Jonathan Katz, and Lior Malka. Faster secure two-party computation using garbled circuits. In *20th USENIX Security Symposium*, San Francisco, CA, August 8–12, 2011.

88. IARPA. Broad agency announcement IARPA-BAA-11-01: Security and privacy assurance research (SPAR) program. https://www.fbo.gov/notices/c55e38dbde30cb668f 687897d8f01e69, February 2011.

89. Internet Engineering Task Force. Request for comments 4301. Security architecture for the internet protocol. https://tools.ietf.org/html/rfc4301. Accessed: July 09, 2015.

90. Internet Engineering Task Force. Request for comments 5246. The Transport Layer Security (TLS) protocol version 1.2. https://tools.ietf.org/html/rfc5246. Accessed: July 09, 2015.

91. Alexander Iliev and Sean W Smith. Private information storage with logarithm-space secure hardware. In Yves Deswarte, Frédéric Cuppens, Sushil Jajodia, and Lingyu Wang, editors, *Information Security Management, Education and Privacy, IFIP 18th World Computer Congress, TC11 19th International Information Security Workshops*, Toulouse, France, August 22–27, 2004, volume 148 of *IFIP*, pp. 199–214. Kluwer.

92. Yuval Ishai, Eyal Kushilevitz, and Rafail Ostrovsky. Efficient arguments without short PCPs. In *Proceedings of the Twenty-Second Annual IEEE Conference on Computational Complexity, CCC '07*, Washington, DC, 2007, pp. 278–291. IEEE Computer Society.

93. Stanislaw Jarecki, Charanjit S Jutla, Hugo Krawczyk, Marcel-Catalin Rosu, and Michael Steiner. Outsourced symmetric private information retrieval. In Ahmad-Reza Sadeghi, Virgil D. Gligor, and Moti Yung, editors, *2013 ACM SIGSAC Conference on Computer and Communications Security, CCS'13*, Berlin, Germany, November 4–8, 2013, pp. 875–888. ACM.

94. David Kahn. *The Codebreakers: The Comprehensive History of Secret Communication from Ancient Times to the Internet.* Scribner, 2nd Revised edition, December 1996.

95. Seny Kamara and Charalampos Papamanthou. Parallel and dynamic searchable symmetric encryption. In *Financial Cryptography and Data Security—17th International Conference, FC 2013, Revised Selected Papers*, Okinawa, Japan, April 1–5, 2013, pp. 258–274.

96. Seny Kamara, Charalampos Papamanthou, and Tom Roeder. Dynamic searchable symmetric encryption. In *the ACM Conference on Computer and Communications Security, CCS'12*, Raleigh, NC, October 16–18, 2012, pp. 965–976.

97. Jonathan Katz and Yehuda Lindell. *Introduction to Modern Cryptography.* Chapman and Hall/CRC Press, 2007.

98. Jonathan Katz and Arkady Yerukhimovich. On black-box constructions of predicate encryption from trapdoor permutations. In *Advances in Cryptology—ASIACRYPT 2009, Proceedings of the 15th International Conference on the Theory and Application of Cryptology and Information Security*, Tokyo, Japan, December 6–10, 2009, pp. 197–213.

99. Roger I Khazan and Dan Utin. Lincoln open cryptographic key management architecture. https://www.ll.mit.edu/publications/technotes/TechNote_LOCKMA.pdf. Accessed: June 2015, 26.

100. Joud Khoury, Gregory Lauer, Partha P Pal, Bishal Thapa, and Joseph P Loyall. Efficient private publish-subscribe systems. In *17th IEEE International Symposium on Object/Component/Service-Oriented Real-Time Distributed Computing, ISORC 2014*, Reno, NV, June 10–12, 2014, pp. 64–71, 2014.

101. Joe Kilian. A note on efficient zero-knowledge proofs and arguments (extended abstract). In *Proceedings of the Twenty-fourth Annual ACM Symposium on Theory of Computing, STOC '92*, New York, pp. 723–732, 1992. ACM.

102. Vladimir Kolesnikov, Payman Mohassel, and Mike Rosulek. FleXOR: Flexible garbling for XOR gates that beats free-XOR. In *Advances in Cryptology - CRYPTO 2014, 34th Annual Cryptology Conference*, Santa Barbara, CA, August 17–21, 2014, Part II, pp. 440–457.

103. Vladimir Kolesnikov and Thomas Schneider. Improved garbled circuit: Free XOR gates and applications. In *Automata, Languages, and Programming, 35th International Colloquium, ICALP 2008*, Reykjavik, Iceland, July 7–11, 2008, Part II, pp. 486–498.

104. Ben Kreuter, Benjamin Mood, Abhi Shelat, and Kevin Butler. PCF: A portable circuit format for scalable two-party secure computation. In *Proceedings of the 22th USENIX Security Symposium*, Washington, DC, August 14–16, 2013, pp. 321–336.

105. Benjamin Kreuter, Abhi Shelat, and Chih-Hao Shen. Billion-gate secure computation with malicious adversaries. In *Proceedings of the 21th USENIX Security Symposium*, Bellevue, WA, August 8–10, 2012, pp. 285–300.

106. Eyal Kushilevitz and Rafail Ostrovsky. Replication is NOT needed: SINGLE database, computationally-private information retrieval. In *38th Annual Symposium on Foundations of Computer Science, FOCS '97*, Miami Beach, FL, October 19–22, 1997, pp. 364–373. IEEE Computer Society.

107. Eyal Kushilevitz and Rafail Ostrovsky. One-way trapdoor permutations are sufficient for non-trivial single-server private information retrieval. In Bart Preneel, editor, *Advances in Cryptology—EUROCRYPT 2000, Proceeding of the International Conference on the Theory and Application of Cryptographic Techniques*, Bruges, Belgium, May 14–18, 2000, volume 1807 of *Lecture Notes in Computer Science*, pp. 104–121. Springer.

108. RSA Laboratories. Public-Key Cryptography Standards (PKCS) #1: RSA Cryptography Specifications Version 2.1. https://tools.ietf.org/html/rfc3447. Accessed: June 26, 2015.

109. Leslie Lamport, Robert Shostak, and Marshall Pease. The byzantine generals problem. *ACM Trans. Program. Lang. Syst.*, 4(3):382–401, July 1982.

110. Thomas A Limoncelli, Strata R Chalup, and Christina J Hogan. *The Practice of Cloud System Administration: Designing and Operating Large Distributed Systems*, volume 2, 1st edition. Addison-Wesley Professional, September 2014.

111. Yehuda Lindell and Benny Pinkas. Secure two-party computation via cut-and-choose oblivious transfer. In *Theory of Cryptography Conference - 8th Theory of Cryptography Conference, TCC 2011*, Providence, RI, March 28–30, 2011, pp. 329–346.

112. Yehuda Lindell, Benny Pinkas, and Nigel P Smart. Implementing two-party computation efficiently with security against malicious adversaries. In *Security and Cryptography for Networks, 6th International Conference, SCN 2008*, Amalfi, Italy, September 10–12, 2008, pp. 2–20.

113. Chang Liu, Yan Huang, Elaine Shi, Jonathan Katz, and Michael W Hicks. Automating efficient RAM-model secure computation. In *2014 IEEE Symposium on Security and Privacy, SP 2014*, Berkeley, CA, May 18–21, 2014, pp. 623–638.

114. Chang Liu, Xiao S Wang, Kartik Nayak, Yan Huang, and Elaine Shi. ObliVM: A programming framework for secure computation. In *IEEE Symposium on Security and Privacy, SP 2015*, San Jose, CA, May 17–21, 2015, pp. 359–376.

115. Ben Lynn. PBC: The pairing-based crypto library. http://crypto.stanford.edu/pbc/. Accessed: June 26, 2015.

116. Dahlia Malkhi, Noam Nisan, Benny Pinkas, and Yaron Sella. Fairplay—A secure two-party computation system. In *Proceedings of the 13th USENIX Security Symposium*, San Diego, CA, August 9–13, 2004, pp. 287–302.

117. Mandiant. http://intelreport.mandiant.com/Mandiant_APT1_Report.pdf, Feb 2013.

118. Charalampos Mavroforakis, Nathan Chenette, Adam O'Neill, George Kollios, and Ran Canetti. Modular order-preserving encryption, revisited. In Timos Sellis, Susan B Davidson, and Zachary G Ives, editors, *Proceedings of the 2015 ACM SIGMOD International Conference on Management of Data*, Melbourne, Victoria, Australia, May 31–June 4, 2015 pp. 763–777. ACM.

119. MIT Lincoln Laboratory. HEtest. https://www.ll.mit.edu/mission/cybersec/softwaretools/ hetest/hetest.html, February 2011.

120. Tarik Moataz, Travis Mayberry, and Erik-Oliver Blass. Resizable tree-based oblivious RAM. IACR Cryptology ePrint Archive, Report 2014/732, 2014.

121. Tarik Moataz, Travis Mayberry, and Erik-Oliver Blass. Constant communication oblivious RAM. IACR Cryptology ePrint Archive, Report 2015/570, 2015.

122. Marti Motoyama, Damon McCoy, Kirill Levchenko, Stefan Savage, and Geoffrey M Voelker. An analysis of underground forums. In *Proceedings of the 2011 ACM SIG-COMM Conference on Internet Measurement Conference, IMC '11*, New York, 2011, pp. 71–80. ACM.

123. Dalit Naor, Moni Naor, and Jeffery Lotspiech. Revocation and tracing schemes for stateless receivers. In *Advances in Cryptology—CRYPTO 2001, Proceedings of the 21st Annual International Cryptology Conference*, Santa Barbara, CA, August 19–23, 2001, pp. 41–62.

124. Moni Naor, Benny Pinkas, and Reuban Sumner. Privacy preserving auctions and mechanism design. In *ACM Conference on Electronic Commerce (EC)*, Denver, CO, 1999, pp. 129–139.

125. National Security Agency. Suite B cryptography. https://www.nsa.gov/ia/programs/ suiteb_cryptography/. Accessed: June 26, 2015.

126. Muhammad Naveed, Shashank Agrawal, Manoj Prabhakaran, XiaoFeng Wang, Erman Ayday, Jean-Pierre Hubaux, and Carl A Gunter. Controlled functional encryption. In *Proceedings of the 2014 ACM SIGSAC Conference on Computer and Communications Security*, Scottsdale, AZ, November 3–7, 2014, pp. 1280–1291.

127. National Institute of Standards and Technology. Cryptographic toolkit. http://csrc. nist.gov/groups/ST/toolkit/index.html. Accessed: June 26, 2015.

128. National Institute of Standards and Technology. Federal information processing standards publication 197: Announcing the advanced encryption standard (AES). http://csrc.nist. gov/publications/fips/fips197/fips-197.pdf. Accessed: July 09, 2015.

129. OASIS. Key management interoperability protocol specification version 1.2. http:// docs.oasis-open.org/kmip/spec/v1.2/cos01/kmip-spec-v1.2-cos01.html. Accessed: June 26, 2015.

130. Adam O'Neill. Definitional issues in functional encryption. IACR Cryptology ePrint Archive, Report 2010/556, 2010.

131. Oracle. Oracle label security with oracle database 12c. http://www.oracle.com/ technetwork/database/options/label-security/label-security-wp-12c-1896140.pdf? ssSourceSiteId=ocomen, June 2013.

132. Pascal Paillier. Public-key cryptosystems based on composite degree residuosity classes. In *Advances in Cryptology - EUROCRYPT '99, International Conference on the Theory and Application of Cryptographic Techniques*, Prague, Czech Republic, May 2–6, 1999, pp. 223–238.

133. Vasilis Pappas, Fernando Krell, Binh Vo, Vladimir Kolesnikov, Tal Malkin, Seung Geol Choi, Wesley George, Angelos D Keromytis, and Steve Bellovin. Blind Seer: A scalable private DBMS. In *2014 IEEE Symposium on Security and Privacy, SP 2014*, Berkeley, CA, May 18–21, 2014, pp. 359–374. IEEE Computer Society.

134. Hyun-A Park, Bum H Kim, Dong H Lee, Yon D Chung, and Justin Zhan. Secure similarity search. In *2007 IEEE International Conference on Granular Computing, GrC 2007*, San Jose, CA, November 2–4, 2007, p. 598. IEEE.

135. Bryan Parno, Jon Howell, Craig Gentry, and Mariana Raykova. Pinocchio: Nearly practical verifiable computation. In *Proceedings of the 2013 IEEE Symposium on Security and Privacy, SP '13*, Washington, DC, 2013, pp. 238–252. IEEE Computer Society.

136. Benny Pinkas, Thomas Schneider, Nigel P Smart, and Stephen C Williams. Secure two-party computation is practical. In *Advances in Cryptology - ASIACRYPT 2009, 15th International Conference on the Theory and Application of Cryptology and Information Security*, Tokyo, Japan, December 6–10, 2009, pp. 250–267.

137. Matthew Pirretti, Patrick Traynor, Patrick McDaniel, and Brent Waters. Secure attribute-based systems. In *Proceedings of the 13th ACM Conference on Computer and Communications Security, CCS 2006*, Alexandria, VA, October 30–November 3, 2006, pp. 99–112.

138. Raluca A Popa, Frank H Li, and Nickolai Zeldovich. An ideal-security protocol for order-preserving encoding. In *2013 IEEE Symposium on Security and Privacy, SP 2013* Berkeley, CA, May 19–22, 2013, pp. 463–477. IEEE Computer Society.

139. Raluca A Popa, Catherine MS Redfield, Nickolai Zeldovich, and Hari Balakrishnan. CryptDB: Protecting confidentiality with encrypted query processing. In *Proceedings of the 23rd ACM Symposium on Operating Systems Principles, SOSP 2011*, Cascais, Portugal, October 23–26, 2011, pp. 85–100.

140. Thomas Pöppelmann, Michael Naehrig, Andrew Putnam, and Adrian Macias. Accelerating homomorphic evaluation on reconfigurable hardware. Cryptology ePrint Archive, Report 2015/631, 2015. http://eprint.iacr.org/.

141. Michael O Rabin. Digitalized signatures and public-key functions as intractable as factorization. MIT Laboratory for Computer Science. http://publications.csail.mit.edu/lcs/pubs/pdf/MIT-LCS-TR-212.pdf, January 1979.

142. Michael O Rabin. How to exchange secrets with oblivious transfer. IACR Cryptology ePrint Archive, Report 2005/187, 2005.

143. Tal Rabin and Michael Ben-Or. Verifiable secret sharing and multiparty protocols with honest majority. (extended abstract). In *Proceedings of the 21st Annual ACM Symposium on Theory of Computing*, Seattle, WA, May 14–17, 1989, pp. 73–85.

144. Aseem Rastogi, Matthew A Hammer, and Michael Hicks. Wysteria: A programming language for generic, mixed-mode multiparty computations. In *IEEE Symposium on Security and Privacy, SP 2014*, Berkeley, CA, May 18–21, 2014, pp. 655–670.

145. Oded Regev. On lattices, learning with errors, random linear codes, and cryptography. In *Proceedings of the Thirty-seventh Annual ACM Symposium on Theory of Computing, STOC '05*, New York, 2005, pp. 84–93. ACM.

146. Ling Ren, Christopher W Fletcher, Albert Kwon, Emil Stefanov, Elaine Shi, Marten van Dijk, and Srinivas Devadas. Ring ORAM: Closing the gap between small and large client storage oblivious RAM. IACR Cryptology ePrint Archive, Report 2014/997, 2014.

147. Ronald L Rivest, Len Adleman, and Michael L Dertouzos. On data banks and privacy homomorphisms. *Foundations of Secure Computation*, pp. 169–180, 1978. Academic Press, New York.

148. Ronald L Rivest, Adi Shamir, and Leonard M Adleman. A method for obtaining digital signatures and public-key cryptosystems. *Commun. ACM*, 21(2):120–126, 1978.

149. Ahmad-Reza Sadeghi, Virgil D Gligor, and Moti Yung, editors, In *2013 ACM SIGSAC Conference on Computer and Communications Security, CCS'13*, Berlin, Germany, November 4–8, 2013. ACM.

150. Amit Sahai and Hakan Seyalioglu. Worry-free encryption: Functional encryption with public keys. In *Proceedings of the 17th ACM Conference on Computer and Communications Security, CCS 2010*, Chicago, IL, October 4–8, 2010, pp. 463–472.

151. Amit Sahai and Brent Waters. Fuzzy identity-based encryption. In *Advances in Cryptology—EUROCRYPT 2005, Proceedings of the 24th Annual International Conference on the Theory and Applications of Cryptographic Techniques*, Aarhus, Denmark, May 22–26, 2005, pp. 457–473.

152. Felix Schuster, Manuel Costa, Cédric Fournet, Christos Gkantsidis, Marcus Peinado, Gloria Mainar-Ruiz, and Mark Russinovich. VC3: Trustworthy data analytics in the cloud using SGX. In *2015 IEEE Symposium on Security and Privacy, SP 2015*, San Jose, CA, May 17–21, 2015, pp. 38–54.

153. Adi Shamir. How to share a secret. In *Communications of the ACM*, 1979.

154. Abhi Shelat and Chih-Hao Shen. Fast two-party secure computation with minimal assumptions. In Ahmad-Reza Sadeghi, Virgil D Gligor, and Moti Yung, editors, *2013 ACM SIGSAC Conference on Computer and Communications Security, CCS'13*, Berlin, Germany, November 4–8, 2013, pp. 523–534.

155. Emily Shen, Elaine Shi, and Brent Waters. Predicate privacy in encryption systems. In Omer Reingold, editor, *Theory of Cryptography, Proceedings of the 6th Theory of Cryptography Conference, TCC 2009*, San Francisco, CA, March 15–17, 2009, volume 5444 of *Lecture Notes in Computer Science*, pp. 457–473. Springer.

156. New York Times. 9 recent cyberattacks against big businesses. http://www.nytimes.com/interactive/2015/02/05/technology/recent-cyberattacks.html, February 5, 2015. Accessed: July 09, 2015.

157. New York Times. Hacking linked to China exposes millions of U.S. workers. http://www.nytimes.com/2015/06/05/us/breach-in-a-federal-computer-system-exposes-personnel-data.html, June 4, 2015. Accessed: July 09, 2015.

158. Nigel P Smart and Frederik Vercauteren. Fully homomorphic encryption with relatively small key and ciphertext sizes. In *Public Key Cryptography – PKC 2010*, volume 6056 of *LNCS*, pages 420–443, 2010.

159. Nigel P Smart and Frederik Vercauteren. Fully homomorphic SIMD operations. In *Designs, Codes and Cryptography*, 71(1):57–81, 2014.

160. Sean W Smith and David Safford. Practical server privacy with secure coprocessors. *IBM Syst. J.*, 40(3):683–695, 2001.

161. Dawn X Song, David Wagner, and Adrian Perrig. Practical techniques for searches on encrypted data. In *2000 IEEE Symposium on Security and Privacy*, Berkeley, CA, May 14–17, 2000, pp. 44–55. IEEE Computer Society.

162. Ebrahim M Songhori, Siam U Hussain, Ahmad-Reza Sadeghi, Thomas Schneider, and Farinaz Koushanfar. TinyGarble: Highly compressed and scalable sequential garbled circuits. In *IEEE Symposium on Security and Privacy, SP 2015*, San Jose, CA, May 17–21, 2015, pp. 411–428.

163. Emil Stefanov and Elaine Shi. ObliviStore: High performance oblivious cloud storage. In *2013 IEEE Symposium on Security and Privacy, SP 2013*, Berkeley, CA, May 19–22, 2013, pp. 253–267. IEEE Computer Society.

164. Emil Stefanov, Marten van Dijk, Elaine Shi, Christopher W Fletcher, Ling Ren, Xiangyao Yu, and Srinivas Devadas. Path ORAM: An extremely simple oblivious RAM protocol. In Ahmad-Reza Sadeghi, Virgil D Gligor, and Moti Yung, editors, *2013 ACM SIGSAC Conference on Computer and Communications Security, CCS'13*, Berlin, Germany, November 4–8, 2013, pp. 299–310.

165. The Apache Software Foundation. Accumulo. https://accumulo.apache.org/. Accessed: July 09, 2015.

166. Stephen Tu, M. Frans Kaashoek, Samuel Madden, and Nickolai Zeldovich. Processing analytical queries over encrypted data. *Proc. VLDB Endow.*, 6(5):289–300, March 2013.

167. Peter van Liesdonk, Saeed Sedghi, Jeroen Doumen, Pieter H Hartel, and Willem Jonker. Computationally efficient searchable symmetric encryption. In *Secure Data Management, Proceedings of the 7th VLDB Workshop, SDM 2010*, Singapore, September 17, 2010, pp. 87–100.

168. Mayank Varia, Benjamin Price, Nicholas Hwang, Ariel Hamlin, Jonathan Herzog, Jill Poland, Michael Reschly, Sophia Yakoubov, and Robert K Cunningham. Automated assessment of secure search systems. *Operat. Syst. Rev.*, 49(1):22–30, 2015.

169. Mayank Varia, Sophia Yakoubov, and Yang Yang. HEtest: A homomorphic encryption testing framework. In *Financial Cryptography and Data Security – FC 2015 International Workshops, BITCOIN, WAHC, and Wearable*, San Juan, PR, January 30, 2015, Revised selected papers, pp. 213–230.

170. Michael Walfish and Andrew J Blumberg. Verifying computations without reexecuting them. *Commun. ACM*, 58(2):74–84, 2015.

171. Guojun Wang, Qin Liu, and Jie Wu. Hierarchical attribute-based encryption for fine-grained access control in cloud storage services. In *Proceedings of the 17th ACM Conference on Computer and Communications Security, CCS 2010*, Chicago, IL, October 4–8, 2010, pp. 735–737, 2010.

172. Andrew Chi-Chih Yao. Protocols for secure computations (extended abstract). In *23rd Annual Symposium on Foundations of Computer Science*, Chicago, IL, November 3–5, 1982, pp. 160–164.

173. Samee Zahur, Mike Rosulek, and David Evans. Two halves make a whole: Reducing data transfer in garbled circuits using half gates. In *Advances in Cryptology – EUROCRYPT 2015, 34th Annual International Conference on the Theory and Applications of Cryptographic Techniques*, Sofia, Bulgaria, April 26–30, 2015, Part II, pp. 220–250.

174. Yihua Zhang, Aaron Steele, and Marina Blanton. PICCO: A general-purpose compiler for private distributed computation. In *ACM SIGSAC Conference on Computer and Communications Security, CCS'13*, Berlin, Germany, November 4–8, 2013, pp. 813–826.

Chapter 11

Some Issues of Privacy in a World of Big Data and Data Mining

Daniel E. O'Leary

CONTENTS

Abstract

This chapter examines privacy issues in the context of big data and potential data mining of that data. Issues are analyzed based on five emerging unique characterizations associated with big data: the five Vs, the Big Data Lake, "thing" data, the quantified self, repurposed data, and the generation of knowledge from unstructured communication data, e.g., Twitter Tweets. Each of those sets of emerging issues is analyzed in detail for their potential impact on privacy.

As part of this analysis, this chapter identifies "unevenness" (of volume, velocity, variety and veracity) as potential concerns of privacy and equity. Further, this analysis notes that in some cases systems developed to "defend" individuals and organizations from privacy infringements, ironically can infringe on the privacy of "adjacent others."

This analysis also notes the importance of governance in emerging structures such as the Big Data Lake and repurposed data. Finally, this chapter questions the ethical use of some data used to create knowledge from the equivalence of digital conversations.

11.1 Introduction

Big data has been developed, in part, because of the availability of technology, such as Hadoop, the increase in the volumes of data, emerging new sources of data (Internet of Things, Twitter, and others), the increase in the velocity of other data (e.g., real-time data and log data), and the approaches being used to assess the veracity of that data (e.g., continuous monitoring) with all of that data being put in juxtaposition with transaction processing and a range of other more traditional enterprise databases. With these new technologies and data sources, new issues can be investigated to support decision making.

However, although big data provides even more data, it is still data, and data mining still is necessary to troll through the data to find key relationships, whether descriptive, diagnostic, predictive, or prescriptive. Unfortunately, with more data and different data, there are more opportunities to violate personal and organizational privacy. As a result, it is probably not surprising that Michael and Miller (2013) call privacy one of the biggest potential challenges of big data.

11.1.1 Privacy

Privacy has long been a concern in data mining, virtually since the first international meeting addressing data mining (e.g., O'Leary 1991). These concerns have led to notions of so-called privacy preservation (e.g., Verykios et al. 2004a) within databases to protect privacy, which has become a field of its own.

Google.com defines privacy as "the state or condition of being free from being observed or disturbed by other people" and "the state of being free from public attention." Thus, the key to privacy seems to be related to observation and monitoring, unwanted attention, and disturbing.

A recent Google search (June 29, 2015) found 5,370,000,000 results related to privacy. From 1900 to 2008, the use of the term "privacy" in books has increased roughly 750%.* Just from 1985 to 2008, there was an increase of 150% of the occurrences of privacy in books. These results suggest that over time "privacy" is becoming more of an issue, even independent of big data.

Databases capture information about a wide range of human behavior, including spending activity (credit and debit card activity), health-care activity, location activity (phone movement), and computer browser activity. With big data, the potential for being "observed" or "monitored," as manifested in all of these databases, can substantially increase. In particular, individual behavior as captured in data trails left behind by individuals as part of general living can be linked together. Thus, behavior in different segments of a person's life can be stitched together inferring or telling a story of potentially private events.

11.2 Selected Previous Research

There has been some previous research on privacy in big data.

Wong (2012) raised the issue of the implications of storing data on a single site, as opposed to multiple sites. Cavoukian and Jonas (2012) discuss privacy by design as an approach to the needs of this situation. They make seven assertions that would facilitate designing in privacy, largely aimed at the record level. For example, there is the assertion of the need for full attribution, which refers to capturing where every record has come from and when it was acquired. Cavoukian and Jonas (2012) also generate what they call information transfer accounting, in which a log of every data item is captured for any transfer to other systems or reports. Cormonde and Srivastava (2009) note that the two primary controls designed to facilitate privacy are limiting access and anonymizing the data. Wu et al. (2014) note that many privacy concerns focus on not allowing third-party data miners to access the data. Tene and Polonetsky (2012) examine some of the big data and privacy issues from a legal perspective. In addition, they examine big data privacy in terms of particular industries.

The importance of privacy in databases, in general, provides an important background. As an example, Verykios et al. (2004b) surveys the privacy preservation literature.

11.3 Five Vs and Privacy

Big data has been characterized by the five Vs (e.g., Zikopoulos et al. 2012, 2013): volume, velocity, variety, veracity, and value. This section examines how the five Vs relate to challenges

---

*See https://books.google.com/ngrams/graph?year_start=1800&year_end=2008&corpus=15&smoothing=7& case_insensitive=on&content=privacy&direct_url=t4%3B%2Cprivacy%3B%2Cc0%3B%2Cs0%3B%3Bprivacy%3B% 2Cc0%3B%3BPrivacy%3B%2Cc0.

in big data privacy. Analysis of privacy using these characteristics leads to two key insights: First, there is "unevenness" in each of the Vs; second, systems developed by those concerned about privacy violations are likely to affect the privacy of "adjacent others."

11.3.1 Volume

A key component of big data is more data, that is, increased volumes of data. As more and more data are being captured, whether from the Internet of Things (SRA 2009), communication data (Twitter) or even transaction data, the potential supplies of data are increasing. With larger amounts of data, there is an increasing "coverage" in a digital format of people's lives. This broader coverage suggests more digital observing, more digital monitoring, more attention, and potentially more "disturbing" the privacy of individuals.

11.3.1.1 Uneven volume

However, volume is not necessarily uniform across or even within different databases. For example, more volume will be generated about some individuals than other individuals, more information can be generated about some products than others, and some event types occur much more frequently than other events. There are a number of factors that affect the "evenness" of the supply of data about some object. For example, if a product has some failings, then there is likely to be news media or social media discussions about the product, captured in a social media or external news database. As another example, if a manager does something unusually good or bad, then there is likely to be social media or news media about that manager. Similarly, internal activities can generate additional transaction data or operational data. For example, a machine that is making product "out of specifications" likely is generating substantial monitoring data.

11.3.1.2 Unevenness—Example

Unevenness can be further illustrated with a search of the Internet for different people. For one person, there may be pictures and multiple Google entries, whereas for another, there may be virtually nothing. Unevenness may be the result of a conscious effort, or it may result because of activity in different domains. For example, universities and faculty generally have a substantial Internet presence, whereas similar people in other domains have more limited presence.

11.3.1.3 Unevenness—More observations, more variables

Unevenness can result in either more observations over the same number of variables or more variables or both. More observations can suggest that better estimates are available for the variables under consideration. More variables can suggest a better (or at least more depth) description of the object or individual – more context. However, in either case, "more" suggests that "more" observing is being done. As a result, uneven data suggests uneven monitoring, uneven analysis, and uneven privacy for the objects and agents in the databases.

11.3.1.4 What does unevenness imply?

What does the existence of a lack of evenness of data tell us or imply? For those activities with more data, it is easier to do "more" analysis. As a result, there may be an inappropriate amount of analysis of some individuals that could ultimately lead to prejudicial behavior toward those individuals: more data suggests that more observation can be or has been done. As a result, the lack of evenness suggests uneven observation and monitoring.

Similarly, there are implications associated with less data on some set of objects. If there is less data, then the analysis is likely to put more weight on fewer items potentially overweighing those items as part of the monitoring process. Although this may not be a direct privacy issue, it could result in a lack of equity and comparably inappropriate monitoring.

11.3.1.5 Unevenness and the demand for more data

Evenness or unevenness of the data can affect additional demand for more data. The existence of data for some object, but not another, can lead the organization to identify gaps in the data and lead to a search for additional data for those recognized gaps. However, ironically, the search for gap information only occurs because of the existence of other information.

11.3.1.6 Organizational monitoring of data flows—"Adjacent" others

Finally, because of potential analysis of data that could impact their privacy, individuals and organizations have begun to try to monitor existing flows of data. Unfortunately, increasing volumes of data can make keeping up with privacy invading information very difficult. As a result, increasingly, systems are being built to monitor those data flows (e.g., Ziegler and Skubacz 2006). Ironically, such systems can result in a loss of privacy for those adjacent to those developing the systems ("adjacent others"). For example, systems that have been built typically do not just search for information about specific individuals and organizations, but they also search for information about benchmark individuals providing another prying source for those adjacent others. In the case of a chief executive officer (CEO) of some company, such a system is likely to look not just for information about the specific CEO but for information about other CEOs from other companies in the same industry.

11.3.2 Velocity

It has been reported that roughly 90% of the world's data has been generated in the past 2 years.* As a result, it is clear that increasing velocity has had a major effect on the development of large volumes of data.†

E-commerce sites are set for real-time processing of transactions, generating substantial velocity. In addition, computer-based systems and logs associated with those systems provide high velocity that generates big data. Data velocity also is related to feedback. For example, one tweet might be retreated and further retweeted, with many more retweets than the original message. The more retweets the higher the velocity.

Further, not only do objects generate data but as information about those objects is generated, there is response to that information and those responses lead to additional information, cascading on itself. As an example, tags on one photo could cascade to other photos that, whether appropriate or not, potentially impinge on the privacy of some person.

11.3.2.1 Uneven velocity

As with volume, data velocity and feedback also are uneven across and within databases. For example, for most Twitter tweets, there are limited retweets. However, for celebrities and sports figures, there can be a substantial number of retweets and those retweets increase velocity.

Uneven velocity can lead to disproportionate data being available for some individuals. Because of that increased availability, this can lead to additional analysis of some objects

---

* http://www.sciencedaily.com/releases/2013/05/130522085217.htm.
† http://www.scaledb.com/high-velocity-data.php.

or agents in comparison with other objects, potentially reducing the privacy related to those objects or agents. Further, a deeper analysis of those objects can lead to inequitable analysis and potentially inequitable treatment of those objects and individuals.

11.3.2.2 Organizational monitoring of data flows—"Adjacent" others

For those individuals or organizations that monitor their digital footprint, as data velocity increases, it becomes more difficult to keep up with data flows. As a result, because of increasing volume, systems must be built to account for that velocity by capturing information at increasing rates. This in turn potentially generates additional feedback on existing information, which further speeds the velocity for the specific agent, and any benchmark agents about whom the system gathers data.

11.3.3 Variety

Variety refers to the notion that there can be multiple different types of data and databases.

In a world of relational databases, structured numeric data was typically the focus. However, with variety that changes. Data from "things" on the Internet of Things, or unstructured data, such as Twitter tweets are now expected to be used in conjunction with each other and with structured transaction data. For example, a company may be interested in analyzing the impact of Twitter message activity of a certain sentiment on sales transactions.

11.3.3.1 Uneven variety

Uneven variety can derive from a number of reasons. Available databases may reflect the commercial systems that have been built. For example, there are a number of devices that allow individuals to capture the number of steps that they walk every day and their heartbeat. Although these provide important health information, they are only a few of the many potential metrics available concerning individual health. However, these two are among the more likely available, illustrating that the variety may be uneven.

11.3.3.2 Impact of uneven variety

Uneven database variety is likely to primarily affect the number of variables available to analyze the particular object or individual, and the observations for those variables. As a result, this suggests that the inferences that can be drawn also will be uneven. Accordingly, the monitoring and observation that can be done is also likely to be uneven.

Further, any gaps in databases suggest that additional databases be generated for those not included in the specified databases. Although this will tend to make the data "even," it also can increase the potential for privacy invasion.

11.3.3.3 Law of requisite variety

As noted by Ashby (1956, p. 207), "only variety can destroy variety." As a result, in order to handle the broad variety of databases will require a number of methodologies. Because different databases will necessarily be treated differently, this suggests that in some cases, there will be a potential for the lack of equity. Further, some approaches will work better than others, potentially influencing manifested privacy.

11.3.4 Veracity

A potential problem with big data is that the quality of the data may not be the same as historically may have been the case. In particular, different databases are likely to differ in their veracity (uneven veracity). For example, accounting and financial databases are typically audited, by both internal and external auditors to ensure that the data is complete, correct, and accurate. However, few other databases can make such a claim. As an example, operations data is likely to be of high quality, but there is not necessarily an audit to make sure that the data is correct.

This unevenness can cast doubt on implications generated from the integration of multiple databases. Unfortunately, there are few indicators in data capturing the veracity.

However, systems can be developed to continuously monitor data to see if it is reasonable, meets expectations, and so on. Ironically, such monitoring can result in potential privacy violations. For example, it is not unusual for continuous monitoring systems to build "profiles" to monitor different objects and their behavior. Such profiles typically are based on generating expectations and then matching the actual behavior to expectations. Unfortunately, profiles require data that ultimately is used for the continuous observation of the object, which is in direct opposition to the very definition of privacy.

11.3.5 Value

Unfortunately, the potential value of data can be a problem. In the technology bust of 2000, a number of companies sought to create (any) value by actually selling their data. Similarly, today, if an organization does not create enough value, then it may need to take to drastic measures. In an effort to create value from their data, organizations may end up selling the data to third parties. In such situations, privacy can be a potential concern

11.4 Big Data Lake and Privacy

The Big Data Lake is an object-based storage capability that holds virtually all of the databases of an organization in their original format (e.g., O'Leary 2014). As a result, the data would all be up to date, and it would be available for analysis and comparison to or with other data in the organization's Big Data Lake.

Although the notion of the Big Data Lake has been criticized, the concept is very intuitive and very appealing. Typically, the Big Data Lake would include independent databases that usually stand alone. Further, the Big Data Lake would hold virtually all of an organization's databases with the purpose of allowing users to analyze and generate many relationships between data in different databases. As a result, a knowledgeable user might expect to find unique relationships that had not yet been investigated. Accordingly, the availability of such data could provide a clear value creation capability.

However, such a colocation of many independent databases clearly could cause a number of potential privacy breeches depending on the data colocated in the lake and who has access. It is easy to imagine that someone interested in productivity studies would be interested in payroll data in order to fully understand the costs of particular activities. However, it is also easy to imagine that making payroll data available for general use in the Big Data Lake potentially would lead to a number of privacy violations as payroll data became public. As a result, structures such as the Big Data Lake need appropriate governance and access limitations so that privacy concerns are minimized. In particular, access controls would be necessary to mitigate against potential privacy violations and concerns.

11.5 Thing Data

"Things" relate to the increasing number of objects that are becoming a part of the "Internet of Things." Refrigerators, air conditioners, heart monitors, dish washers, radio-frequency identification (RFID) tags, mobile phones, and other devices are being hooked to the Internet. Those things generate data about their processes and send the information to some source that may gather and analyze the data.

Because devices are hooked to the Internet, agents may communicate with the things generating additional data. For example, the owner of an air conditioner can set the air conditioner to turn on or off at particular times, and the owner may set the temperature for the device. Similarly, the device can send information over the Internet, such as how much power it is using, the temperature of the environment where it is located, and a range of other information. Thus, there are two-way flows of information.

Information about air conditioners, such as when it is scheduled to turn on or off, could provide information about when the inhabitant is home. Such information could be particularly useful to intruders. Further, in the case of refrigerators, potential information about the content could lead to privacy concerns. Perhaps the refrigerator contains medicines that could be used to identify diseases with which the owner is afflicted. Heart monitors or sleep apnea devices might be manipulated to the detriment of the owner. For example, in the case of heart monitors, there has been concern that hackers would be able to access the device and potentially devastate the owner. Reportedly, the physicians of Dick Cheney, former Vice President of the United States did not implement the wireless capability on his heart monitoring system (Peterson 2013). Accordingly, thing data provides potential privacy, security, and safety limitations.

As another example of thing data, O'Leary (2013) examined the City of Boston's mobile phone app called StreetBump. StreetBump is an app for mobile phones that allows the user to help the City of Boston by using his or her phone to identify potholes. After the app is downloaded, the driver of a car sets his or her phone in the car and starts the app. The accelerometer captures when the car goes over a bump and notes the location. At the end of the trip, the user tells the app to upload the information, and potentially the city has real-time information about the existence of potholes.

Unfortunately, there are many potential privacy issues associated with such thing data. In the case of StreetBump, there was concern that the city would be able to "know" where users were whenever they used the app. There was also a concern that where people had been and what they had done might be derivable from the information that was uploaded. After all, trip location and time information could provide deep insights into the app user's visits. In this setting, the City of Boston opted to not take anything but independent bump information in order to preserve privacy.

This provides a generic approach to organizations preserving the privacy of those from whom they have gathered data. In particular, data needs to be organized in a manner from which only the minimal level of information can be derived. In the case of the City of Boston, they only need the bump location. As a result, "only" the bump location is captured. If more than the minimal level is gathered, then the database needs to be placed at its minimal level by removing all potential identifying information.

11.6 "Quantified Self" Data

Closely related to thing data is that increasingly individuals are self-capturing digital data about themselves and storing that data in a wide range of databases. For example, there are devices that measure sleep quality, heart rate, activity (such as the number of steps walked), psychological state and traits, environmental variables (e.g., location), and others. These efforts and technologies have resulted in the so-called quantified self (e.g., Swan 2013). Of course, this generates an extreme state of "self-inflicted" observation and monitoring, but it is self-observation and self-monitoring, and would only cause privacy issues if someone else were to gain access.

11.6.1 Quantified self as big data

As a stand-alone set of data, the quantified self could be high velocity, high variety, and high volume, becoming big data, in and of itself. If it stands alone and there is no outside access, then privacy is not an issue. However, individuals may not understand security well enough to keep their data private. In addition, with such data, it is likely that either the patient or medical facilities would be interested in the information and gaining access to the information, introducing potential privacy issues.

11.6.2 Sharing quantified self data

This sharing of data already has manifested itself in a number of sites, including "Patients-likeme" where shared data allows people to share personal health records with others so that they can compare "treatments, symptoms, and experiences." Unfortunately, these comparisons provide a number of potential opportunities for privacy violations.

11.7 Repurposed Data

Recently, it was declared that "all data is credit data".* In that setting, all available data (internally and externally) apparently is being used to analyze different individuals' credit. Data originally generated for one purpose has been "repurposed" for credit analysis. Repurposed data is data that can be used for purposes other than those for which it was originally intended. Recently, observers have commented that organizations can create a substantial value by repurposing data (Shacklett 2014). As an another example, "thing" data generated to monitor the behavior of things can be used to do more than just monitor. Thing activity data can be analyzed at the time of events to determine changes in the data stream, in order to better understand the impact of events. One particularly important set of events are "failure events." In particular, if failure of the thing can be predicted, then in anticipation of that event, personnel can repair the device, rather than have the device fail.

Unfortunately, repurposed data may open up additional privacy concerns. For example, credit card expenditures by a person are accumulated and used for credit card billings. A repurpose is to use the data to find what kinds of products are bought together, or understand how much people will pay for particular items.

Unfortunately, repurposing data typically brings an alternative frame of analysis, a different group of people, and a different set of governance rules. Credit card expenditures initially provide input to some accounting function, specifically designed to create a bill for the customer. Typically accounting functions are run relatively independent of line or marketing management and designed to preserve the privacy of the individuals. In addition, there are typically a number of audit functions and personnel in place to ensure appropriate use of the data. In particular, accounting functions generally are designed to maintain privacy of their partners.

---

* http://www.zestfinance.com/how-we-do-it.html.

However, when the credit card data is repurposed, for example, to determine what products are bought together, the frame of reference and the responsibility for the data change. Such repurposes typically are under the purview of line management and have the express purpose of generating information—the more information the better. Unfortunately, such repurposing analyzes, by their very nature, may result in privacy violations. If a database is repurposed, it is important that the reuses maintain the same governance and access controls as the original purpose.

11.8   Unstructured Communication Data

Unstructured communication data is being used to create knowledge. As an example, Twitter data is being mined to create a broad range of insights, such as predicting the future (Asur and Huberman 2010). Clearly analyzing communication information can provide insight, but is it ethical? Does it invade privacy? Does access to the data make it ethical to try and generate meaning out of conversation data?

In order to address this issue, two perspectives are employed: First, this analysis asks if the original data is directly or indirectly related to its expected use. If the data was not designed for what it is being used and the inferences are indirect, there is some question as to how ethical it is to use such conversation data (from Twitter or Yelp or any social media) in order to make these inferences. Second, this chapter asks how comfortable would we be if the equivalent verbal conversations were captured and analyzed for alternative information?

11.8.1   Direct data inferences

Knowledge acquisition will often "directly" ask an expert or multiple experts about some process in order to acquire knowledge to embed in a system. For example, *The Wall Street Journal* chronicled how knowledge was gathered from a dam expert (Rose 1988) and tax experts (Kneale 1986). Exit interviews of voters are conducted to determine how they voted. The Delphi prediction method generates multiple rounds of feedback from experts in order to determine what they directly expect the world to be like in future. Information from multiple experts is fed back to the experts, typically through three or more rounds in order to gain some consensus. In each of these settings, the specific sources (experts and voters) were directly addressed and the purpose of the analysis was clear—information or knowledge was being directly captured for some specific purposes. In each case, the person being addressed or questioned knows that their information is being used to answer a specific question, build a system, and so on.

11.8.2   Indirect data inferences

However, in contrast to directly soliciting knowledge or information, increasingly, unstructured information generated for communication purposes is being used to infer what different agents or groups of agents think or do. This is an "indirect" approach to generating knowledge or information.

Typically digital media, such as Twitter, are used to communicate with others or to express an opinion or to gather information directly from others. Similarly, other sources, such as Yelp, allow people the opportunity to provide their opinion about a product or service.

However, increasingly that communication information is being used for other purposes. For example, Twitter data has been found to be useful to make certain predictions and knowledge discovery opportunities, and Twitter tweets have been found to be related to music purchases. However, it is unlikely that many Twitter users know that their Twitter tweets are analyzed and are used to make inferences about information contained in the messages. Although those communications are made for others to read, it is unlikely that the originators planned to have their comments aggregated and analyzed for knowledge discovery.

11.8.3 Conversations as data

In the case of verbal conversation, it is reasonable to assume that a conversation will not be recorded and used without the participants' permissions. Yet in many ways, Twitter and other conversations are very similar to verbal conversations.

Is it ethical for others to use such "conversations" as data, particularly if the data provides insight to the Tweeter's behavior? Should people have some privacy assurance that what they say will not be used to generate either aggregate or individual profiles?

Further, increasingly, there is more being captured than just the Twitter conversation. In some settings, Twitter insights are matched with location information.

How would people feel if information about their mobile phone conversations was used to create knowledge, either directly or indirectly? The number of phones in a given location, at some point in time, could provide information about potential store locations. In addition, demographic information could be coupled with the location information to generate the appropriate demographics. Such location information could be used to determine where stores might be located to maximize revenues from particular economic groups. Would people think that use of such data was ethical?

11.9 Continuous Monitoring and Intrusion Detection Systems

Continuous monitoring systems (e.g., Vasarhelyi et al. 2004) and intrusion detection systems (e.g., O'Leary 1992) repurpose data, so that it probably could be said that "all data potentially is continuous monitoring data" or "all data potentially is intrusion detection data." As with other repurposed data, such repurposing can cause potential privacy intrusions for the individuals whose behavior forms the basis of the ability of these systems to make inferences.

In particular, these kinds of systems use a wide range of data generated by the users to develop inferences as to whether an event is anomalous or expected. For example, profiles are made of users that could include information that captures various dimensions of the users' behavior such as where they access their computer resources? When they access their computing resources? What computing resources they access?. In addition, such systems can use information gathered from digital sources, such as calendars, in order to draw additional inferences about the information that is generated. If the behavior differs from the expected behavior, then a potential anomaly that needs explanation is generated.

Although the information used in such systems generally would be innocuous, potentially if it were linked to adjacent monitoring, intrusion-detection data, or other types of information, it may provide potential privacy concerns. As an example, web page visits adjacent to calendared absences could be analyzed to determine that a particular user has potential health concerns. In addition, using streams of data gathered from personal settings might be used to identify specific individuals outside of a work setting.

11.10 Summary—Contributions and Extensions

This chapter investigated multiple characteristics of big data and examined how each of those characteristics relates to different potential privacy concerns. In particular, this chapter focused on the five Vs, the Big Data Lake, the thing data, the quantified self, repurposed data, unstructured communications data and continuous monitoring and intrusion detection systems as providing big data or unique characteristics of big data that potentially could influence privacy.

11.10.1 Contributions

This chapter has a number of contributions. First, this chapter provides a view of privacy based on the five Vs. One of the issues associated with the five Vs was the unevenness of volume, variety, velocity, veracity, and value, and the effects of that unevenness on privacy. Another emerging issue identified here was the impact on "adjacent others" of systems developed to monitor the emerging data available about organizations and individuals. Second, this chapter used the notion of the "Big Data Lake" to address the issue of privacy concerns associated with putting arbitrary, historically independent databases in juxtaposition to each other. Unfortunately, to date there has been little attention given to the potential issues associated with development in the Big Data Lake. However, clearly governance and access issues need to be mapped out. Third, this chapter investigated how thing data could facilitate potential privacy limitations. Fourth, it introduced the notion of the "quantified self" and potential concerns of privacy and also this paper analyzed how systems designed to defend an organization or system could generate information that violates the privacy of the potential users. Finally, it analyzed how communication information could be analyzed to provide indirect information at odds with ethical and privacy concerns.

11.10.2 Extensions

There are a number of potential extensions to this discussion. First, this chapter has identified relatively unique characteristics of big data and used those characteristics to elicit potential privacy issues. Accordingly, additional unique characteristics of big data, beyond those in this chapter, could be examined. Second, additional issues associated with concerns generated here also could be further analyzed. For example, in the case of the Internet of Things, many other appliances are being placed on line, beyond refrigerators. As a result, specific issues related to each of those appliances could be further investigated. Third, "unevenness" was identified as an important issue potentially affecting privacy. Accordingly, there are a number of important issues, including the following:

■ What analytics describe the extent of that unevenness?

■ What steps can be taken to make the data (volume, velocity, variety, veracity) even?

Fourth, this chapter suggests that the privacy of "adjacent others" could be influenced by systems designed to search available information for potential privacy intrusions. Such systems could be referred to as "privacy intrusion detection" systems and be the source of future development (e.g., Venter et al. 2004). Fifth, this chapter specified the importance of governance rules, particularly with the Big Data Lake and repurposed data. Further research could investigate specific governance issues for those settings.

References

Ashby, R.W. 1956, *An Introduction to Cybernetics*. London: Methuen.

Asur, S. and Huberman, B. 2010, Predicting the future with social media, in *Proceedings of the 2010 IEEE/WIC/ACM International Conference on Web Intelligence and Intelligent Agent Technology*, Vol. 1. IEEE, Toronto, Canada. http://www.hpl.hp.com/research/scl/papers/socialmedia/socialmedia.pdf.

Cavoukian, A. and Jonas, J. Privacy by design in the age of big data, 2012, https://privacybydesign.ca/content/uploads/2012/06/pbd-big_data.pdf.

Cormode, G. and Srivastava, D. 2009, Anonymized data: Generation, models, usage, in *Proceedings of SIGMOD*, Providence, RI, 2009. pp. 1015–1018.

Kneale, D. How Coopers & Lybrand put expertise into its computers, *The Wall Street Journal*, November 14, 1986.

Michael, K. and Miller, K. Big Data: New opportunities and new, *Computer* 46.6 (2013): 22–24.

O'Leary, D. E. Knowledge discovery as a threat to database security, *Knowledge Discovery in Databases* 9 (1991): 507–516.

O'Leary, D. E. Intrusion-detection systems, *Journal of Information Systems* 6(1) (1992): 63–74.

O'Leary, D. E. Exploiting Big Data from mobile device sensor-based apps: Challenges and benefits, *MIS Quarterly Executive* 12.4 (2013): 179–187.

O'Leary, D. E. Embedding AI and crowdsourcing in the Big Data lake, *IEEE Intelligent Systems* 29.5 (2014): 70–73.

Peterson, A. Yes, Terrorists could have hacked Dick Cheney's heart, *The Washington Post*, October 21, 2013, http://www.washingtonpost.com/blogs/the-switch/wp/2013/10/21/yes-terrorists-could-have-hacked-dick-cheneys-heart/.

Rose, F. An "Electronic" clone of a skilled engineer is very hard to create, *The Wall Street Journal*, August 12, 1988.

Shacklett, M. Repurpose your data to get more analytics bang for your buck, http://www.techrepublic.com/article/repurpose-big-data-to-get-more-analytics-bang-for-your-bucks/.

SRA. 2009, The internet of things strategic road map, http://sintef.biz/upload/IKT/9022/CERP-IoT% 20SRA_IoT_v11_pdf.pdf.

Swan, M. The quantified self, *Big Data* 1.2 (2013): 85–99.

Tene, O. and Polonetsky, J. Big data for all: Privacy and user control in the age of analytics. *Northwestern Journal of Technology and Intellectual Property* 11 (2012): 240–273.

Vasarhelyi, M. A., Alles, M. G., and Kogan, A. Principles of analytic monitoring for continuous assurance. *Journal of Emerging Technologies in Accounting* 1(1) (2004): 1–21.

Venter, H. S., Olivier, M. S., and Eloff, J. H. P. PIDS: A privacy intrusion detection system. *Internet Research* 14.5 (2004): 360–365.

Verykios, V., Elmagarmid, A., Dasseni, E., Bertino, E., and Saygin, Y. Association rule hiding, *IEEE Transactions on Knowledge and Data Engineering* 16.4 (2004a): 434–447.

Verykios, V. S., Bertino, E., Fovino, I. N., Provenza, L. P., Saygin, Y., and Theodoridis, Y. State-of-the-art in privacy preserving data mining. *ACM Sigmod Record* 33.1 (2004b): 50–57.

Wong, R., Big data privacy, *Information Technology & Software Engineering*, 2.5 (2012): e114.

Wu, X., Zhu, X., Wu, G.-Q., and Ding, W. Data mining with big data. *IEEE Transactions on Knowledge and Data Engineering* 26.1 (2014): 97–107.

Ziegler, C. N. and Skubacz, M. 2006, Towards automated reputation and brand monitoring on the Web, in *Proceedings of the IEEE/WIC/ACM International Conference on Web Intelligence*, pp. 1066–1072.

Zikopoulos, P., DeRoos, D., Parasuraman, K., Deutsch, T., Corrigan, D., and Giles, J. 2013, *Harness the Power of Big Data*. McGraw-Hill, New York.

Zikopoulos, P., Eaton, C., DeRoos, D., Deutsch, T., and Lapis, G. 2012, *Understanding Big Data: Analytics for Enterprise Class Hadoop and Streaming Data*. McGraw-Hill, New York.

Chapter 12

Privacy in Big Data

Benjamin Habegger

Omar Hasan

Thomas Cerqueus

Lionel Brunie

Nadia Bennani

Harald Kosch

Ernesto Damiani

CONTENTS

12.1   Introduction

Personalization consists of adapting outputs to a particular context and user. It may rely on user profile attributes such as the geographical location, academic and professional background, membership in groups, interests, preferences, opinions, and so on. Personalization is used by a variety of Web-based services for different purposes. A common form of personalization is the recommendation of items, elements, or general information that a user has not yet considered but may find useful.

General-purpose social networks such as Facebook.com use personalization techniques to find potential friends based on the existing relationships and group memberships of the user. Professional social networks such as LinkedIn.com exploit the skills and professional background information available in a user profile to recommend potential employees. Search engines such as Google.com use the history of user searches to personalize the current searches of the user.

Big data analysis techniques are a collection of various techniques that can be used to discover knowledge in high volume, highly dynamic, and highly varied data. Big data techniques offer opportunities for personalization that can result in the collection of very comprehensive user profiles. Big data analysis techniques have two strengths in particular that enable collecting accurate and rich information for personalization: (1) Big data analysis techniques process unstructured data as well as structured data. Unstructured data of different varieties generated by users is growing in volume with high velocity and contains lots of useful information about the users. (2) Big data analysis techniques can process high volume data from multiple sources. This enables linking user attribute data from different sources and aggregating them into a single user profile. Moreover, user information from different sources can be correlated to validate or invalidate the information discovered from one source.

On one hand, user profiling with big data techniques is advantageous for providing better services as we have discussed above. On the other hand, user profiling poses a significant threat to user privacy. One can assume that an ethical and trustworthy service would use the information collected for personalization purposes with the user's explicit consent and only for the benefit of the user. However, services that are less inclined toward protecting user privacy may use personalization data for a number of purposes which may not be approved by the user and which may result in loss of private information. One example is the utilization of personalization data for targeted advertising [15]. Another example is the selling of private information in user profiles to third parties for a profit. The third parties may then use the private information for commercial or even malicious purposes [35]. Privacy breaches may occur even when a service is willing to protect a user's privacy [24].

In this chapter, which is in continuation to our article [12] previously presented on the topic, we highlight some of the privacy issues related to personalization using big data techniques. We also present the use case of the Enhancing Europe's eXchange in Cultural Educational and Scientific resources (EEXCESS) research project,* which aims to personalize user recommendations by making intensive use of user profiling and therefore collecting detailed information about users. In this chapter, we supplement the description of the EEXCESS project with an up-to-date account of real experiences gained from the project on implementing personalization with privacy.

The rest of the chapter is organized as follows. Section 12.2 discusses the objectives behind personalization and how it can be achieved through different big data techniques. Section 12.3 presents an introduction to privacy and discusses a user scenario where privacy becomes an issue in the context of personalization. Section 12.4 discusses privacy challenges that may appear in systems relying on information about users and in particular, personalization systems. Section 12.5 recalls the role of the user in relationship to personalization and privacy. Section 12.6 gives an overview of the current state of privacy solutions. Section 12.7 describes the EEXCESS project, its goal of providing personalized content recommendation and the impacts considering privacy. We conclude in Section 12.8.

12.2 Personalization with Big Data Techniques

12.2.1 What is personalization?

The goal of personalization is to provide the most adapted response to a user's current need with the least amount of explicit information provided by him/her. Many existing systems provide some form of personalization. Google search personalizes search results using information such as the user's geo-location, IP address, search history, and result click-thru. Facebook provides *friend* recommendations based on a user's social network already known by the service. Many location-based services, at a very minimum, use a user's geo-location to provide results near the user's current position. Personalized advertisements and marketing solutions attempt to better understand buying habits in order to propose advertisements to users for products they could likely be interested in.

Personalization is not limited to online services. For example, medical analysis systems try to build patient profiles which are as fine-grained as possible (e.g., taking into account genetic

*The presented work was developed within the EEXCESS project funded by the EU Seventh Framework Program, grant agreement number 600601.

information) in order to propose the most adapted treatment to the patient. Personalization even reaches industrial processes, for example, the industrial process of printing. Many printing firms offer the possibility to personalize statement documents such as bank statements, with adapted advertisements and offers. With the arrival of technologies such as 3D printers, it is very likely that the near future makes even more room for personalization.

There are clear advantages to personalization. A typical example is the utilization of user profile data for targeted advertising [15]. This way, users only receive advertisements that they have the most interest in and are not overwhelmed with advertisements for products they wouldn't even consider buying. Another example is filtering out spam emails. Personalization also improves the impact of a given service. In search systems, it allows users to more quickly find the information they are looking for. More generally it relieves users from the information overload they face every day by letting the systems dig through the massive amounts of data on their behalf and letting them find the relevant data for the users.

12.2.2 How big data techniques allow for personalization?

Big data techniques offer opportunities for personalization that can result in very comprehensive user profiles. Big data techniques have two strengths in particular that enable collecting accurate and rich information for user profiles: (1) Big data techniques process unstructured data as well as structured data. Unstructured data of different varieties generated by users is growing in volume with high velocity and contains lots of useful information about the users. (2) Big data techniques can process high volume data from multiple sources. This enables linking user attribute data from different sources and aggregating them into a single user profile. Moreover, user information from different sources can be correlated to validate or invalidate the information discovered from one source.

We list some of the big data analyses techniques below that can be used for collecting information about a user and building a user profile. An extended list of big data techniques that can be used for personalization can be found in [21].

> *Network analysis*: These algorithms are used to discover relationships between the nodes in a graph or a network. Network analysis is particularly useful in the context of social networks where important information about the user such as his/her friends, co-workers, relatives, and so on can be discovered. Social network analysis can also reveal central users in the network, that is, users who exert the most influence over other users. This information can be used to populate the attributes of social and environmental contexts, individual characteristics, and so on in a user profile.

> *Sentiment analysis*: It analysis is a natural language processing technique that aims to determine the opinion and subjectivity of reviewers. The Internet is replete with reviews, comments, and ratings due to the growing popularity of websites such as Amazon.com, Ebay.com, and Epinion.com where users provide their opinion on other users and items. Moreover, micro blogging sites such as Twitter.com and social network sites such as Facebook.com also hold a large amount of user opinions. The goal of sentiment analysis is to classify user opinions. The classification may be a simple polarity classification, that is, negative or positive, or a more complex one, for example, multiple ratings. Sentiment analysis can be used to process unstructured text written by a user to discover their interests, opinions, preferences, and so on to be included into their profile.

Trust and reputation management: It is a set of algorithms and protocols for determining the trustworthiness of a previously unknown user in the context of his/her reliability in performing some action. For example, a reputation management system could be used for computing the trustworthiness of a online vendor who may or may not deliver the promised product once he/she receives payment. The reputation of a user is computed as an aggregate of the feedback provided by other users in the system. Trust and reputation information can be an important part of a user profile. It can convey the user's trust in other users as well as his/her own reputation in various contexts. This information can be subsequently used as a basis for recommending trustworthy users and avoiding those who are untrustworthy. Trust and reputation management systems can function in conjunction with sentiment analysis for obtaining user opinions and then computing trustworthiness and reputation.

Machine learning: It is a sub field of artificial intelligence that aims to build algorithms that can make decisions not based on explicit programming but instead based on historical empirical data. An example often cited is the algorithmic classification of email into spam and non spam messages without user intervention. In the context of personalization, machine learning can be used for learning user behavior by identifying patterns. Topics in machine learning include supervised learning approaches, for example, neural networks, parametric/nonparametric algorithms, support vector machines, and so on, and unsupervised learning approaches, for example, cluster analysis, reduction of dimensionality, and so on.

Cluster analysis: It is the process of classifying users (or any other objects) into smaller subgroups called clusters given a large single set of users. The clusters are formed based on the similarity of the users in that cluster in some aspect. Cluster analysis can be applied for discovering communities, learning membership of users in groups, and so on. Cluster analysis can be considered as a sub topic of machine learning.

12.3 Privacy

On one hand, personalization with big data techniques is advantageous for providing better services as we have discussed above. On the other hand, big data poses a significant threat to user privacy. One can assume that an ethical and trustworthy service providing personalization would use the information collected about users with their explicit consent. However, services that are less inclined toward protecting user privacy, may use such data for a number of purposes which may not be approved by the user and which may result in loss of private information. An example is the selling of private information to third parties for a profit. The third parties may then use the private information of the users for commercial or even malicious purposes [35].

12.3.1 What is privacy?

Depending on the application and the targeted privacy requirement we can have different levels of information disclosure. Let's take privacy preserving reputation systems (e.g., [14]) as an example. We can have five different levels for privacy depending on whether identities, votes, and aggregated reputation score are disclosed and linked or not. For example, in the context of calculating the reputation of a user Alice by three other users Bob, Carol, and David, which respectively have the votes +1, +1, and −1, the reputation system may disclose the following information to Alice.

> *Full disclosure*: All tuples (Bob,+1), (Carol,+1), (David,−1) as well as the aggregated score (+1 if sum is used) are known by Alice.

> *Permuted disclosure*: All voters Bob, Carol, and David are known by Alice as well as the scores but permuted so Alice cannot determine who voted what.

> *Identity disclosure*: All voters Bob, Carol, and David are known by Alice, however, individual votes are hidden and only the aggregated score is known by Alice.

> *Votes disclosure*: All votes are known by Alice but the voters are hidden.

> *Result disclosure*: No details are disclosed except the aggregated score.

> *No disclosure*: An aggregated score for Alice is calculated but she does not have access to it.

More generally, we can subdivide privacy objectives in two, elaborated as follows

> *User anonymity*: The first objective is preserving user anonymity. In this setting, untrusted peers should not be able to link the identity of the user to the requests that they receive. For example, if Bob is navigating the Web, any request that a content provider receives should not be linkable to the real user Bob. Information such as his IP address, email identifiers, or any other such information which may help identify Bob should not be made available.

> *Disclosure of private information about known users*: The second objective is preventing the disclosure of private information. Let's take the same example of Bob searching on the Web but desiring his age to be kept private. However, let's suppose that he/she does not mind the origin of his query being revealed. In this case, privacy preservation does not necessarily require anonymity but rather providing guarantees that Bob's age will not be disclosed.

Ideally, the user would like to obtain quality and personalized recommendations without revealing anything about himself. Attaining such an objective means ensuring that a user remains anonymous with respect to the peers he/she considers non trustworthy. Works have shown that in some restricted cases anonymization is possible [6,10,11]. This, however, often comes at the cost of quality or utility of the disclosed information [4,19,24,31]. It may also be the case that users do not necessarily require anonymity (e.g., in social networks), but rather have control over what is disclosed or not disclosed.

12.3.2 How can privacy be breached?

Depending on the definition of privacy, different techniques can be used to breach privacy even within systems which intend to protect it. We identify two types of privacy attacks: (1) *protocol* attacks are those relying on protocol exchanges between peers, in particular using connection information (IP address, cookies, etc.), to identify users or information about them and (2) *statistical* attacks are those relying on statistical techniques (in particular statistical machine learning) to analyze flows of information reaching a peer and using automated reasoning techniques to deduce user identity or private characteristics.

Protocol attacks: These attacks rely on the fact that since a user wants to obtain an information from a peer, then the peer will have to be contacted by some means. For example, a user wanting to access a Web page on *looms* will have his/her browser making a request to the hosting server. Having been contacted, the server has a trace of the user's IP and knows that this IP has requested the page on looms. Protection from such attacks can be obtained by using proxies but this just moves the problem of trust from the content provider to the proxy provider. It is then the proxy which must be trusted. This very basic example gives an initial intuition on the fact that protecting from protocol attacks can get complex.

Statistical attacks: These attacks rely on the information which legitimately flows to a given peer. Even if users are protected by a privacy-preserving protocol, the data which ends in the hands of a potentially malicious or curious peer may be used to break this anonymity. For example, to be able to find interesting documents for a user, a search engine must be provided with a search query. This query in itself provides information about the user from which it originates (be it only that he/she is interested in the topic of the query). By correlating together the information that an untrusted peer has collected and linked together about a user, it can become possible to de-anonymize the user [27].

12.3.3 User scenario

Let us consider a user scenario in which a recommender system pushes information to a user in order to help his/her with his/her work. Alice is an economist employed by a consulting firm. She is currently working on a business plan for one of her customers on a market which is new to her. She uses a search engine that integrates a recommender system to investigate on the different actors of the market and in particular the potential competitors for her client. The recommender system component of the search engine pushes relevant content from an economic database to Alice. This content includes detailed descriptions of companies found in the target market of her client and strategic economic data.

In this scenario, the recommender system will have collected significant information about Alice: her interests (economic information), some comprehension of her goal (writing a business plan), her knowledge (expert in economics), and her context of work (information about her customer, the target market, the information she has already collected, etc.). Knowing as much as possible about Alice and her customer will allow the recommender system to provide her with adapted recommendations. For example, instead of presenting general-purpose information about the market, the system will propose more detailed technical data which Alice needs and understands.

However, Alice requires that a high level of privacy is ensured by the system. In fact, she is legally- tied by a non disclosure policy with her customer. In particular, it should not be learned that Alice's customer is taking a move toward the new market.

It would be unacceptable to Alice that any information about herself or her customer leak out of the system. Alice's project may even be so sensitive that simply the fact that *someone* (without particularly knowing who) is setting up a business plan on the target market may be an unacceptable leak. Such a disclosure could lead to competitors taking strategic moves. This emphasizes the fact that preserving only anonymity may not be sufficient in some cases.

12.4 Privacy Challenges

Providing users with quality recommendations using big data techniques is a seemingly conflicting objective with the equally important goal of privacy preservation. Even a small amount of personal information may lead to identifying a user with high probability in the presence of side channel external data [24].

Currently, systems that provide personalization function as a black box from the user's perspective. Users do not know what is really collected about them, what is inferred about them by the system, with which other data sources their private data may be combined, what are their benefits of disclosure. Furthermore, faced with the multitude and growing number of external data sources, even limited disclosure of information to a given system may reveal enough about them for the same system to be able to infer knowledge they would have otherwise preferred to remain private. We list below (Sections 12.4.1 through 12.4.11) some of the main categories of challenges that users face concerning their privacy in the existing systems using big data techniques for personalization.

12.4.1 Transparency

Users are often unable to monitor and follow precisely what information about them the system has collected. For example, it is common knowledge that different services, such as Google, Facebook, Amazon, and so on, use big data analytics to provide personalization in many of their services. However, it is not always transparent to users what information has been collected, inferred, and how it is used by whom. Even if these services wish to provide more transparency it is often technically challenging to provide tools to visualize complex processing and manipulation (and in particular aggregation) of user information.

12.4.2 Control

Users are often unable to express their private information disclosure preferences. This can either be due to the unavailability of such options, the complexity of the provided tools, or even their unawareness of privacy issues. They should be able to specify what is disclosed and how detailed the disclosure should be as well as to whom it is disclosed. A big challenge for control is that the more fine-grained privacy settings are the more complex and time consuming it becomes for users to set them. Furthermore, not all users have the same level of requirements, some desire such fine-grained control, whereas others would be satisfied with simpler high level control.

Another related issue is that users have different views on what should be private and give privacy varying importance. Some may prefer having very good personalization whereas others favor privacy. Privacy preservation is already a challenge in itself. Taking user privacy preferences into account requires that privacy algorithms should be able to dynamically adapt to the user's preferences.

12.4.3 Feedback

Users often have difficulties understanding the impacts of disclosing or not disclosing certain pieces of information on personalization. Personalization is impacted by the type, quantity, and quality of information users provide. It is difficult for users to clearly perceive how their inputs impact personalization. This is amplified by the fact that often, these impacts are differed in time and their effects come only later. Also, in many cases, when they do perceive the advantages or lack of value of providing some piece of information, it is long after they have provided it. To make things worse, once the information is released, it is hard for it to be completely retracted.

12.4.4 Re-identification

Because of big data techniques, such as machine learning, very few discriminant data allow to (re)identify the user at the origin of a request. For example, it is possible for a search engine to re-identify some queries sent by a single user among all the queries. This is true even if the user is connected to the search engine via an anonymous network such as TOR [9]. This is done by using the content of the messages (rather than who they are coming from) and using classification techniques to re-identify their likely origin. This suggests that anonymous networks or query shuffling to guarantee unlinkability between users and their requests may not be enough. Therefore, within the context of personalization we are faced with a paradox: on one hand we want to adapt results to specific users, which requires discriminating the user from the others, and on the other hand, to preserve user privacy we should rather not discriminate them.

12.4.5 Discovery

Big data techniques can be utilized for discovering previously unknown information about a given individual. For example, through statistical reasoning, having access to the list of visited websites may reveal the gender of the users even if they have not given them explicitly.

12.4.6 Privacy and utility balance

On one hand, personalization pushes toward providing discriminant data (the more the better) about users whereas privacy pushes to have non discriminant data (the less the better). However, many personalization techniques rely on using data from similar users. If groups of similar users are sufficiently wide, it becomes difficult to distinguish users among these groups.

Ideally, privacy-preservation mechanisms should not impact the quality of personalization obtained from the user. However, this is likely not easily achievable. A less restrictive requirement is that the privacy-preservation mechanisms should minimize the impacts of privacy-preservation on the quality of personalization. This implies, of course, being capable of measuring such quality which in itself could be a challenge.

12.4.7 Collusion

Collusion between peers is another risk for privacy. Indeed, the information which may not be individually discoverable through two uncombined sources of information, when combined through collusion, could lead to new discoveries and therefore privacy breaches.

12.4.8 Providing privacy guarantees

At all levels within the system, user privacy guarantees should be provided. This is most likely one of the hardest tasks. Indeed, as soon as information flows out of a system, sensitive information leaks become a risk. Solutions which may seem trivial, such as anonymization, have been shown to be ineffective or inefficient. A well-known example showing that simple anonymization is insufficient to protect privacy is the de-anonymization of the data of the Netflix contest [24]. Furthermore, Dwork [10] has shown that the published results of a statistical database may lead to privacy breaches even for users who are not originally part of the database. These examples show the difficulties which will have to be overcome in order to provide a privacy-safe system. Furthermore, these works show that research on privacy has shifted from totally preventing privacy breaches to minimizing privacy risks. One of the difficulties to overcome is to ensure that the collection of information flowing out of the system to potentially malicious peers, limits the risks in breaching any of the users' policies. It goes without saying that the attackers themselves very likely have access to big data techniques and that this aspect should be taken into account.

12.4.9 Flexible privacy policies

Users are different, in particular with respect to privacy. Some may not have any privacy concerns at all whereas others may not want to disclose a single piece of information about themselves. For example, in one hypothesis, our user Alice may simply wish to remain anonymous. In another hypothesis, Alice may not be concerned by her identity being revealed, but wish that some information about her be kept private (e.g., she may wish to keep private that she is affected by a particular disease). One big challenge will be to define a policy model which allows for such flexibility and at the same time allows to ensure the policy is respected. Preventing direct disclosure of information marked private is quite straightforward. However, a real challenge is preventing the disclosure of the same information *indirectly*. Indeed, leaking other nonprivate information of a user's profile can lead, through inference, to unwanted disclosures.

12.4.10 Evaluating trust and reputation

What user profile information is disclosed, or at which granularity it is disclosed, may depend on the trust (with respect to privacy concerns) that the user has in a recommender system or a content provider. Calculating an entity's reputation and trustworthiness in a privacy-preserving manner is thus an issue.

12.4.11 Performance

Supplementing big data techniques for personalization with privacy-preserving functionalities often entails added complexity and consequently an increase in the computational resources required. It is imaginable that the raise in computational requirements may be prohibitive enough for a provider that they are forced to sacrifice privacy of users for the sake of practicality. This could be the case even when the provider has the will to implement privacy-preserving features. Thus, a principal challenge in personalization with privacy is to limit the demand on computing resources while maintaining an acceptable level of performance.

12.5 User-Specific Considerations for Personalization with Privacy

Personalization with privacy aims to provide users a service that better fits their needs. Users therefore need to be implied in the process. In particular, users play an important role in information disclosure, which, of course, has an impact on personalization with privacy.

12.5.1   Information disclosure

Even though privacy is often considered as a technical issue, it is also important to understand the user's perspective. In a study on user behavior, [17] has shown that user's globally tend to disclose less information when users are faced with a system explicitly talking about privacy. The interpretation given is that when privacy issues are put in the focus, users tend to become more suspicious and therefore leak less information. This is quite a paradox as a system willing to be transparent about privacy finds itself disadvantaged with respect to one not mentioning privacy at all. However, the same work studies how to improve disclosure (compared to a system not mentioning privacy issues). Giving the same explanations to everyone will lead to the tendency of users disclosing less because of the invocation of privacy. However, adapting explanations to the users can allow to improve disclosure. For example, within the test groups of [17], giving an explanation to men about what the data will be used for, and giving information to women about the percentage of users the data will be disclosed to, tended to globally improve disclosure.

12.5.2   Impact of including users in the process

A system aiming to have its users disclose information willingly and at the same time respect their privacy must have solutions which adapt to their needs. Furthermore, giving high and precise control to users can on one hand show a will for transparency from the service provider. However, on the other hand, this may make the system seem too complex. Therefore, users should be provided with a system allowing them to set their privacy settings *simply* but without losing *flexibility*. To this effect, users should be able to specify their privacy concerns at a high level, but also be allowed more fine-grained settings.

Another important aspect to consider is providing users with elements to understand the effects of disclosing information. As discussed previously, this involves providing the appropriate explanations to the appropriate users. In our user scenario, the objective of user information disclosure is mainly to improve the quality of the recommendations for the user. This can, for example, be obtained through a tool allowing to compare results using different privacy settings.

Given a user's preferences it is then necessary to have a system capable of optimizing the use of the disclosed information. In the user scenario, this means that the quality of the recommendations should be maintained as close as possible to those that the user could have expected with a more detailed profile. Furthermore, providing recommendation quality will also rely on user profiling. Such deep user profiling entails many privacy concerns. Indeed, while users are likely to be interested in having very precise recommendations, they may not at the same time be willing that a third party collects private information about them.

12.6   Privacy Solutions

There is a significant amount of research currently in progress to achieve the goal of preserving user privacy while collecting personal information. Big data techniques offer excellent opportunities for more accurate personalization. However, privacy is an issue that can hinder acceptance by users of personalization with big data techniques. Therefore, there is a need to develop big data techniques that can collect information for user profiles while respecting the privacy of the users. Privacy-preserving big data techniques for personalization would raise the confidence of users toward allowing services to collect data for personalization purposes. In Section 12.6.1 through 12.6.6, we list some of the works on privacy preservation in domains related to big data.

12.6.1 Differential privacy

In the domain of statistical databases, a major shift occurred with the work of Dwork and the introduction of differential privacy [10,11,25]. Through a theoretical framework, the authors demonstrate that, as soon as we consider external knowledge, privacy breaches can occur even for people who do not participate in a statistical database. This has introduced a shift in the way privacy is perceived. The objective is no longer to preserve privacy in an absolute manner, but rather limit the risk of increasing the privacy breach for a participant of a statistical database. To this effect, differentially private mechanisms are those that ensure that the statistical outputs of two databases which are only different by a single participant return similar statistical results. This most often consists in adding sufficient noise to the outputs. Even though there are situations in which differential privacy is attainable, in particular count queries, there are many constraints imposed by differential privacy [4,31]. In particular, in situations which should allow multiple queries, noise must be augmented proportionally to the number of queries to prevent noise reduction techniques to be applied. However, adding too much noise can deprive the outputs of the system of any utility. Therefore much research is ongoing to evaluate the trade-offs between privacy and utility [30]. However, in practice, differential privacy can render some subsets of the randomized data less useful while poorly preserving the privacy of specific individuals. This has been demonstrated for instance in [31]. Thus, differential privacy-preserving techniques still have much to achieve in order to render personal information of users truly private.

12.6.2 k-anonymity

Recommender systems need to gather massive instances of past user interactions and their ratings about objects that they were concerned with. This allows them to propose a selection of predicted objects to a current user, based on profile similarity analysis with the current user, using techniques like collaborative filtering. While this allows having good recommendation quality, it also creates user privacy concerns. *k*-anonymity is one of the well-known techniques to preserve user privacy. The recommender in this case should ensure that each selected object has been selected by at least *k* users and that each object has been rated similarly by at least *k* users. This allows avoiding structure-based and label-based attacks [7]. Several methods have been proposed to ensure *k*-anonymity. We can cite [1,7,20,29,33]. Many solutions are aimed at resolving *k*-anonymity problems in databases [1,29,33]. References [7,20] both proposed using *k*-anonymity for privacy-preserving recommenders. In both, past user ratings are represented using a bi partite graph, where nodes are subdivided into user nodes and object nodes. A graph edge represents the rated selection of an object by a user. Projecting the graph on a single user gives the knowledge that the system has about that user's ratings and selections. The *k*-anonymity is obtained then by padding the graph such that user clustering with less recommendation accuracy can be obtained. Whereas, most solutions proposed for recommenders are based on a centralized gathering of user rating [20], propose a user-centric distributed and anonymous solution to gather useful information to make recommendations. Interestingly, recent work has shown that it can be linked with differential privacy under certain circumstances [19].

12.6.3 Anonymization protocols

Reference [8] introduced a routing protocol allowing the anonymization of communications between two peers by shuffling messages and therefore disabling a server from knowing where a given message comes from. The onion router [9] improves anonymity by using cryptography. A client message is encrypted multiple times with the keys of the peers of the routing path.

This protocol preserves the target server from knowing the address of the client as long as the intermediate peers do not collude. However, it is often possible to still identify the original user through information provided within the message itself. This is typically the case of Web cookies and/or protocol headers. Solutions exist through browser extensions such as FoxTor or TorButton cookies and headers. However, the body of the message itself (e.g., a search query) which is required for the target server to provide a response (e.g., search results) itself reveals information about the user which in some cases may lead to user re-identification [27].

12.6.4 Data obfuscation

To tackle attacks based on the content of the message, works in the literature have proposed to rely on data obfuscation. Different works have suggested such an approach in the case of Web search [22,26]. In the case of search queries, seen as a collection of terms, the idea is to build an obfuscated query by adding extra decoy terms to the query. The obfuscated query is sent to the search engine which can therefore not know what the original query was. Search results are then filtered by the client in order to restore the accuracy of the original request.

12.6.5 Privacy preserving reputation management

A privacy preserving reputation management system operates such that the opinions used to compute a reputation score remain private and only the reputation score is made public. This approach allows users to give frank opinions about other users without the fear of rendering their opinions public or the fear of retaliation from the target user. Privacy preserving reputation management systems for centralized environments include those by Kerschbaum [16] and by Bethencourt et al. [2]. The system by Kerschbaum introduces the requirement of authorizability, which implies that only the users who have had a transaction with a ratee are allowed to rate him/her even though rating is done anonymously. Bethencourt's et al. system lets a user verify that the reputation of a target user is composed of feedback provided by distinct feedback providers (implying no collusion) even when users are anonymous. Hasan et al. [13,14] propose privacy preserving reputation management systems for environments where the existence of centralized entities and trusted third parties cannot be assumed. Current privacy preserving reputation management systems still face a number of open issues. These include attacks such as self-promotion and slandering, in which a user either submits unjustified good opinions about himself or unwarranted bad opinions about a competitor.

12.6.6 User groups

Based on the ideas of [6], their exist many works relying on providing personalization for groups of similar users rather than individual users themselves. For example, [23,32], propose aggregating data of multiple users belonging to similar interest groups. A group profile is built anonymously using distributed and cryptographic techniques.

12.7 The EEXCESS Use Case

12.7.1 What is EEXCESS?

EEXCESS (eexcess.eu) is a European Union FP7 research project that commenced in February 2013. The project consortium comprises of INSA Lyon (insa-lyon.fr), Joanneum Research (joanneum.at), University of Passau (uni-passau.de), Know-Center (know-center.tugraz.at),

ZBW (zbw.eu), Bit media (bit.at), Archäologie und Museum Baselland (archaeologie.bl.ch), Collections Trust (collectionstrust.org.uk), Mendeley (mendeley.com), and Wissenmedia (wissenmedia.de). In this section we present the EEXCESS project to illustrate how user profiling can benefit recommender systems particularly with the use of big data techniques. We also discuss the associated privacy issues and the approaches currently being considered in the project for tackling them.

The main objective of EEXCESS is to promote the content of existing rich data sources available throughout Europe. While user context is more and more present, the current response of Web search engines and recommendation engines to the massive amount of data found on the Web has been to order query results based on popularity. Obviously, the introduction of the PageRank algorithm [5] in search engines has changed the landscape of online searching. However, this has made more difficult to access some valuable content, as it gets buried in the mass of data. This unseen data is sometimes referred to as *long-tail content* in reference to the long-tail of a power-law distribution, which in many cases characterizes the distribution of user interest. This type of long tail content is provided by the EEXCESS partners. This content includes precise and rich content such as museum object descriptions, scientific articles, and business articles. Currently, this high-quality content is not very visible, even though it would be invaluable in the appropriate contexts where fine-grained and precise information is sought for.

The aim of EEXCESS is to push such content made available by the partners to users when it is relevant for them. However, this relies on having a precise understanding of a given user's interests and his/her current context. Different levels of user profiling can help to characterize a user's interests. In EEXCESS, basic, yet informative, user profiles are collected to improve the recommendation results.

12.7.2 Architecture

Figure 12.1 presents a sketch of the current architecture of the EEXCESS system. From this perspective, EEXCESS is made of four components: (1) A plugin added to the user's client whose role is to collect and transfer the user's context, trigger recommendation requests, and render them through rich visualizations, (2) a privacy proxy which collects the user's privacy policy and ensures that it is respected, (3) a usage mining component allowing to identify common usage patterns and enrich user profiles accordingly, and (4) a federated recommender service composed of individual data sources hosting a specific data collection. The circled numbers on the figure give the information flow when content is being recommended.

A privacy proxy is part of the system, as preserving users' privacy is a crucial concern in the project. More specifically, no information about a user profile data should leak out of the system without the user's consent. As it is generally the case when considering privacy in data management systems, the mechanisms implemented to guarantee a certain level of privacy jeopardize some other features of the system (e.g., accuracy and performance).

12.7.3 Personalization

One of the major objectives of EEXCESS is to provide users with relevant and personalized content from the EEXCESS partner data sources. To achieve this goal, accurate user-profiling is an important part of the project and consists in collecting sensitive data about users. An important usage-mining component is responsible for collection or enrichment of user profiles

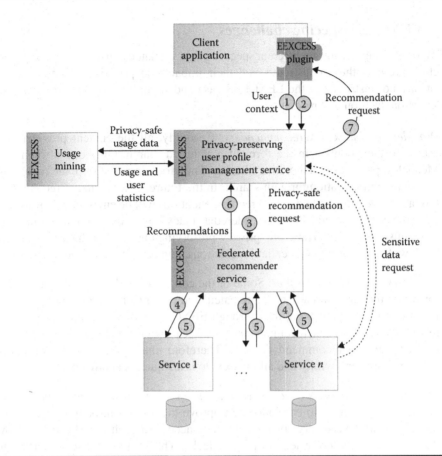

Figure 12.1: Architecture of the EEXCESS system.

using big data techniques as those described in Section 12.2. Some basic information (e.g., age range, location, and interests) is explicitly given by users. These *individual character-istics* form his/her profile. User interactions with the system are monitored to keep track of a user's *behavior*. Among the partners of EEXCESS, BitMedia is an e-learning platform. In this case, it is clear that the user's learning *goals* and current *knowledge* (e.g., in the form of courses already taken) will be part of the user's profile. In EEXCESS, the user's *context* consists of information such as his/her current geo-location, the document, or Web page (both URL and content) he/she is working on and the navigation page which lead to the current page.

To capture a more comprehensive understanding of the user, different big data techniques can be applied to further enrich his/her profile. For example, usage mining aims at identifying usage trend, as well as information about the user's implicit goals and knowledge. On-the-fly analysis of user interests, context, and expectations is also planned. Clustering techniques may be used to identify communities within EEXCESS users. This profiling and better under-standing of users only aims at providing them with personalized services, and in particular *personalized recommendations*. Indeed, the content of the EEXCESS partners being very specific (i.e., being in the long tail of documents when ordered by popularity), having a fine-grained understanding of EEXCESS users is essential to link the correct users to the correct content.

12.7.4 EEXCESS-specific challenges

The EEXCESS context brings in different specific privacy-related constraints and requirements. This section presents these constraints and their impacts on privacy, as similar *real world* constraints are not restricted to the EEXCESS case and should be considered when trying to reconcile personalization and privacy.

Providers as black box recommenders: Currently, all the content providers give an access to their content in the form of a standard search. Only one of them, namely Mendeley, provides collaborative filtering. Therefore, in a first step, the focus has been put on recommendation through search. In the future, other forms of recommendation (collaborative filtering and hybrid recommendation) will be envisaged. However, in any case, the content providers are considered as black boxes; the internal recommendation mechanism is hidden. Therefore, privacy-preserving mechanisms have to be modular to work with different recommendation solutions, and cannot be limited to one form.

Providers with existing profiles: Some of the content providers already have user bases for which they may already have pre-calculated recommendations available. If privacy is limited to anonymization, then through EEXCESS, users will loose access to those recommendations as the recommenders of these providers will not be aware of the users they are sending recommendations to. Therefore, the privacy solutions proposed go beyond simple anonymity and allow users to issue queries anonymously.

Provider recommenders needing feedback to quality: An important objective of recommender systems is to continuously improve themselves through user feedback. To this effect, it is necessary for them to have access to such user feedback. However, this feedback should not lead to privacy leaks. This is a challenge as many attempts toward anonymizing recommendation data have failed in that the data could be de-anonymized [24].

Let us consider that a user wishes to remain anonymous to all the recommenders. In this case, the attacker could be one of the content providers trying to collect information about the user that it receives queries from. The EEXCESS privacy requirements for such a user would include

Content anonymity: To guarantee privacy, the attacker should not be able to identify the user from the provided data. Therefore, the system should ensure that an attacker cannot infer from the content of a request who issued it.

Request unlinkability: If multiple queries can be linked together, even while having content-anonymity for each individual query, the combination of the two could reveal information about the user. Therefore, it is required that the protocols guarantee that two independent requests coming from the same user are unlinkable.

Origin unlinkability: This should be feasible by anonymizing the origin of the request, but under the condition that the origin is not revealed by the application-level protocols. Therefore, we also need to guarantee that the application-level protocols are privacy-preserving (i.e., an attacker cannot link a given request to the user who issued it).

Respecting these three constraints is an ideal goal which requires to limit the information transmitted in each request. Such limitations have a high impact on the utility of the profile information disclosed. Thus the challenge is rather to find a balance between privacy and utility than to ensure complete privacy.

In information systems (such as recommender systems and statistical databases), the main goal of privacy preservation is not to reveal sensitive information about a single entity within the underlying data. This has been shown to be a difficult goal [10,24]. In a survey on privacy in social networks, Zheleva and Getoor [36] describe some of the common approaches for preserving privacy: *differential privacy* and k-*anonymity*. In the context of recommender systems using collaborative filtering, an approach is to use big data techniques such as clustering to group users together in order to provide privacy [3,6,18] with the theory of k-anonymity.

In our particular setting, we are faced with a federated recommender system in which trusted and untrusted peers may exchange information. This requires that both the protocols for exchanging information and the content disclosed are privacy-safe. Furthermore, recommendations may not always be limited to a single recommendation technique among the peers. Each content source may use its own approach. In the context of EEXCESS, few hypotheses can be made on the computational capacities or the background knowledge that an untrusted peer may have access to.

Our work in the EEXCESS project includes the development of mechanisms to allow the definition of flexible user privacy policies, guarantees based on the user privacy policies for non disclosure of private information, quantification of the risk of disclosing private information, mechanisms for exchange of information based on the reputation, and trustworthiness of partners, as well as the definition of the relationship between the amount of information revealed and the quality of recommendations.

12.7.5   Impacts within the EEXCESS use case

Privacy has multiple impacts on the design of systems heavily relying on personalization. Much depends on the trustworthiness of the peers, but most of all, the legal entities running these peers. In the case of EEXCESS, the architecture of the system and the recommendation algorithms are highly dependent on the trust put in the legal entity which will host EEXCESS software. If the federated recommender component could be hosted by possibly untrustworthy peers, then it could be required that the component be distributed and/or make use of cryptographic solutions.

The impacts of privacy mechanisms on personalization within the EEXCESS system can be summarized as follows:

> *Adapting personalization algorithms*: Providing privacy-preserving personalization implies adapting existing personalization algorithms. Many approaches include cryptographic mechanisms, distribution over a network of peers, working with partial and incomplete data, working with groups of users, or pseudonyms.

> *Architectural impacts*: In general, privacy-preservation is not limited to inventing a new version of an algorithm. It has impacts on the global architecture of the privacy-preserving system. Indeed, many privacy-preservation mechanisms rely on the fact that all the data does not reside on a single peer. This is particularly true to allow relaxing trustworthiness assumptions on some or part of the peers. Figure 12.2 gives different options of the trustworthiness assumptions within the EEXCESS architecture. Depending on the chosen trust scenarios, the privacy-preserving strategies need to be adapted.

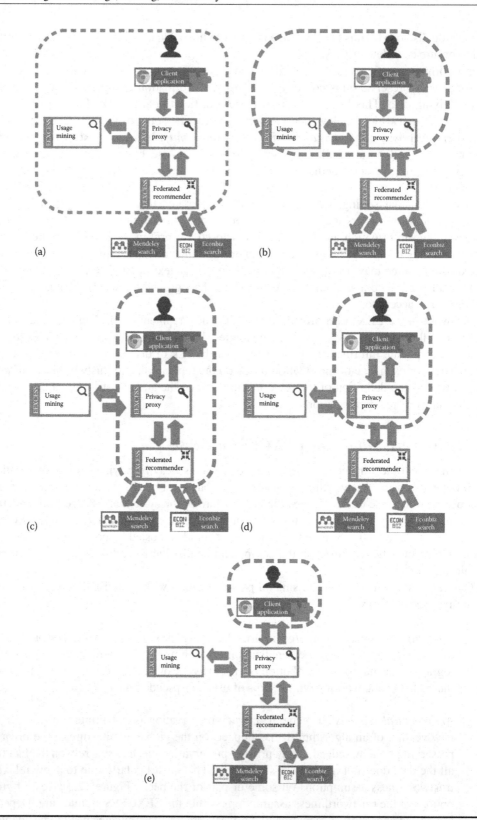

Figure 12.2: EEXCESS trustworthiness scenarios. (a) All trusted, (b) untrusted federator, (c) untrusted mining, (d) only proxy trusted, and (e) untrusted proxy.

For example, if the privacy proxy cannot be trusted (scenario [e] of Figure 12.2), then it requires to distribute the proxy component over multiple non-colluding authorities to ensure that a single authority does not have all the information.

Making privacy-preservation dynamic: Finally, taking user preference into the privacy-preservation mechanisms requires that the personalization algorithms dynamically adapt to each user. In particular, the information provided for two similar users but with different privacy preferences implies that the data available for each of those users be not as detailed. For example, users may provide some information at different levels of granularity. One user may allow providing a complete birth date whereas another may only allow revealing his/her age range.

Performance: The introduction of mechanisms to ensure privacy comes at a price. For instance, encryption mechanisms aiming at hiding users' identity when using a system add computation complexity. Similarly, query obfuscation techniques which create fake queries to prevent attackers to construct accurate user profiles generate network traffic and tend to overload the systems they target. TrackMeNot in an example of such a system [34]. The challenge system designers have to face is then to offer the best trade-off between privacy preservation and performance (i.e., computation cost, resource consumption, and response time). The approach envisioned in the context of EEXCESS is to raise users' awareness regarding their privacy. In other words, it consists in informing users of the private data they divulge by using the system. Then, users are equipped to tune the privacy settings according to their expectations, and therefore achieve a satisfactory trade-off between privacy preservation and performance.

12.7.6 Implementation

In this section, we focus on the mechanisms used to preserve users' privacy in the EEXCESS recommender system. These mechanisms are based on the PEAS protocol [28]. PEAS aims at protecting users privacy in the context of Web search. It is composed of two protocols to ensure users privacy: an unlinkability protocol and an indistinguishability protocol. They are respectively implemented on a privacy proxy and the client application.

Privacy proxy: It is located between the client application (where the queries are issued) and the federated recommender. The privacy proxy allows hiding the identity of the user who sent the query. The privacy proxy itself is composed of two components: the receiver and the issuer. By using an encryption protocol, PEAS assures that the receiver knows the user who sent the query, but not the content of the query; and the issuer knows the content of the query, but not the identity of the user. This protocol is made available through Web services.

Client application: The main application developed in the EEXCESS project is a Google Chrome extension. When a user is browsing the Web, queries are sent automatically (according to the page content) to the federated recommender and resources are suggested. In order to prevent the federated recommender to retrieve the identity of users by analyzing the content of the queries, PEAS offers a technique to obfuscate queries and filter the results. This technique uses a group profile, which is built by the issuer, and generates fake queries that are added to the original query. To remove irrelevant results introduced by the obfuscation technique, a filtering technique is implemented. The protocol is provided through a Javascript component.

12.8 Conclusion

In this chapter, we discussed the challenges raised when building systems, which simultaneously require a deep level of personalization as well as a high level of user privacy. Big data analysis techniques play an important role in making such personalization possible. On the other hand, this raises the issue of respecting a given user's privacy. Big data may even increase this risk by providing attackers the means of circumventing privacy-protective actions. We illustrated these issues by introducing the challenges raised by EEXCESS, a concrete project aiming both to provide high quality recommendations and to respect user privacy.

References

1. Claudio A Ardagna, Giovanni Livraga, and Pierangela Samarati. Protecting privacy of user information in continuous location-based services. In *IEEE 15th International Conference on Computational Science and Engineering*, pp. 162–169, Nicosia, Cyprus. IEEE, December 2012.

2. John Bethencourt, Elaine Shi, and Dawn Song. Signatures of reputation: Towards trust without identity. In *14th International Conference on Financial Cryptography and Data Security*, pp. 400–407, Barbados, 2010.

3. Antoine Boutet, Davide Frey, Arnaud Jegou, and Anne-marie Kermarrec. Privacy-preserving distributed collaborative filtering. Technical Report, INRIA, Rennes, France, February 2013.

4. Hai Brenner and Kobbi Nissim. Impossibility of differentially private universally optimal mechanisms. In *51st Annual IEEE Symposium on Foundations of Computer Science*, pp. 71–80, Las Vegas, NV, 2010.

5. Sergey Brin and Lawrence Page. The anatomy of a large-scale hypertextual web search engine. *Computer Networks*, 30(1–7):107–117, 1998.

6. John F Canny. Collaborative filtering with privacy. In *IEEE Symposium on Security and Privacy*, pp. 45–57, Oakland, CA, 2002.

7. Chih-Cheng Chang, Brian Thompson, Hui (Wendy) Wang, and Danfeng Yao. Towards publishing recommendation data with predictive anonymization. In *5th ACM Symposium on Information, Computer and Communications Security*, p. 24. ACM, New York, April 2010.

8. David Chaum. Untraceable electronic mail, return addresses, and digital pseudonyms. *Communications of the ACM*, 24(2):84–90, February 1981.

9. Roger Dingledine, Nick Mathewson, and Paul F Syverson. Tor: The second-generation onion router. In *13th USENIX Security Symposium*, pp. 303–320, San Diego, CA, 2004.

10. Cynthia Dwork. Differential privacy. In *33rd International Colloquium on Automata, Languages and Programming*, Venice, Italy, volume 4052 of *Lecture Notes in Computer Science*, pp. 1–12. Springer-Verlag, Berlin, Germany, July 2006.

11. Cynthia Dwork. Differential privacy: A survey of results. In *Theory and Applications of Models of Computation*, T-H. Hubert Chan, Lap Chi Lau, and Luca Trevisan (eds.), volume 4978 of *Lecture Notes in Computer Science*, pp. 1–19. Springer-Verlag, Berlin, Germany, April 2008.

12. Benjamin Habegger, Omar Hasan, Lionel Brunie, Nadia Bennani, Harald Kosch, and Ernesto Damiani. Personalization vs. privacy in big data analysis. *International Journal of Big Data*, 1:25–35, 2014.

13. Omar Hasan, Lionel Brunie, and Elisa Bertino. Preserving privacy of feedback providers in decentralized reputation systems. *Computers & Security*, 31(7):816–826, October 2012.

14. Omar Hasan, Lionel Brunie, Elisa Bertino, and Ning Shang. A decentralized privacy preserving reputation protocol for the malicious adversarial model. *IEEE Transactions on Information Forensics and Security*, 8(6):949–962, 2013.

15. Paul Jessup. Big data and targeted advertising. http://www.unleashed-technologies.com/blog/2012/06/28/big-data-and-targeted-advertising, June 2012.

16. Florian Kerschbaum. A verifiable, centralized, coercion-free reputation system. In *8th ACM Workshop on Privacy in the E-Society*, pp. 61–70. ACM, New York, 2009.

17. Alfred Kobsa. Privacy-enhanced personalization. *Communications of the ACM*, 50(8): 24–33, 2007.

18. Dongsheng Li, Qin Lv, Huanhuan Xia, Li Shang, Tun Lu, and Ning Gu. Pistis: A privacy-preserving content recommender system for online social communities. In *IEEE/WIC/ACM International Conferences on Web Intelligence and Intelligent Agent Technology*, Lyon, France, pp. 79–86, 2011.

19. Ninghui Li, Wahbeh Qardaji, and Dong Su. On sampling, anonymization, and differential privacy or, k-anonymization meets differential privacy. In *7th ACM Symposium on Information, Computer and Communications Security*, Seoul, Korea, pp. 32–33, 2012.

20. Zhifeng Luo, Shuhong Chen, and Yutian Li. A distributed anonymization scheme for privacy-preserving recommendation systems. In *IEEE 4th International Conference on Software Engineering and Service Science*, pp. 491–494, IEEE, Beijing, China May 2013.

21. James Manyika, Michael Chui, Brad Brown, Jacques Bughin, Richard Dobbs, Charles Roxburgh, and Angela H Byers. Big data: The next frontier for innovation, competition, and productivity. Technical report, The McKinsey Global Institute, Charlotte, NC, May 2011.

22. Mummoorthy Murugesan. Privacy through deniable search. PhD thesis, Purdue University, West Lafayette, IN, January 2010.

23. Animesh Nandi, Armen Aghasaryan, and Makram Bouzid. P3: A privacy preserving personalization middleware for recommendation-based services. In *Proceedings of 4th Hot Topics in Privacy Enhancing Technologies Symposium*, Waterloo, Canada, pp. 1–12, July 27–29, 2011.

24. Arvind Narayanan and Vitaly Shmatikov. Robust de-anonymization of large sparse datasets. In *2008 IEEE Symposium on Security and Privacy*, pp. 111–125, Oakland, CA, 2008.

25. Kobbi Nissim. Private data analysis via output perturbation—A Rigorous Approach to Constructing Sanitizers and Privacy Preserving Algorithms. In *Privacy-Preserving Data Mining—Models and Algorithms*, Charu C. Aggarwal and Philip S. Yu (eds.), pp. 383–414. Springer, New York, 2008.

26. HweeHwa Pang, Xuhua Ding, and Xiaokui Xiao. Embellishing text search queries to protect user privacy. *Proceedings of the VLDB Endowment*, 3(1):598–607, 2010.

27. Sai T Peddinti and Nitesh Saxena. On the limitations of query obfuscation techniques for location privacy. In *UbiComp'11*, p. 187. ACM, New York, 2011.

28. Albin Petit, Thomas Cerqueus, Sonia B Mokhtar, Lionel Brunie, and Harald Kosch. PEAS: Private, efficient and accurate web search. In *14th IEEE International Conference on Trust, Security and Privacy in Computing and Communications*, Helsinki, Finland, pp. 111–123, August 20–22, 2015.

29. Pierangela Samarati. Protecting respondents identities in microdata release. *IEEE Transactions on Knowledge and Data Engineering*, 13(6):1010–1027, 2001.

30. Lalitha Sankar, Siva Raj Rajagopalan, and Vincent H. Poor. Utility-privacy tradeoffs in databases: An information-theoretic approach. *IEEE Transactions on Information Forensics and Security*, 8(6):838–852, June 2013.

31. Rathindra Sarathy and Krishnamurty Muralidhar. Evaluating laplace noise addition to satisfy differential privacy for numeric data. *Transactions on Data Privacy*, 4(1):1–17, April 2011.

32. Shang Shang, Yuk Hui, Pan Hui, Paul Cuff, and Sanjeev Kulkarni. Privacy preserving recommendation system based on groups. arXiv preprint arXiv:1305.0540, pp. 1–28, 2013.

33. Latanya Sweeney. k-Anonymity: A model for protecting privacy. *International Journal of Uncertainty, Fuzziness and Knowledge-Based Systems*, 10(05):557–570, October 2002.

34. Vincent Toubiana, Lakshminarayanan Subramanian, and Helen Nissenbaum. Trackmenot: Enhancing the privacy of web search. arXiv preprint arXiv:1109.4677, 2011.

35. Jamie Yap. User profiling fears real but paranoia unnecessary. http://www.zdnet.com/user-profiling-fears-real-but-paranoia-unnecessary-2062302030/, September 2011.

36. Elena Zheleva and Lise Getoor. Privacy in social networks: A survey. In *Social Network Data Analytics*, pp. 277–306. Springer, 2011.

Chapter 13

Privacy and Integrity of Outsourced Data Storage and Processing

Dongxi Liu

Shenlu Wang

John Zic

CONTENTS

The organizations and people in our digital society are collecting more and more data in their daily business and life. For an individual company or person, they may not have enough computation and storage capability to manage their data. To address this problem, they can outsource their data to a cloud, which provides a powerful platform to store and process data for its clients. However, with this outsourcing solution, the clients might concern whether their data is managed properly in the cloud, in terms of both data privacy and integrity, since they no longer have the physical control over their data.

In this chapter, we describe an architecture that uses a lightweight homomorphic encryption scheme to ensure the privacy and integrity of outsourced data. With homomorphic encryption, a client can encrypt its data for privacy and the encrypted data can still be directly processed in the cloud. Moreover, the client can verify the integrity of processing results in this architecture, and if the data is archived for a long period of time in the cloud, the client can periodically check whether the encrypted data can be fully retrieved without downloading the whole data.

13.1   Overview

In this architecture, a client is assumed to have a set of numerical values, such as his blood pressures or salaries. For this set of values, the user wants to outsource to a public cloud for storage and processing. Each value is attached with a unique index, such that a value can be selected with its index. To access or process particular values, the user sends corresponding indexes to the cloud, which sends back these values or their processing results. A user may change the stored values or store more new values into the cloud.

The cloud in this architecture is assumed to be honest-but-curious [15]. In this security model, the cloud is honest to store and process data, but might be curious to know what the data is about. In particular, we assume an honest-but-curious cloud only derives information from the encrypted data through deterministic calculation steps, and it does not perform brute-force attacks to randomly guess unknown values, such as the corresponding plaintext of a ciphertext. Note that an honest cloud might not store or process data correctly due to accidental mistakes, so the architecture is useful for clients to be able to check the integrity of outsourced data and processing results.

13.2   Lightweight Homomorphic Encryption

In this section, we describe a homomorphic encryption scheme [7] for concretely defining our secure data storage and processing method. This encryption scheme supports both addition and multiplication of encrypted data, and is efficient for practical applications with good usability features. This encryption scheme is secure only for cyphertext-only attacks, but it is suitable for encrypting data to be stored in the honest-but-curious cloud.

13.2.1 Key generation

A key $K(m)$ in this encryption scheme is a tuple of three components (Γ, Θ, Φ). The component Γ is a list $[(k_1, s_1, t_1), ..., (k_m, s_m, t_m)]$, where k_i, s_i, and t_i are random real numbers. All real numbers in this scheme are supposed with finite precision, such that random real numbers can be uniformly sampled. To ensure correctness, this scheme requires $m \geq 4$, $k_i \neq 0$ for $1 \leq i \leq m - 1$, $k_m + s_m + t_m \neq 0$, and only one $t_i \neq 0$ for $1 \leq i \leq m - 1$.

The second component Θ consists of two random real numbers θ_1 and θ_2. The third component Φ is defined as $\Phi = (\Phi_1, \Phi_2, \Phi_3)$, where Φ_j is a m dimensional vector $(\phi_{j1}, ..., \phi_{jm})$ for $1 \leq j \leq 3$. The following is the definition of each ϕ_{ji}, where θ_1 and θ_2 are specified in Θ and $\theta_3 = 1$.

- Uniformly sample m random real numbers $r_{j1}, ..., r_{jm}$, which can be arbitrarily large

- Compute $\phi_{j1} = k_1 * t_1 * \theta_j + s_1 * r_{jm} + k_1 * (r_{j1} - r_{j(m-1)})$

- Compute $\phi_{ji} = k_i * t_i * \theta_j + s_i * r_{jm} + k_i * (r_{ji} - r_{j(i-1)})$ for $2 \leq i \leq m - 1$

- Compute $\phi_{jm} = (k_m + s_m + t_m) * r_{jm}$

13.2.2 Encryption and decryption

Given the key $K(m)$, let $\Phi = (\Phi_1, \Phi_2, \Phi_3)$ in the key, where Φ_j is a m dimensional vector $(\phi_{j1}, ..., \phi_{jm})$ for $1 \leq j \leq 3$. Then, the algorithm Enc encrypts a real number v into the ciphertext $(e_1, ..., e_m)$, denoted by $\text{Enc}(K(m), v) = (e_1, ..., e_m)$, by using the following steps:

- Uniformly sample 2 arbitrarily large random real numbers ru_1 and ru_2

- Compute $ru_3 = v - ru_1 * \theta_1 - ru_2 * \theta_2$

- Compute $e_i = ru_1 * \phi_{1i} + ru_2 * \phi_{2i} + ru_3 * \phi_{3i}$ for $1 \leq i \leq m$

Since ru_1 and ru_2 are random numbers, this encryption scheme is probabilistic. That is, the same value can be encrypted into different ciphertexts. Given a ciphertext $(e_1, ..., e_m)$ and the key $K(m)$, the decryption algorithm Dec returns the plaintext v by the following steps:

- $T = \sum_{i=1}^{m-1} t_i$

- $S = e_m / (k_m + s_m + t_m)$

- $v = (\sum_{i=1}^{m-1} (e_i - S * s_i) / k_i) / T$

The decryption operation is denoted as $\text{Dec}(K(m), (e_1, ..., e_m)) = v$. The conditions $k_m + s_m + t_m \neq 0$, $k_i \neq 0$ for $1 \leq i \leq m - 1$, and $t_i \neq 0$ for some i ensure the validity of the above decryption steps.

When an adversary is allowed to choose one or two plaintexts to encrypt for an encryption key, this scheme is semantically secure when the size of plaintexts is bounded (i.e., not allowing arbitrarily large plaintexts). Let e be a ciphertext which encrypts v or v'. If an encryption scheme is semantically secure, the adversary cannot distinguish whether e encrypts v or v' with a probability non-negligibly higher than $\frac{1}{2}$ [6]. Since the random numbers in this scheme can be arbitrarily large, the value of e is dominated by the value of random numbers. Hence, the advantage of distinguishing whether e encrypts v or v' is asymptotically equal to distinguishing the value of random numbers used in e. Suppose each random number has p decimal digits. Then, the probability of distinguishing a particular value of random numbers is $O(\frac{1}{10^p})$ due to uniform sampling. The function $\frac{1}{10^p}$ is negligible with respect to the variable p, because it is the inverse of the exponential function 10^p.

13.2.3 An example of encryption and decryption

The encryption scheme is illustrated with an example in this section. In this example, we let $m = 4$ in $K(m)$, so the ciphertexts generated with this key will be four-dimensional.

13.2.3.1 An instance of $K(4)$

For a key $K(4)$, suppose we have $\Gamma = [(3.2, 2.7, 0), (9.1, 3.1, 1.5), (3.6, 7.9, 0), (2.1, 8.8, 7.9)]$ and $\Theta = (34.7, 51.3)$. Then, Φ_1, Φ_2, and Φ_3 in the key component Φ are generated as follows.

- Generation of $\Phi_1 = (\phi_{11}, \phi_{12}, \phi_{13}, \phi_{14})$

  - Let 5689.23, 375623145.2, -8523.87, and 24231.47 be the random numbers for generating Φ_1

  - $\phi_{11} = 3.2 * 0 * 34.7 + 2.7 * 24231.47 + 3.2 * (5689.23 - (-8523.87)) = 110906.889$

  - $\phi_{12} = 9.1 * 1.5 * 34.7 + 3.1 * 24231.47 + 9.1 * (375623145.2 - 5689.23) = 3418194440.539$

  - $\phi_{13} = 3.6 * 0 * 34.7 + 7.9 * 24231.47 + 3.6 * (-8523.87 - 375623145.2) = -1352082580.039$

  - $\phi_{14} = (2.1 + 8.8 + 7.9) * 24231.47 = 455551.636$

- Generation of $\Phi_2 = (\phi_{21}, \phi_{22}, \phi_{23}, \phi_{24})$

  - The random numbers for generating Φ_2 are supposed to be 2378.48, -52378342.4, 2347.9, and 3478.71

  - $\phi_{21} = 3.2 * 0 * 51.3 + 2.7 * 3478.71 + 3.2 * (2378.48 - 2347.9) = 9490.373$

  - $\phi_{22} = 9.1 * 1.5 * 51.3 + 3.1 * 3478.71 + 9.1 * ((-52378342.4) - 2378.48) = -476653075.762$

  - $\phi_{23} = 3.6 * 0 * 51.3 + 7.9 * 3478.71 + 3.6 * (2347.9 - (-52378342.4)) = 188597966.889$

  - $\phi_{24} = (2.1 + 8.8 + 7.9) * 3478.71 = 65399.748$

- Generation of $\Phi_3 = (\phi_{31}, \phi_{32}, \phi_{33}, \phi_{34})$

  - Assume the random numbers for generating Φ_3 are 23789.23, 834458.32, 67280.67, and 7666.43

  - $\phi_{31} = 3.2 * 0 * 1 + 2.7 * 7666.43 + 3.2 * (23789.23 - 67280.67) = -118473.247$

  - $\phi_{32} = 9.1 * 1.5 * 1 + 3.1 * 7666.43 + 9.1 * (834458.32 - 23789.23) = 7400868.302$

  - $\phi_{33} = 3.6 * 0 * 1 + 7.9 * 7666.43 + 3.6 * (67280.67 - 834458.32) = -2701274.743$

  - $\phi_{34} = (2.1 + 8.8 + 7.9) * 7666.43 = 144128.884$

13.2.3.2   Example of encryption and decryption

Suppose the plaintext value is 307.1 and the random numbers $rv_1 = 2378.2$ and $rv_2 = -790234.7$. Then, we have $rv_3 = 307.1 - 34.7*2378.2 - 51.3*(-790234.7) = 40456823.67$. The following is the encryption and decryption of 307.1 with the above key instance $K(4)$. We can see the plaintext value is correctly recovered.

- Encryption $\text{Enc}(K(4), 307.1) = (e_1, e_2, e_3, e_4)$

  - $e_1 = 2378.2*110906.889 + (-790234.7)*9490.373 + 40456823.67*(-118473.247) = -4800287126788.47979$

  - $e_2 = 2378.2*3418194440.539 + (-790234.7)*(-476653075.762) + 40456823.67*7400868.302 = 684212574246257.49954$

  - $e_3 = 2378.2*(-1352082580.039) + (-790234.7)*(188597966.889) + 40456823.67*(-2701274.743) = -261537176538763.16491$

  - $e_4 = 2378.2*455551.636 + (-790234.7)*65399.748 + 40456823.67*(144128.884) = 5780399088401.76388$

- Decryption $\text{Dec}(K(4), (e_1, e_2, e_3, e_4)) = 307.1$

  - $T = 0 + 1.5 + 0 = 1.5$

  - $S = 5780399088401.76388/(2.1 + 8.8 + 7.9) = 307468036617.1151$

  - $(e_1 - 2.7*S)/3.2 = -1759515883017.0908$

  - $(e_2 - 3.1*S)/9.1 = 75083453113488.4003$

  - $(e_3 - 7.9*S)/3.6 = -73323937230010.6595$

  - $(-1759515883017.0908 + 75083453113488.4003 + (-73323937230010.6595))/T = 307.1$

13.3   Processing of Encrypted Data

For the encrypted data stored in the cloud, it is desirable that the cloud can process the data directly. Thus, a client does not need to download, decrypt, and then process the data by themselves. The encryption scheme introduced in the last section is homomorphic, hence allowing the cloud to process the data directly without knowing encryption keys.

The encryption scheme supports both addition and multiplication over ciphertexts. As a result, the cloud can add or multiply ciphertexts and the resulting ciphertexts can then be decrypted by the client to get back the sum or product of the corresponding plaintexts. In the following, we describe the homomorphic operations of the encryption scheme (i.e., the addition or multiplication of ciphertexts).

13.3.1   Addition and multiplication of ciphertexts

Suppose two plaintext values v and v' are encrypted with the key $K(m)$:

$$\text{Enc}(K(m), v) = (e_1, ..., e_m)$$
$$\text{Enc}(K(m), v') = (e'_1, ..., e'_m).$$

Then, the encryption scheme has the additively homomorphic property by ensuring

$$\text{Dec}(K(m), \text{Enc}(K(m), v) \oplus \text{Enc}(K(m), v')) = v + v',$$

where $\oplus$ indicates vector addition, that is,

$$\text{Enc}(K(m), v) \oplus \text{Enc}(K(m), v') = (e_1 + e'_1, ..., e_m + e'_m).$$

Let b be a real number. As an extension of the additively homomorphic property, we have

$$\text{Dec}(K(m), b \odot \text{Enc}(K(m), v)) = b * v,$$

where $b \odot \text{Enc}(K(m), v) = (b * e_1, ..., b * e_m)$. Combined with the $\oplus$ operation, we can get

$$\text{Dec}(K(m), \text{Enc}(K(m), v) \oplus (b \odot \text{Enc}(K(m), v'))) = v + b * v'.$$

When $b = -1$, we can calculate $v - v'$ from the two ciphertexts.

The multiplicative homomorphism is about calculating $v * v'$ from the multiplication of ciphertexts $(e_1, ..., e_m)$ and $(e'_1, ..., e'_m)$. The ciphertext multiplication is represented by $\otimes$ and generates the resulting ciphertext, as shown below, which is their outer product.

$$(e_1 * e'_1, ..., e_1 * e'_m,$$
$$...,$$
$$e_m * e'_1, ..., e_m * e'_m)$$

Then, $v * v'$ is obtained by decrypting this ciphertext in two steps:

■ Calculate $\text{Dec}(K(m), (e_i * e'_1, ..., e_i * e'_m)) = e_i * v'$ for $1 \le i \le m$

■ Calculate $\text{Dec}(K(m), (e_1 * v', ..., e_m * v')) = v * v'$

Note that the above decryption steps for multiplied ciphertexts do not require v and v' be encrypted in the same key. That is, when multiplying their ciphertexts, v and v' can be encrypted with different keys.

13.3.1.1 Example of homomorphic operations

We explain the homomorphic operations with an example. In addition to the ciphertext of 307.1 encrypted in the last section, our examples need the encryption of another value, say 53.71. The ciphertext for the value 53.71 is shown below, encrypted with the same key, and its correctness can be verified by decryption.

$$(13892023092.168459, 10220726367973.636436,$$
$$-4068687196870.903229, -14983269964.005508)$$

In this example, we want to calculate $0.3 * 307.1 + 0.7 * 53.71$, which is the weighted average of 307.1 and 53.71, from their ciphertexts. For this purpose, we add the two ciphertexts in the way shown below.

$$(0.3 * (-4800287126788.47979) + 0.7 * 13892023092.168459,$$
$$0.3 * 684212574246257.49954 + 0.7 * 10220726367973.636436,$$
$$0.3 * (-261537176538763.16491) + 0.7 * (-4068687196870.903229),$$
$$0.3 * 5780399088401.76388 + 0.7 * (-14983269964.005508))$$

That is, the resulting ciphertext is

$$(-1430361721872.0260157, 212418280731458.7953672,$$
$$-81309233999438.5817333, 1723631437545.7253084)$$

As shown by the following steps, the above ciphertext is decrypted into 129.727, which is the correct result of $0.3 * 307.1 + 0.7 * 53.71$.

- $T = 0 + 1.5 + 0 = 1.5$

- $S = 1723631437545.7253084/(2.1 + 8.8 + 7.9) = 91682523273.708793$

- $((-1430361721872.0260157) - 2.7 * S)/3.2 = -524345167097.199924$

- $(212418280731458.7953672 - 3.1 * S)/9.1 = 23311435704319.812979$

- $((-81309233999438.5817333) - 7.9 * S)/3.6 = -22787090537028.022555$

- $((-524345167097.199924) + 23311435704319.812979 +$
  $(-22787090537028.022555))/T = 129.727$

13.3.2 Evaluation of multivariate polynomials

Based on the additive and multiplicative homomorphism, this encryption scheme supports the evaluation of multivariate polynomials over encrypted data. We explain such calculations conceptually with the following example. Suppose six real numbers v_i $(1 \le i \le 6)$ are encrypted into the ciphertexts $\text{Enc}(K(m), v_i)$. From these ciphertexts, we want to calculate $v_1 + v_2 * v_3 + v_4 * v_5 * v_6$. This is done by evaluating the following expression over the six ciphertexts:

$$(\text{Enc}(K(m), v_1) \otimes \text{Enc}(K(m), 1) \otimes \text{Enc}(K(m), 1)) \oplus$$
$$(\text{Enc}(K(m), v_2) \otimes \text{Enc}(K(m), v_3) \otimes \text{Enc}(K(m), 1)) \oplus$$
$$(\text{Enc}(K(m), v_4) \otimes \text{Enc}(K(m), v_5) \otimes \text{Enc}(K(m), v_6)).$$

In the above expression, the three encryptions of one, $\text{Enc}(K(m), 1)$, are used to ensure the three intermediate ciphertexts, which encrypt $v_1 * 1 * 1$, $v_2 * v_3 * 1$, and $v_4 * v_5 * v_6$, have the same number of subciphertexts, so that they can be added to produce the ciphertext for $v_1 * 1 * 1 + v_2 * v_3 * 1 + v_4 * v_5 * v_6$.

The multiplication of two matrices and inner product of two vectors, widely used in practical applications, are instances of multivariate polynomials. Hence, their calculations can be done over encrypted data with this encryption scheme.

13.3.3 Comparison of ciphertexts

In addition to addition and multiplication, we also need to compare data in many data processing applications. However, comparing encrypted values are not directly allowed by the homomorphic encryption scheme in this chapter. The reason is that during encryption plaintext values are usually randomized and even the same value is encrypted into different ciphertexts. There are order-preserving encryption schemes [1,3], but they are not homomorphic and hence cannot support additions and multiplications over ciphertexts.

The homomorphic encryption scheme supports only ciphertext additions and multiplications. Hence, to compare two plaintext values through their ciphertexts, the comparison operations should be expressed as additions and multiplications over ciphertexts. This can be achieved by representing plaintext values in the binary format.

Let $V = b_1 b_2 ... b_N$ and $V' = b'_1 b'_2 ... b'_N$ be the binary representation of V and V'. Based on the binary representation, the expression below implements the comparison $V > V'$, by returning 1 if $V > V'$, and 0 otherwise.

$$(b_1 \oplus b'_1) * b_1 + (b_1 \oplus b'_1 \oplus 1) * (b_2 \oplus b'_2) * b_2 + \cdots +$$
$$(b_1 \oplus b'_1 \oplus 1) * \cdots * (b_{N-1} \oplus b'_{N-1} \oplus 1) * (b_N \oplus b'_N) * b_N$$

In the above expression, $\oplus$ is overloaded to represent the XOR operation and defined as $b_i \oplus b'_i = b_i + b'_i - 2 * b_i * b'_i$. Similarly, the comparison $V < V'$ can be implemented with the following expression, which returns 1 if $V < V'$, and 0 otherwise.

$$(b_1 \oplus b'_1) * b'_1 + (b_1 \oplus b'_1 \oplus 1) * (b_2 \oplus b'_2) * b'_2 + \cdots +$$
$$(b_1 \oplus b'_1 \oplus 1) * \cdots * (b_{N-1} \oplus b'_{N-1} \oplus 1) * (b_N \oplus b'_N) * b'_N$$

For the comparison $V = V'$, it is defined by the expression $(1 - (V > V')) * (1 - (V < V'))$, which returns 1 if $V = V'$, and 0 otherwise.

Thus, if we encrypt V and V' bit by bit, then the comparison of V and V' can be done with the additions and multiplications over the ciphertexts of their encrypted bits. The decrypted result will be 1 or 0, reflecting the comparison result.

13.3.4   Usability features

We can use other homomorphic encryption schemes, such as [4], to encrypt and process data. However, the homomorphic encryption scheme described in this chapter has two good usability features.

First, this scheme directly supports real numbers and negative values for both encryption and homomorphic operations. Thus, such types of data commonly found in business applications can be directly encrypted by the client and processed by the cloud. For example, the average salaries of staffs in a company, which are supposed to be a real number, can be directly calculated in the cloud.

Second, this scheme does not rely on modulo operations in decryption, and hence it does not bound the results computed from ciphertexts. The current homomorphic encryption schemes usually bound the results of homomorphic operations with a modulus. The processing results from ciphertexts cannot be decrypted correctly [9], if the results are bigger than the modulus used in decryption.

13.4   Retrievability of Encrypted Data

A client may use the cloud to archive his data. That is, such data will not be accessed frequently. For this case, the client may wonder whether his data is stored properly and can be fully retrieved after a long period of time. This concern is reasonable, since some storage errors may not even be known by cloud administrators. In this section, we describe a proof of retrievability (PoR) protocol proposed in [8], which allows the client to check the integrity of data archived in the cloud without downloading the whole data. This PoR protocol is built on the homomorphic encryption scheme introduced in this chapter.

The basic idea of this PoR protocol is that an encrypted value is attached with a message authenticator, which provides the redundant information to check the integrity of the encrypted value. However, to ensure the efficiency of integrity checking, the message authenticators should be homomorphic. Thus, the message authenticators of a set of encrypted values can be added up to check the integrity of the sum of these encrypted values in one proof. Moreover, the message authenticators should be probabilistic, such that the statistical information of the encrypted values is not leaked by their message authenticators.

13.4.1   Encrypted values with message authenticators

To use this PoR protocol, a client needs to generate three different homomorphic encryption keys $K(m)$, $K_R(m)$, and $K_I(m)$, and a pseudorandom function key k. For a value v_i, the client stores into the cloud the following tuple:

$$(i, \text{Enc}(K(m), v_i), \text{PoR}(K_R(m), K_I(m), k, i, v_i))$$

In this tuple, i is the unique index of the value v_i, $\text{Enc}(K(m), v_i)$ is the homomorphic encryption of v_i, and $\text{PoR}(K_R(m), K_I(m), k, i, v_i)$ generates the message authenticator of v_i, which is a pair (R, I), as defined below.

- $R = \text{Enc}(K_R(m), v_i)$

- let $R = (e_1^R, ..., e_m^R)$

- $I = \text{EncFixed}(K_I(m), f_k(i), e_m^R)$

In the above definition, f is a pseudorandom function with the key k. The element I is generated with the function $\text{EncFixed}(K_I(m), v, w)$. It encrypts v in the same way as Enc, but the last element of the ciphertext from EncFixed is fixed to be the argument w. That is, let $\text{EncFixed}(K_I(m), f_k(i), e_m^R) = (e_1^I, ..., e_m^I)$ in the above definition. Then, EncFixed ensures $e_m^R = e_m^I$. The following is the definition of $\text{EncFixed}(K_I(m), v, w)$, which is built over the Enc algorithm.

- Uniformly sample an arbitrarily large random real numbers ru_1

- Compute $ru_2 = (w - ru_1 * (\phi_{1m} - \theta_1 * \phi_{3m}) - v * \phi_{3m}) / (\phi_{2m} - \theta_2 * \phi_{3m})$

- Compute $ru_3 = v - ru_1 * \theta_1 - ru_2 * \theta_2$

- Compute $e_i = ru_1 * \phi_{1i} + ru_2 * \phi_{2i} + ru_3 * \phi_{3i}$ for $1 \leq i \leq m$

We prove that the last element of $\text{EncFixed}(K_I(m), v, w)$ is equal to w. According to the definition of EncFixed, we have

$$e_m = ru_1 * \phi_{1m} + ru_2 * \phi_{2m} + ru_3 * \phi_{3m}$$

Replacing ru_3 and ru_2 with their definitions, respectively, we get

$$
\begin{aligned}
e_m^I &= ru_1 * \phi_{1m} + ru_2 * \phi_{2m} \\
&\quad + (v - ru_1 * \theta_1 - ru_2 * \theta_2) * \phi_{3m} \\
&= ru_1 * (\phi_{1m} - \theta_1 * \phi_{3m}) \\
&\quad + ru_2 * (\phi_{2m} - \theta_2 * \phi_{3m}) + v * \phi_{3m} \\
&= ru_1 * (\phi_{1m} - \theta_1 * \phi_{3m}) \\
&\quad + (w - ru_1 * (\phi_{1m} - \theta_1 * \phi_{3m}) - v * \phi_{3m}) + v * \phi_{3m} \\
&= w
\end{aligned}
$$

Note that in the function $\text{EncFixed}(K_I(m), v, w)$, the term $\phi_{2m} - \theta_2 * \phi_{3m}$ appears as a divisor. It needs to have a multiplicative inverse with finite precision, such that the quotient of the division can be precisely represented. This can be achieved by choosing appropriate parameters for $K_I(m)$. By letting $k_m + s_m + t_m = 1$ in $K_I(m)$, we have

$$\phi_{2m} - \theta_2 * \phi_{3m} = r_{2m} - \theta_2 * r_{3m},$$

since $\phi_{2m} = (k_m + s_m + t_m) * r_{2m}$ and $\phi_{3m} = (k_m + s_m + t_m) * r_{3m}$. To make $r_{2m} - \theta_2 * r_{3m}$ have a multiplicative inverse with finite precision, we randomly sample r_{3m} and then define $r_{2m} = \theta_2 * r_{3m} + g$, where g is uniformly sampled from a secret set that contains invertible real numbers, such as the set $\{0.1, 0.2, 0.5, 0.4, 0.8, 1.0, 1.6, 2.0, 2.5, 3.2\}$. Thus, $r_{2m} - \theta_2 * r_{3m} = g$ and g has a multiplicative inverse with finite precision.

13.4.2 Proof of retrievability

Let the client have the values $(i, C_i, (R_i, I_i))$ in the cloud, where $C_i = \text{Enc}(K(m), v_i)$ and $(R_i, I_i) = \text{PoR}(K_R(m), K_I(m), k, i, v_i)$, for $1 \leq i \leq N$. Then, the client can verify the integrity of stored values with the following protocol.

Client $\longrightarrow$ Cloud: A set Q of pairs (i, r_i), where i is a random integer from 1 to N and is unique in Q, and r_i is a random real number.

Cloud $\longrightarrow$ Client: A tuple (C, R, I), with each component defined below.

- $C = \oplus_{(i,r_i) \in Q} (r_i \odot C_i)$
- $R = \oplus_{(i,r_i) \in Q} (r_i \odot R_i)$
- $I = \oplus_{(i,r_i) \in Q} (r_i \odot I_i)$

User: Successful integrity check if the following conditions hold. Let $I = (e_1^I, ..., e_m^I)$ and $R = (e_1^R, ..., e_m^R)$.

- $e_m^I = e_m^R$
- $\text{Dec}(K_I(m), I) = \Sigma_{(i,r_i) \in Q} r_i * f_k(i)$
- $\text{Dec}(K(m), C) = \text{Dec}(K_R(m), R)$

From the protocol, we can see R_i in a message authenticator (R_i, I_i) is used as a redundant copy to ensure the correctness of the value v_i, while I_i is used to ensure that the value taken by the cloud in the protocol is really the value with the index i. Thus, the retrievability proof constructed by the cloud really takes into account the value v_i, when its index i is included in Q.

The proof (C, R, I) returned from the cloud always consists of $3 * m$ real numbers, regardless of the size of Q and the number of outsourced values. Hence, the communication cost between the user and the cloud is much lower than downloading $|Q|$ values for the user to check. The notation Q means the size of Q. Hence, our PoR is compact. Note that a message authenticator (R, I) is homomorphic and probabilistic, since they are the ciphertexts of a probabilistic homomorphic encryption scheme.

An execution of this protocol only ensures the retrievability of the selected values in Q. If Q selects all outsourced values, it might cause a burden for the cloud to construct the proof. Another way is that a client chooses a small Q for each proof, but runs this protocol periodically to cover all outsourced values.

13.4.3 Integrity of outsourced data processing

The homomorphically encrypted data can be directly processed by the cloud. Ideally, the processing results from the cloud can be verified by the client, such that the client is confident that the processing results are correctly computed from the data specified by him and the input data in the processing is not tampered with. Using only homomorphic encryption cannot ensure that the cloud faithfully carries out the computation specified by the client. For example, the computation specified by the client is to calculate the sum of all values, but the cloud may accidentally choose two wrong ciphertexts and return their sum. At this case, the processing result is not integrated.

The check of processing integrity is a specific usage of the PoR protocol. Hence, as in the retrievability proof, we still let a value v_i be stored as $(i, C_i, (R_i, I_i))$ in the cloud, where $C_i = \text{Enc}(K(m), v_i)$ and $(R_i, I_i) = \text{PoR}(K_R(m), K_I(m), k, i, v_i)$, for $1 \leq i \leq N$. In the following, we first introduce how to check the integrity of processing results obtained with additions of ciphertexts.

According to the additively homomorphic property of the encryption scheme, the client can ask the cloud to do the following computation over ciphertexts, where a_i is a real number provided by the client to specify the input data. If $a_i = 0$, an input data v_i is not selected for processing.

$$C = \oplus_{i=1}^{N}(a_i \odot C_i) \tag{13.1}$$

From the decryption of C with $K(m)$, the client gets $\Sigma_{i=1}^{N} a_i * v_i$. Changing a_i can result in different calculations. For example, if $a_i = 1$, then the above expression calculates the sum of all outsourced values; if $a_i = 1/N$, the average of outsourced values is calculated.

To prove the integrity of the computation $\oplus_{i=1}^{N}(a_i \odot C_i)$, we ask the cloud to calculate the following two additional values, as in the PoR protocol.

$$
\begin{aligned}
R &= \oplus_{i=1}^{N}(a_i \odot R_i) \\
I &= \oplus_{i=1}^{N}(a_i \odot I_i)
\end{aligned}
$$

That is, for the check of processing integrity, the user sends $Q = \{(1, a_1), ..., (N, a_N)\}$ to the cloud. Then, the integrity of encrypted data processing is checked in the same way as the PoR.

Using the multiplicative homomorphism, the cloud can do more complex computations. We take the multiplication of two encrypted matrices as the example to illustrate how the client verifies the integrity of complex computations. Suppose the client has two $N * N$ matrices M and M'. The ith row of M and the jth column of M' are represented by the vectors $(v_{i1}, ..., v_{iN})$ and $(v'_{1j}, ..., v'_{Nj})$, respectively. Then, the element in the ith row and the jth column of the resulting matrix is calculated over ciphertexts by the expression

$$C_{ij} = \sum_{l=1}^{N} C_{il} \otimes C'_{lj}$$

where C_{il} and C'_{lj} encrypt the plaintext values v_{il} in M and v'_{lj} in M', respectively. To verify the integrity of C_{ij}, the client sends $Q_i = \{(1, a_{i1}), ..., (N, a_{iN})\}$ and $Q'_j = \{(1, a'_{1j}), ..., (N, a'_{Nj})\}$, and the cloud needs to calculate R_{ij} and I_{ij}.

$$
\begin{aligned}
R_{ij} &= \oplus_{l=1}^{N}((a_{il} * a'_{lj}) \odot (R_{il} \otimes R_{lj})) \\
I_{ij} &= \oplus_{l=1}^{N}((a_{il} * a'_{lj}) \odot (I_{il} \otimes I_{lj}))
\end{aligned}
$$

Note that if they are the same for the last elements R_{il} and I_{il}, and the last elements of R_{lj} and I_{lj}, then the last elements of $R_{il} \otimes R_{kl}$ and $I_{il} \otimes I_{lj}$ are also the same. The verification steps for C_{ij} are the same as in the PoR protocol, except that we require

$$\text{Dec}(K_I(m), I_{ij}) = \sum_{l=1}^{N} (a_{il} * a'_{lj} * f_k(il) * f_k(lj)),$$

where the parameters of the function f_k, il and lj, represent the indexes of v_{il} and v'_{lj}, respectively.

13.5 Discussion of Security

In this section, we discuss the security of the PoR protocol in two types of attacks: deletion attack and modification attack. The security of the PoR protocol is dependent on the property of the pseudorandom function.

13.5.1 *Property of pseudorandom function*

The pseudorandom function f is used in the PoR protocol with a secret key k. Given a value i, $f_k(i)$ returns a pseudorandom number, which is hard to be distinguished from a real random number. Hence, the results of linear combination of $f_k(i)$ and $f_k(j)$, such as $f_k(i) + f_k(j)$ and $f_k(i) - f_k(j)$, are also pseudorandom numbers and hard to be guessed by the cloud, even if the indexes i and j are known.

Given two stored values $(i, C_i, (R_i, I_i))$ and $(j, C_j, (R_j, I_j))$, the cloud cannot apply the homomorphic operations to two existing valid encryptions of indexes I_i and I_j, and get a valid encryption of a specific index. For example, if the cloud knows I_1, it cannot get a valid I_2 by applying the homomorphic addition $I_1 \oplus I_1$, because $f_k(1)$ is encrypted in I_1 and we do not have $f_k(1) + f_k(1) = f_k(2)$.

13.5.2 *Deletion attacks*

In a deletion attack, some outsourced values are supposed to be deleted by the cloud or they are actually not stored after the cloud receives them. As described above, a client stores the tuple $(i, C_i, (R_i, I_i))$ into the cloud. We discuss three cases of deletion attacks. In all cases, we assume the index i is specified in Q in an execution of the PoR protocol.

■ I_i *is deleted*: In this case, the cloud does not have I_i to calculate a correct I that can be decrypted to $\Sigma_{(i,r_i) \in Q} r_i * f_k(i)$. Thus, the second condition of the PoR protocol cannot be satisfied. Moreover, the cloud cannot generate I_i with the combination of other undeleted I_j, where $i \neq j$, according to the property of the pseudorandom function f_k.

■ R_i *is deleted*: Since I_i cannot be deleted (i.e., it is used when calculating I), then the deletion of R_i leads to two ciphertext vectors I and R with their last elements different. Thus, the first condition of the PoR protocol is not satisfied.

■ C_i *is deleted*: This deletion will cause the cloud to return C that is decrypted into a value different from the decryption of R, since R_i cannot be deleted as discussed above, making the third condition of the PoR protocol unsatisfied.

13.5.3 Modification attacks

In this type of attacks, some stored values are supposed to be modified by the cloud. That is, one or more elements in the ciphertexts C_i, R_i, and I_i are modified. If a ciphertext C_i is modified by using homomorphic operations (e.g., a ciphertext is replaced by the sum of another two ciphertexts), then a successful modification attack requires that the modified and original ciphertexts encrypt different plaintext values; otherwise, semantically it is not a modification attack, because the plaintext value is not modified.

If a ciphertext is modified by directly changing a ciphertext element (i.e., not by using homomorphic operations), then the decryption algorithm of the homomorphic encryption scheme returns a random result, since the random numbers in the ciphertext modified in this way cannot be completely removed. In the discussion below, we still assume the index i is specified in Q of the PoR protocol.

■ I_i *is modified*: Then, I cannot be decrypted into the same value as $\Sigma_{(i,r_i) \in Q} r_i * f_k(i)$, regardless of whether it is modified homomorphically or not.

■ R_i *is modified*: Since I_i cannot be modified, the last element of R_i cannot be modified; otherwise, the last elements of I and R will not be equal. Moreover, the modification of any element of R_i causes R to be decrypted into a random value and hence not equal to the decryption of C_i.

■ C_i *is modified*: The modified C_i will be decrypted into a random number. Thus, it cannot be equal to the decryption of R_i, which might also be a random number, if R_i is modified, too.

13.6 Related Works

The architecture in this chapter is designed for the scenario, in which a client may not only use the cloud as a big storage, but also as a powerful data processing platform to discover knowledge from their data (i.e., outsourcing of data analysis). There have been a number of PoR schemes, such as [2,5,12–14]. However, these scheme might not be used to verify the integrity of all types of outsourced computations, such as the integrity of multiplication of two encrypted matrices.

There are also other specific schemes to verify the integrity of outsourced computations, such as [10,11]. Compared with them, our scheme is more generic, since it is built over homomorphic encryption, and all computations allowed by the homomorphic encryption schemes can be verified. For example, the comparisons of encrypted values can be verified in our architecture, since they are supported by the homomorphic encryption scheme. In addition, our architecture provides a unified framework for the preservation of data privacy, proof of retreivability, and verification of computation integrity.

In this architecture, a client can choose his favorite homomorphic encryption scheme, such as the homomorphic encryption scheme proposed in [4]. In [4], a message m is encrypted into a ciphertext with two elements (c_0, c_1), where $c_0 = a*s+t*e+m$ and $c_1 = -a$, where s and t are secret values, and a and e are random numbers. With this encryption scheme, an authenticator (R, I) in our PoR protocol can be defined in the following way.

$$
\begin{aligned}
R &= (c_{R0}, c_1) \\
c_{R0} &= a*s_R + t_R*e + m \\
I &= (c_{I0}, c_1) \\
c_{I0} &= a*s_I + t_I*e' + f_k(i) \\
c_1 &= -a
\end{aligned}
$$

The PoR protocol in our architecture does not need to be changed for different choices of homomorphic encryption schemes. However, as discussed before, the homomorphic encryption scheme used in this chapter has good usability features for real numbers and negative values in business applications.

References

1. Rakesh Agrawal, Jerry Kiernan, Ramakrishnan Srikant, and Yirong Xu. Order preserving encryption for numeric data. In *Proceedings of the ACM SIGMOD International Conference on Management of Data*, Paris, France, pp. 563–574, 2004.

2. Giuseppe Ateniese, Randal Burns, Reza Curtmola, Joseph Herring, Lea Kissner, Zachary Peterson, and Dawn Song. Provable data possession at untrusted stores. In *Proceedings of the 14th ACM Conference on Computer and Communications Security*, Alexandria, VA, pp. 598–609, 2007.

3. Alexandra Boldyreva, Nathan Chenette, Younho Lee, and Adam O'Neill. Order-preserving symmetric encryption. In *Proceedings of the 28th Annual International Conference on Advances in Cryptology*, Cologne, Germany, pp. 224–241, 2009.

4. Zvika Brakerski and Vinod Vaikuntanathan. Fully homomorphic encryption from ring-lwe and security for key dependent messages. In *CRYPTO*, Santa Barbara, CA, pp. 505–524, 2011.

5. Ari Juels and Burton S Kaliski, Jr. Pors: Proofs of retrievability for large files. In *Proceedings of the 14th ACM Conference on Computer and Communications Security*, Alexandria, VA, pp. 584–597, 2007.

6. Jonathan Katz and Yehuda Lindell. *Introduction to Modern Cryptography*. Chapman & Hall/CRC Press Cryptography and Network Security Series. Chapman & Hall/CRC Press, Boca Raton, FL, 2007.

7. Dongxi Liu. Homomorphic encryption for database querying. Inernational Patent Application No.: PCT/AU2013/000674 Accessible via http://patentscope.wipo.int/search/en/WO2013188929, 2013.

8. Dongxi Liu and John Zic. Proofs of encrypted data retrievability with probabilistic and homomorphic message authenticators. In *IEEE International Symposium on Recent Advances of Trust, Security and Privacy in Computing and Communications*, Helsinki, Finland, pp. 897–904, 2015.

9. Michael Naehrig, Kristin Lauter, and Vinod Vaikuntanathan. Can homomorphic encryption be practical? In *Proceedings of the 3rd ACM Workshop on Cloud Computing Security Workshop*, Chicago, IL, pp. 113–124, 2011.

10. Charalampos Papamanthou, Elaine Shi, and Roberto Tamassia. Signatures of correct computation. In *Proceedings of the 10th Theory of Cryptography Conference on Theory of Cryptography*, pp. 222–242. Springer-Verlag, Berlin, Germany, 2013.

11. Bryan Parno, Jon Howell, Craig Gentry, and Mariana Raykova. Pinocchio: Nearly practical verifiable computation. In *IEEE Symposium on Security and Privacy*, pp. 238–252, Berkeley, CA, May 19–22, 2013.

12. Hovav Shacham and Brent Waters. Compact proofs of retrievability. In *Proceedings of the 14th International Conference on the Theory and Application of Cryptology and Information Security: Advances in Cryptology*, pp. 90–107. Springer-Verlag, Berlin, Germany, 2008.

13. Elaine Shi, Emil Stefanov, and Charalampos Papamanthou. Practical dynamic proofs of retrievability. In *Proceedings of the ACM SIGSAC Conference on Computer and Communications Security*, Berlin, Germany, pp. 325–336, 2013.

14. Jia Xu and Ee-Chien Chang. Towards efficient proofs of retrievability. In *Proceedings of the 7th ACM Symposium on Information, Computer and Communications Security*, Seoul, Korea, pp. 79–80, 2012.

15. Andrew Chi-Chih Yao. How to generate and exchange secrets. In *27th Annual Symposium on Foundations of Computer Science*, Toronto, Canada, pp. 162–167, 1986.

Chapter 14

Privacy and Accountability Concerns in the Age of Big Data

Manik Lal Das

CONTENTS

Abstract

In the age of big data, information leakage through (un)intentional exposures and accountability of data usage pose serious threats to organizations and individuals. With the proliferation of Internet technology in modern applications, enormous volumes of data are generated every minute such as health records, airlines information, social networking, online shopping, weather forecasting, and so on. It has been observed that most of the users do not pay much attention to their data, for example, who stores their data, who else can access their data, who is liable if their data are misused, and so on? Importantly, the uprising usage of social networking and smartphones have made the situation worse. Any individual could disclose his friend's personal data to others, even a user gives his own sensitive data to other application service providers for getting some free services temporarily. Furthermore, a (malicious) legitimate user could publish some sensitive information of others in social networking sites, which not only disclose the private information of organization/individual, but also harm others' privacy. Many applications that share individuals' identifying information with dozens of advertising and Internet tracking companies either do not have a proper accountability of data access mechanism in place or they depend on a chaining outsourcing mechanism, which raises the privacy concern of individual. In this chapter, we discuss about the privacy and accountability concerns in big data applications, review the privacy and accountability issues in big data applications, and present a generic construction for addressing these concerns. The proposed construction allows consumers to access data from service provider in a controlled privacy-preserving framework. The service provider will be able to identify the source of data leakages with the help of the proposed construction. The proposed construction can be extended further to other security features based on the applications' requirements.

14.1 Introduction

Big data promises a better decision-making process, a better growth of an organization, and a better society. Big data applications refer to large volume of datasets and the tools/algorithms used to analyze them, train them, and measure the behavior of the users and associated actions provided to them. Undoubtedly, big data have been considered as a powerful computational and knowledge building tool for managing applications in an efficient way for increased growth of the applications and for better human life [1,2]. Nowadays, computing paradigms are pervasive in nature, that is to say that in every hour massive amount of data are being generated, transferred, and stored in different devices for various applications such as weather forecasting, health care, space applications, hotel and travel tickets booking, supply chain, online trading, and so on. On top of these, a large volume of data are being posted to online social networking medium that includes Twitter, YouTube, Facebook, Instagram, and other social networking sites [3]. While operating on large-scale data can help organizations in getting better services, privacy of individual as well as data of big data applications pose a major concern than handling data in traditional applications. Intentional or unintentional leakage of sensitive data of organization and/or individual is undoubtedly one of the most severe privacy (and security as well) threats that organizations face in the modern digital age. This is because the personal information is available in social networks and service providers can outsource other third-party storage and computing platform to serve its users. Although users create their individual profile while using many online applications, which they make private or public, the service provider could use tools and techniques by which they can get full or partial information of the users. Furthermore, once an individual share his/her data with other authorized users then how to control his/her data from unauthorized leakages. In such scenarios, some common questions arise—who owns the data? who is to be blamed for data leakages? how to control the data free-flow mechanism? how to measure the accountability of data access? and so on—all these questions require reasonable assurance in big data applications to ensure data privacy that may belong to organization and/or individual. Therefore, orienting application with big data would certainly give better world, but at the same time it requires substantial efforts for addressing data privacy, data leakage, data ownership, regulatory norms, and ethical practices. It has been observed that the chance of getting caught for data leakages is primarily lying on the service provider. However, if the service provider outsourced storage and computing platform from other party(ies) then the problem becomes complex if they do not have a proper audit and accountability mechanism in place.

Data leakages through social networking: It has been seen that many third-party applications of the online social networking sites leak sensitive information about their users to advertising companies. One may disable the application by detecting it; however, by the time it gets noticed the analytics tool can extract information about the users. Several applications (pretends to be free service provider) ask new users to login to their applications with the social networking credentials, and users do login to the system without taking a moment to check where their credentials go and how it gets validated. With such cross-connectivity among many applications through the same credential, it is sometimes difficult to judge which application is responsible for data leakages that have happened, as many applications have given access to the sensitive data pertaining to the target user.

Data privacy breach through outsourcing: Outsourcing of computing and storage platform offers significant incentives to small- and middle-sized companies including low maintenance cost, less upfront investment, pay-as-you-use option, and so on. However, adopting a poor mechanism for platform outsource could spoil the entire business plan of an organization. It is not new that outsourcing strategy for many applications follows a chain. The outsourcing company that is responsible to handle sensitive data of its users hires a subcontractor that again hires another subcontractor in different locations and delegates each others roles and responsibilities. Eventually when some leakages are noticed, it is not so easy to determine who did the leakages.

Audit and accountability of data access: In modern computing era, third party-dependent resource sharing is a preferred business model because of ubiquitous network access, on-demand service, high return-of-investment, availability of service/data, and efficient way of managing data and users. The service is typically consumed with respect to three types of entities—service provider, data owner, and service consumer. Service provider facilitates resources such as computing, storage, and services; data owner hires the resources from service provider by storing users data in it; and service consumer obtains intended services from the service provider. Typically, data owner hires the third-party storage server to store its data, as well as application. While resource outsourcing provides significant advantages to application providers and service consumers, there are important concerns such as privacy of data, data access, misuse of data, and ownership of data. A proper accountability of data access and usage may provide useful information of the source of data leakages and possible reasons behind the leakages.

In this chapter we discuss the privacy concerns of big data applications and how one can address those concerns with some effective measures. We present a generic construction to address the privacy concerns in big data applications. In particular, the proposed construction aims to address the following concerns in big data applications.

Privacy protected data access: Allowing data access to authorized users is a key success to any applications, irrespective of small or large volumes of data size of the applications. Unauthorized data access and data update (addition, modification, deletion) can not only cause tangible loss to organization, but also compromise the security and privacy of the system and people involved in it. The traditional security solution that works for application provider will not be sufficient for the applications that orient toward big data. By employing data analytics on users data that reside in various third-party servers may lead to additional security threats and importantly, may compromise privacy of users. The proposed construction allows only authorized users to access data from the service provider, where as others will not be able to know what are being accessed in a particular session. With the proposed construction both the user and the service provider will authenticate each other and will know what kinds of data are shared by them. In addition, an adjudicator can resolve the issue of possible data leakages for any complaints that cause user, service provider, or others' privacy breach. We note that the construction does not address privacy-preserving data access. In other words, stored data and access to it by users is known to service provider. However, the proposed construction can be extended further to incorporate this feature with additional steps.

Accountability of data usage: Accountability of user data access on specific application helps in monitoring, controlling, and assessing of data usage by the user for the application. Data loss is the main source of leaking information that may possibly compromise the privacy of individual and/or organization. Therefore, the naive question is: how the data leakages can be controlled and detected? The simple answer to this would be: audit logs and effective measures of data usage. The proposed construction aims to capture this important feature in order to identify the source of data leakages.

The organization of the chapter is as follows. Section 14.2 provides necessary background and important security and privacy challenges in big data applications. Section 14.3 discusses the defense mechanism for data leakages. Section 14.4 presents the generic construction suggested for addressing the above two issues—privacy of data and accountability of data usage. Section 14.5 presents the privacy experiment by which the privacy claims can be validated. We conclude the chapter in Section 14.6.

14.2 Security and Privacy Challenges in Big Data Applications

Security is an integral requirement in applications ranging from consumer electronics, mobile communications, home appliances, and so on to bureaucracy and military application. Nevertheless, the security requirement varies from application to application. With the proliferation of ubiquitous computing and increased social networking trends, privacy issues of data as well as individual become a big concern, even more challenging factor to ensure it in comparison to security requirements in traditional applications. For example, how to ensure the privacy of a mobile phone owner when the crime branch officials want to have law enforcement to track the mobile phone for it whereabouts? how to ensure privacy of a person who puts his wrist watch (a radio-frequency identification [RFID] tag enabled in it) in his bed room? how to ensure the privacy of a user's (un)willingness when business analytics reveal his behavior? and so on. As a result, it is a challenging task to ensure privacy of humans as well as objects in the context of big data applications.

Broadly speaking, security of an organization (assuming it has application(s) to protect from bad people) can be considered as a game between attacker and organization, in which the attacker wants to access information assets from the organization, whereas the organization will put defense mechanism to protect their information assets. Therefore, it is easy to understand that how important the organizational data is, and how data needs to be protected from people who have some malicious intent. The problems become more complex in the context of big data applications. Apart from traditional security and privacy challenges in applications, other important factors that play crucial role in big data applications include *data classification, data storage, data analysis,* and *data leakage*.

14.2.1 Data classification

Classification of data [4,5] is a fascinating art that has got enormous attention from the research community, particularly, in the age of big data. Conventional attribute-based data classification [5,6] is not adequate in modern competitive e-commerce market, as e-retailers like to have data on shopping behavior of users, wish list, browsing trend, and so on, in order to figure out types of products and nature of purchases that users prefer from time to time. Data classification requires effective analysis of business data, filtering data from unstructured data, identification of hot and cold data, and prediction of business trend. As a result, business intelligence, predictive analysis, and expert systems play important roles in classifying data to make data business-centric as well as customer-centric.

14.2.2 Data storage

Data storage and sharing among others require partitioning, multilevel indexing, randomization, and/or encryption based on the nature of applications. It is indeed an uprising business strategy to store application data in third-party server(s) for making data available to all stakeholders, updating data in an efficient ways and finally earning more revenue from the business goal. However, storing large volume data (e.g., exabyte) in partitioning way or semistructured way requires additional efforts for fast data access.

14.2.3 Data analysis

Data analysis [2] has become an extremely important problem area in disciplines like computer science, biology, medicine, finance, and homeland security. The problem is gaining significant momentum in big data applications that involve several aspects. First, large volumes of data must be stored in third-party servers, relying on cleansing and filtering techniques. Second, sophisticated algorithms can be used to analyze the data and extract *useful* information as per application's requirement. Third, various user interfaces and tools can be used to visualize and understand the pattern of data evolved in various sources. Finally, assessing data access and its usage in various inter-dependent applications and taking real-time proactive measures appropriately would be a key success factor.

14.2.4 Data leakage

Data leakages happen due to various reasons—insider leakage, outsider leakage, third-party regulation, proxy agents compromise, and societal and political factors. However, the root cause, to the best of the author's knowledge, of most of these reasons is the lack of a proper accountability of data access provided to various stakeholders of an application, whether it is big or small. In the age of big data, leakages of data are more prone because of third party dependency for storage servers, not having a uniform privacy laws in different countries, data sharing through various channel modes, and so on. Considering these factors of data science, privacy challenges are more prevailing issues in the age of big data than before. Accountability of data access needs to be addressed properly in order to reduce (and control) data leakages. However, enabling strong accountability may also compromise individual/data privacy, in other words, adopting strong privacy feature in applications could relax strong accountability features in the application. Therefore, there is a need to have a controlled balance of these two important features in big data applications.

14.2.5 Privacy challenges

Privacy violations while dealing with big data applications are concerns for human protection, organizational and societal loss, and country's law and order problem, where the primary engines act behind the privacy violations by using various analytics tools and improper use of individuals' credentials in various online media. For example, with a social networking site one can restrict pages to friends, where some pictures posted in the site go against a particular community. Now, if anyone of the friends leaks information of the page then it is not easy to identify the source of the leakages. Furthermore, if the social networking site runs an analytics tool over its databases to extract all the friend's leakages, then who did the leakage of the data and at what level would lead to compromise of individual privacy. In recent trends, application data are mostly stored in third-party servers located in different countries. In that case, third party may run analytics tools and do spying on data access from it by various parties. This will not only compromise individual privacy, but also organization and countries' security. There have been concerns over intelligence over Google and social networking sites [7], where many countries have raised privacy threats of individual in particular and country in general. Ubiquitous computing with extensive usage of mobile devices allows one to track other objects. For example, locating mobile phone users, tracking vehicles, and tracing out an object that has RFID tag embedded in it. All these provide effective monitoring services for better human lives, but at the same time, they may allow someone to compromise other's privacy. Considering these factor into account, big data applications face potential challenges and offer enough space to service providers to address the privacy aspects of individual and organizations that accept the following objectives.

■ *Privacy-preserving data access over protected data*: With increasing demand of data availability, storing application data in third-party server is an efficient business strategy. However, data owner prefers to store data in an encrypted form so that the third-party server cannot learn anything from the stored data. This requires user's query to be processed over encrypted data. Over the years many interesting works [8–10] have been proposed in literature on this direction that allow users to access data while being anonymous and without letting the server know what are being accessed from it.

■ *Audit and accountability*: Controlled data access through authorized entities is a classical approach which has been used in many applications. However, in present ubiquitous computing era, controlling data access is not adequate to prevent data misuse, as it does not control the movement of the data among multiple cross-platform entities. Therefore, proper audit logs (both at application level and system level) and follow-up mechanism should come in force to control and track the free-flow nature of data sharing between entities. By having an acceptable follow-up mechanism, the accountability of data access and its usage would provide the service provider a real picture of data disclosure and its misuse that could ensure the privacy of several stakeholder of an application.

■ *Privacy versus accountability*: As stated before, privacy is a crucial concern in big data applications. And, accountability is required in order to measure culpability and to identify suspicious activities in the system. However, there is an implicit linkage between privacy and accountability. When one measures accountability of data access by a user, what data he/she has accessed from the system, where his/her data have been disseminated, and so on, it provides enough information that leak about the behavior of the user and activities in which the user is involved in. In other words, measuring accountability of a user takes away partially his /her privacy matters. This also implies that preserving strong privacy of user as well as data may partially relax fine-grained audit and thus, allow a strict negotiation of user's accountability.

■ *Multilayer data storage versus data availability*: In the context of big data applications, data classification and storage are two major concerns. Data classification is to be done not only for application data, it also requires concerns of types and mobility of users, frequencies of access, dependency of data between applications, and nature of data. Naturally, data need to be stored in layered protection based on the application requirement. At the same time, data have to be accessible to users from anywhere and anytime. Therefore, storing data in multilayer protection [10,11] and putting defense against denial of service [12] are of potential challenges in big data applications.

■ *Privacy-preserving data mining and analytics*: Companies use modern analytics algorithms and tools for marketing purposes to identify buyers habit and interest and respond to him/her with appropriate services. Simply anonymization of data [13] against analytics is not enough to maintain user privacy. Cryptographic approaches such as homomorphic encryption [14], functional encryption [15] or randomization [16] techniques could provide effective solutions, but with additional computational complexity.

■ *Freedom of expression versus respect to others privacy*: Social networking tools provide a powerful medium of expressing individual (and group) expression. However, these tools need to be used with true spirit for humanitarian assistance and for the betterment of the society. If someone uses it for malicious intent then it spoils the power of social networking objective, instead, invites communal violence, compromises others' privacy, and finally misleads people. Therefore, social networking requires analytics, where analytics require big data and big data collect people expression through social networking, and so on, but this cycle must have proper regulation and norms to ensure privacy of individual and/or organization.

■ *Privacy concerns in distributed computing framework*: Let us consider MapReduce framework [17], where the file is divided and assigned a smaller piece to a Mapper. The Mapper performs intended computation of the smaller piece of the file and outputs a list of key-value pairs. The Reducer aggregates values linked to each unique key and outputs the result. In such scenario, what happens if an untrusted Mapper or Reducer involve in distribution and aggregation process? Is it appropriate to audit neighbors activities in a cooperative monitoring process? How to ensure privacy of data or to identify the source of data leakages? Generally speaking, solution must be emerged based on requirements to address these issues, but one should think beyond the conventional security solution.

■ *Data moderation, sanitization, and dissemination*: Data moderation and proper sanitization not only make big data applications consumer-centric, but also meet the application's security and privacy goals. Inappropriate dissemination of data may leak sensitive information that could harm business objectives. In addition, different countries have different regulatory norms, export/import laws, and varied nature of security policies, therefore, big data applications must adopt acceptable level of data moderation while posting data in public and timely follow-up mechanism if something goes wrong.

14.3 Defense Mechanisms against Data Leakages

Ensuring privacy of personal data in big data applications is a key research problem. In this section we aim to capture privacy of personal as well as organizational data through a generic construction. The construction can be extended to another important requirement that is data availability. However, we focus only on three features in the proposed construction, namely, authorized data access, privacy protection from other users, and accountability of data usage.

14.3.1 Primitives selection

Selection of suitable primitive(s) for security solution of an application is an important criteria. In existing literature many primitives have been used for ensuring privacy of users and data, such as anonymization, randomization, and cryptographic primitives. A brief outline of these primitives is as follows.

- *Anonymization*: Privacy preservation using anonymization technique has been studied extensively in literature. The technique of k-anonymity [13,18] is a classic one. A dataset is said to be k-anonymous ($k \geq 1$) if each record is indistinguishable from at least $k - 1$ other records within the same dataset. The larger the value of k, the better the privacy is protected. However, k-anonymous technique has privacy problems due to lack of diversity in the sensitive attributes and l-diversity ensures better privacy [19]. The notion of l-diversity attempts to solve this problem by requiring that each equivalence class has at least l well-represented value for each sensitive attribute. Although anonymization techniques work for conventional datasets (e.g., relational data), these techniques cannot ensure privacy of user in big data applications such as social networking sites [20], as the adversaries model and user data depend on multiple factors (e.g., nodes, links, and background knowledge) than the conventional datasets.

- *Randomization*: Randomization is a well-known method for preserving privacy of statistical databases [16,21,22]. With this method the original records of the database are perturbed by additional noise so that the behavior of the individual records is masked. However, the de-randomization method should defeat the adversarial capability to know the original dataset from the perturbed dataset.

- *Cryptographic primitives*: Cryptographic primitives such as multiparty computation [23,24], homomorphic encryption [14,25], and attribute-based encryption [9] are suitable for ensuring strong privacy-preserving property but with additional cost in comparison to noncryptographic methods. Although these primitives work toward the privacy goal of applications, additional measures must be taken into consideration while using them in social networking sites or other big data applications.

The proposed construction is inspired by the randomization method and use of standard encryption/decryption algorithms. We note that the proposed construction does not focus on privacy-preserving data mining, instead, it does not allow other users to know what are being transmitted in the current session. The proposed construction allows the service provider to identify the source of data leakages once it is reported to the service provider. Following are the basic assumptions on which the proposed construction works.

- The construction assumes a trusted key generation server (KGS), which generates and manages all keying materials for other principals. The process of keying material generation is assumed to be secure.

- The primitive (e.g., encryption and signature) used in the construction is secure with the standard computational assumptions.

- Adversary can intercept and manipulate data in message exchange.

- Adversary can corrupt a consumer participated into the system and capture the keying materials pertaining to the consumer.

14.4   Proposed Construction

We consider two main principals in our construction—*consumer* and *provider*. Here, *consumer* means common users who use services provided service provider (*provider*). The construction is a tuple (*SysKey, AuthKey, DataAccess, Audit*), and is defined as follows.

- *keying materials* ← *SysKey(k)*: This is a system setup (*SysKey*) phase, which takes the security parameter k as input and outputs all keying materials of all principals of the system. The KGC takes control of this phase and securely gives required keying materials to respective principals in the system.

- *transient key* ← *AuthKey(rand, keying material)*: This phase allows the consumer and the provider to authenticate each other and after successful authentication, they establish a session key for protecting data transmitted in the session. The *AuthKey* takes random number and keying material as inputs and outputs a common transient parameter to gain confidence that they have established a session correctly. With the transient parameter they generate transient key to protect the transmission in the session.

- *data* ← *DataAccess(transient key)*: The *DataAccess* algorithm takes transient key as input and outputs intended data protected under the transient key.

- *source of data leakage* ← *Audit(logs, data)*: This phase takes audit logs and data as inputs and it outputs the possible source of data leakages.

The system architecture is depicted in Figure 14.1.

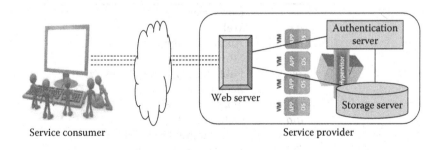

Service consumer   Service provider

Figure 14.1: The system architecture.

14.4.1 System setup

The system might have multiple consumers, service providers, back-end servers, and a KGS. The KGS chooses a security parameter k as per the cryptographic primitives used for the protocol (e.g., for symmetric key cryptographic primitive the security parameter can be 128-bit length [26], and at least 1024-bit for the public key cryptographic primitive [27]). The KGS then generates keying materials (e.g., secret key, public key, and public parameters) for service consumers and service providers. The storage server is assumed to be trusted and must have required level of protection for the data stored in it. For simplifying the trust relationship among the back-end servers, it is advisable to have a separate authentication server, which is responsible for checking the consumers' credentials.

14.4.1.1 Keying material generation

■ *Consumer*: The KGS generates the secret key sk_c and public key pk_c for a consumer who wants to register into the system. A temporary identity tid_c is created on the actual identity aid_c for the consumer and both aid_c and tid_c are mapped to his/her keys. The consumer's credential tuple $< pk_c, tid_c, aid_c, m\_no >$ is stored in the authentication server and storage server, where m_no is the registered mobile number of the consumer. Upon successful registration, the KGS gives $< sk_c, pk_c, tid_c, aid_c >$ to the consumer in a secure manner.

■ *Provider*: The provider is composed of three main servers—storage server, authentication server, and web server. The KGS generates the secret key sk_{ws} and public key pk_{ws} for the web server. It is assumed that the communication channel between the web server, authentication server, and storage server is secure. The KGS stores the public key pk_{ws} of the web server in authentication server and storage server, and the secret key sk_{ws} is kept with the web server. Similarly, the KGS generates the public–private key pair (pk_{as}, sk_{as}) for the authentication server and (pk_{ss}, sk_{ss}) for the storage server. The public key pk_{as} of the authentication server is stored in the web server and storage server, and pk_{ss} of the storage server in the web server and authentication server.

For better readability, we use some symbols in the proposed construction which are mentioned in Table 14.1.

Table 14.1   Notation and meaning

c	Consumer
ws	Web server
as	Authentication server
ss	Storage server
pk_x	Public key of x, $x \in \{c, ws, as\}$
sk_x	Secret key of x, $x \in \{c, ws, as\}$
tid	Temporary identity
aid	Actual identity
$\mathcal{F}$	A cryptographic function
Enc_t	Encryption function under the key t
Sign_t	Signature function under the key t

14.4.2　Data access

Allowing users to access data requires authorization of access; otherwise, unauthorized data access can spoil the business goal. In order to ensure authorized data access, we first discuss how users (service consumer) and servers (service provider) can authenticate each other, and after successful authentication the user is allowed to access his data from the server. Once the user is authorized to access his data, he/she has also established a secure channel between him/her and the server so that transmitted data can be protected from others.

14.4.2.1　Authentication and key agreement

The phase *AuthKey(rand, keying material)* enables the consumer and provider to authenticate each other followed by a transient key agreement by which they can protect data transmitted in the session.

1. Consumer selects a random number *rand* $\in_R Z_p^*$ and computes a challenge $C = \mathcal{F}(rand, tid_c^{(c)}, pk_c)$, then sends $< C, tid_c^{(c)} >$ to the provider. Here, $tid_c^{(c)}$ is the consumer's temporary identity which is valid for the current session.

2. After receiving consumer's request, the web server forwards the challenge C to the authentication server for its validation. The authentication server checks if the $tid_c^{(c)}$ is active in its database. If so, the authentication server generates a new temporary identity $tid_c^{(n)}$ and sends that to the consumer's registered mobile number. Here, $tid_c^{(n)}$ is the consumer's temporary identity which is valid up to the next session. Then the authentication server sends $< \text{Enc}_{pk_{ws}}(tid_c^{(n)}), \text{Sign}_{sk_{as}}(C, tid_c^{(c)}) >$ to the web server as an endorsement of the consumer's request. At the same time the authentication server sends $< \text{Enc}_{pk_{ss}}(tid_c^{(n)}), \text{Sign}_{sk_{as}}(C, tid_c^{(c)}) >$ to the storage server. We note that the tuple from authentication server and storage server may have a validity ticket and timestamp based on the nature of applications.

3. The web server verifies the authentication server's signature on C and if it holds then decrypts the first part of the tuple using the secret key sk_{ws} and gets $tid_c^{(n)}$. Now, the web server computes

$$R = \mathcal{F}(rnd, pk_{ws}), \text{ where } rnd \in_R Z_p^*$$
$$Auth_p = \mathcal{F}_{Skey}(pk_c, tid_c^{(c)}, pk_{ws}, tid_c^{(n)}, C, R)$$

The provider sends $< R, Auth_p >$ to the consumer.

4. Consumer computes $C_p' = \mathcal{F}_{Skey}(pk_c, tid_c^{(c)}, pk_{ws}, tid_c^{(n)}, C, R)$ and checks whether $Auth_p = C_p'$. If it does, then the provider's authentication is confirmed. Now, the consumer computes $Auth_c = \mathcal{F}_{Skey}(pk_{ws}, pk_c, tid_c^{(n)}, R, C)$ and sends $< Auth_c >$ to the provider as a confirmation message.

5. After receiving $< Auth_c >$, the provider (the web server) computes $C_c' = \mathcal{F}_{Skey}(pk_{ws}, pk_c, tid_c^{(n)}, R, C)$ and validates $< Auth_c >$ by checking whether $Auth_c = C_c'$. If it holds, then the consumer's authentication is confirmed.

Note that the authentication code of the provider $Auth_p$ (respectively, $Auth_c$) is computed by a transient secret $Skey = \mathcal{F}(sk_c, pk_c, tid_c^{(n)}, pk_{ws}, sk_{ws})$ generated by the provider (respectively, the consumer). The *Skey* can be used to derive other keys such as *write key*, and *MAC key* for protecting data while transmitting them during the session.

14.4.2.2 Privacy controlled data access

This phase first invokes *AuthKey(rand, keying material)*. Upon successful *AuthKey(rand, keying material)* run, both consumer and provider have established a secure channel between them where all data accessed by the consumer will get protected by the transient key *Skey*. The following steps are required to access data securely:

■ The consumer puts a request to the provider for intended data.

■ The web server sends a signed token *token* to the storage server, where the token is computed as $token = <\text{Sign}_{sk_{ws}}(consumer\ req., pk_{ss}(Skey, tid_c^{(n)}))$. The storage server verifies the received *token* and obtains the *Skey* for the consumer requested data. The storage server verifies $tid_c^{(n)}$ received from the web server with the tuple received from the authentication server in the *AuthKey(rand, keying material)* phase. If they match, then the requested data is provided to the consumer encrypted under *Skey* or by a derived key from it.

14.4.3 Accountability

Once a user has received a document, nothing can prevent him from publishing it, which may cause leakage of others' information. Therefore, there is a need of *check and control* mechanism for data, particularly, for big data applications. The *check and control* mechanism can be effective if proper audit logs and accountability of data access mechanism is followed in the applications. Different techniques can be applied for audit and accountability of data depending on the nature of applications. For example, starting from the classical system logs, database logs, application logs, and so on, and ending with watermarking, and forensic techniques. Importantly, each technique has a cost factor associated with it. We provide a mechanism that allows identifying the source of data leakages.

1. The data that are classified or sensitive will get watermarked with the audit key while accessing the data by a consumer. The audit key is computed as $Akey = \mathcal{F}(tid_c^{(n)}, Skey, sk_{ws})$.

2. The provider maintains audit logs that should have reflection of the reference of the data and watermarked information.

3. For any disputes (e.g., data leakage complaints, misuse of other's data), the provider will be able to resolve it with the audit logs through an adjudicator agreeable to both the consumer and provider. The dispute resolution would identify the temporary identity tid_c used in the watermark/audit key, and the temporary identity would help in identifying the actual identity of the consumer with the help of the authentication server (Figure 14.2).

14.5 Privacy Experiment

We note that claiming privacy or breaking privacy of the system is a context-dependent feature. However, one can consider the privacy feature as a game between adversary and provider. The adversary's goal is to break the privacy of users of the system; whereas, the provider's goal

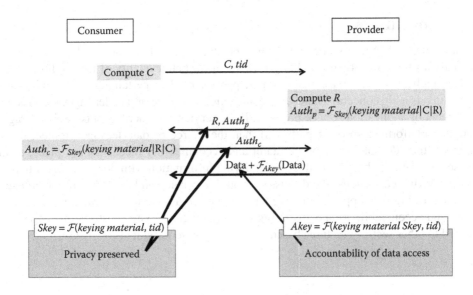

Figure 14.2: Proposed construction for privacy and accountability.

is to preserve privacy of users. If the adversary can prove that he/she knows (by intercepting any number of message exchanges between the consumer and provider, and gathering users information [except the target user]) who is interacting with the provider then the privacy of the users is not preserved to others.

In the proposed construction, if the outputs of any two instances of the protocol run are indistinguishable then the protocol can preserve consumers' privacy while they are interacting to the provider. The experiment consists of a challenger and an adversary. The adversary can control the communication channel between the consumer and provider. The experiment is defined follows.

$\text{Exp}^b_{Adversary}(k)$:

1. $b \in_R \{0,1\}$

2. $\texttt{Provider}(1^k)$, k is the security parameter

3. $m \leftarrow$ *Adversary's capability*

4. Check whether $m = b$

The challenger presents to the adversary the information and asks if the consumer c_i (if $b = 0$) or c_j (if $b = 1$) was active in the *AuthKey(rand, keying material)* phase. We say that the adversary breaks the privacy of the system if $m = b$, that is, if it correctly answers which one of the consumers was active in that session. The advantage Adv of the adversary is defined as

$$\text{Adv}_{Adversary}(k) = \text{Pr}\left[\text{Exp}^0_{Adversary}(k) = 1\right] + \text{Pr}\left[\text{Exp}^1_{Adversary}(k) = 1\right] - 1$$

The proposed construction is said to be privacy preserving from others if no polynomial-time adversary can distinguish a consumer from different instances of the protocol run with non-negligible advantage in security parameter k.

14.6 Conclusion

Data analytics enhance organizational and societal health. Big data application empowers data analytics tools for organizational growths and for better quality of human life. Although big data application can make world smart, privacy of data pertaining to individual and/or organization is a big concern. Therefore, identifying the source of data leakages is an important requirement of application service provider. If the service provider has outsourced storage and computing platform from other party(ies) then the source of data leakage problem becomes complex if they do not have a proper audit and accountability mechanism in place, as many of these applications share individuals' identifying information with dozens of advertising and Internet tracking companies. We discussed about the privacy and accountability challenges in the context of big data applications. We have suggested a generic construction for addressing privacy and accountability concerns in big data applications. The construction can further be extended to other security features based on application requirements.

References

1. M. Milton. *Head First Data Analysis: A Learner's Guide to Big Numbers, Statistics, and Good Decisions.* O'Reilly Media, Sebastopol, CA, 2009.

2. P. K. Janert. *Data Analysis with Open Source Tools.* O'Reilly Media, Sebastopol, CA, 2010.

3. P. K. Ryan. *Social Networking—Digital and Information Literacy.* Rosen Central, New York, 2011.

4. M. S. Chen, J. Han, and P. S. Yu. Data mining: An overview from a database perspective. *IEEE Transactions on Knowledge and Data Engineering*, 8(6):866–883, 1996.

5. W. H. Au, K. C. Chan, A. K. Wong, and Y. Wang. Attribute clustering for grouping, selection, and classification of gene expression data. *IEEE/ACM Transactions on Computational Biology and Bioinformatics*, 2(2):83–101, 2005.

6. A. Almuhareb and M. Poesio. Attribute-based and value-based clustering: An evaluation. In *Proceedings of the Conference on Empirical Methods in Natural Language Processing*, Barcelona, Spain, pp. 158–165, 2004.

7. J. Bamford. The NSA is building the country's biggest Spy center. Retrieved: http://www.wired.com/threatlevel/2012/03/ff_nsadatacenter, 2012.

8. E. Shi, J. Bethencourt, T. H. Chan, D. Song, and A. Perrig. Multi-dimensional Range Query over Encrypted Data. In *Proceedings of IEEE Symposium on Security and Privacy*, Oakland, CA, pp. 350–364, 2007.

9. V. Goyal, O. Pandey, A. Sahai, and B. Waters. Attribute-based encryption for fine-grained access control of encrypted Data. In *Proceedings of the ACM Conference on Computer and Communications Security*, Alexandria, VA, pp. 89–98, 2006.

10. R. A. Popa, C. Redfield, N. Zeldovich, and H. Balakrishnan. CryptDB: Protecting confidentiality with encrypted query processing. In *Proceedings of the ACM Symposium on Operating Systems Principles*, Cascais, Portugal, pp. 85–100, 2011.

11. I. H. Akin, and B. Sunar. On the difficulty of securing web applications using CryptDB. In *Proceedings of the IEEE International Conference on Big Data and Cloud Computing*, Sydney, Australia, pp. 745–752, 2014.

12. X. Wang and M. K. Reiter. Defending against denial-of-service attacks with puzzle auctions. In *Proceedings of the Symposium on Security and Privacy*, Oakland, CA, pp. 78–92, 2003.

13. P. Samarati and L. Sweeney. Generalizing data to provide anonymity when disclosing information. In *Proceedings of the ACM Symposium on Principles of Database Systems*, Seattle, WA, p. 188, 1998.

14. I. Damgard, V. Pastro, N. Smart, and S. Zakarias. Multiparty computation from somewhat homomorphic encryption. In *Proceedings of Advances in Cryptology*, LNCS 7417, Springer, Berlin, Germany, pp. 643–662, 2012.

15. D. Boneh, A. Sahai, and B. Waters. Functional encryption: Definitions and challenges. In *Proceedings of the Theory of Cryptography Conference*, LNCS 6597, Springer, Berlin, Germany, pp. 253–273, 2011.

16. G. T. Duncan and S. Mukherjee. Optimal disclosure limitation strategy in statistical databases: Deterring tracker attacks through additive noise. *Journal of the American Statistical Association*, 95(451):720–729, 2000.

17. J. Dean and S. Ghemawat. MapReduce: A flexible data processing tool. *Communications of the ACM*, 53(1):72–77, 2010.

18. L. Sweeney. *K*-anonymity: A model for protecting privacy. *International Journal on Uncertainty, Fuzziness and Knowledge-Based System*, 10(5):557–570, 2002.

19. A. Machanavajjhala, J. Gehrke, D. Kifer, and M. Venkitasubramaniam. L-diversity: Privacy beyond *k*-Anonymity. In *Proceedings of the IEEE International Conference on Data Engineering*, Atlanta, GA, p. 24, 2006.

20. B. Zhou and J. Pei. Preserving privacy in social networks against neighborhood attacks. In *Proceedings of the IEEE International Conference on Data Engineering*, Cancun, Mexico, pp. 506–515, 2008.

21. T. Evans, L. Zayatz, and J. Slanta. Using noise for disclosure limitation of establishment tabular data. *Journal of Official Statistics*, 14(4):537–551, 1998.

22. V. S. Iyengar. Transforming data to satisfy privacy constraints. In *Proceedings of the ACM SIGKDD International Conference on Knowledge Discovery in Databases and Data Mining*, Edmonton, AB, Canada, pp. 279–288, 2002.

23. Y. Lindell and B. Pinkas. Privacy preserving data mining. *Journal of Cryptology*, 15(3):177–206, 2002.

24. Y. Lindell and B. Pinkas. Secure multiparty computation for privacy-preserving data mining. *Journal of Privacy and Confidentiality*, 1(1):5, 2009.

25. A. Lopez-Alt, E. Tromer, and V. Vaikuntanathan. On-the-fly multiparty computation on the cloud via multikey fully homomorphic encryption. In *Proceedings of the ACM symposium on Theory of Computing*, New York, NY, pp. 1219–1234, 2012.

26. National Institute of Standards and Technology (NIST). FIPS-197: Advanced encryption standard. Retrieved: http://www.itl.nist.gov/fipspubs/, 2001.

27. R. Rivest, A. Shamir, and L. Adleman. A method for obtaining digital signatures and public-key cryptosystems. *Communications of the ACM*, 21(2):120–126, 1978.

Chapter 15

Secure Outsourcing of Data Analysis

Jun Sakuma

CONTENTS

With recent advances in online services, personal data are continually being collected and stored by services for various purposes. Primary data holders need to treat such personal data under prudent control, so personal data are stored in in-house databases hosted by service providers under ordinary circumstances. However, when the size of the personal data is enormous, not only the storage cost but also the computational cost for data analysis is non-negligible. Recently, privacy-enhancing technologies that enable private computation to be outsourced to cloud servers are attracting attention. This chapter provides an overview of technologies for protecting data privacy, including data anonymization, differential privacy, and secure computation, for private computation with the cloud.

15.1 Utilization of Personal Data in the Cloud

15.1.1 Cloud and personal data

Online consumer services generate a huge amount of histories of user activities. Such information can be used for creating novel value and services by combining personal data from different services. The cloud serves as a promising platform for integrating personal data distributed over service providers.

Needless to say, privacy concerns should be carefully considered for utilizing and integrating personal data. Outsourcing to the cloud increases privacy risks, so in-house data processing is preferable in terms of privacy preservation. Nevertheless, we have two motivations to outsource personal data to the cloud. First, in big data analysis, it is quite natural to store data in the cloud and outsource data analysis to the cloud in terms of cost and utility. Second, cloud storage can work as a hub to integrate personal data collected from different sources. If we could resolve privacy concerns that are raised by utilizing personal data in the cloud, diverse types of valuable and promising services would be launched. In most cases, the notion of privacy is related to the identification of de-identified records (re-identification) and the estimation of hidden attribute values.

We refer to hidden attributes as *sensitive attributes*. These attributes do not work as identifiers and should not be published in an identifiable manner in order to be morally and ethically responsible, avoid discrimination, or protect human rights. Sensitive attributes are determined depending on the context of data utilization. For example, arrest records, party affiliation, religious affiliation, and race and ethnicity are almost always treated as sensitive attributes. If a data analyst uses the data for medication research, disease statuses are not recognized as a sensitive attribute even with *personal identifiable information* (PII). However, if the analyst uses the data to make hiring decisions or to perform a credit assessment, then using the disease statuses with PII may be discriminatory. In such a case, disease statuses should be treated as sensitive attributes.

The notations used in this chapter are introduced here. Unless specifically mentioned in this section, a database table is supposedly defined as a collection of records; a record consists of multiple attributes and is associated with a single individual. We call such records *personal data*.

15.1.2 Stakeholder

Several stakeholders appear when outsourcing personal data analysis to the cloud. Individuals that first provide the information are referred to as *data contributors*. The entity that collects personal data from individuals is called the *primary data holder*. The primary data holder deposits the collection of the personal data to the *cloud*. *Data analysts* are individuals or organizations which wish to obtain the results of data analysis with the personal data. We consider two scenarios of data analysis with the data stored in the cloud.

1. Upon requests from a client, the cloud provides a collection of personal data to the client; the client can conduct data analysis with the data.

2. A client can issue a data analysis query to the cloud; the cloud then conducts data analysis and reports the result to the client.

The definition of privacy protection is different in these two scenarios. This will be discussed in Section 15.1.3 in detail.

15.1.3 *Privacy-enhancing technologies for the cloud*

In the two scenarios described above, we can consider two different types of privacy guarantees:

■ The cloud and data analysts do not learn private information contained in personal data provided by the primary data holder (input privacy).

■ The client does not learn private information contained in the data collection from data analysis results provided by the cloud (output privacy).

One straightforward solution for input privacy is to use tamper-resistant devices in the cloud and let the cloud conduct data analysis with the devices. We do not pursue this option because such devices are usually expensive and have a relatively poor computational performance compared to ordinary processors; it is not realistic to use such devices for big data analysis. In this section, it is assumed that the cloud is a collection of computational nodes with ordinary processors and can communicate with other entities with a regular communication channel, namely, the Internet.

We emphasize that preserving privacy in data analysis and access control of data are different notions. Access control is used to distinguish entities that are allowed to access data from entities that are prohibited from accessing the data. Protection of input privacy is done to restrict leaks of the input data throughout the process of data analysis, and protection of output privacy is done to restrict input data from leaking in the result of the analysis.

Table 15.1 summarizes privacy-enhancing technologies for data analysis in the cloud. Input privacy protection can be realized by either stochastic guarantee of privacy or cryptographic guarantee of privacy. For stochastic guarantee, the input data is obfuscated and provided in a plaintext form, so that each record cannot be identified with a high probability. For cryptographic guarantee, the input data is encrypted and provided in a ciphertext form, so that no one without a key for decryption can learn anything from the data. Output privacy, namely, differential privacy, can be realized by a stochastic guarantee of privacy. The analysis results (outputs) are provided in a plaintext form after certain modification or randomization, so that the outputs do not leak too much information about the input data.

Table 15.1 **Privacy protection of data analysis in the cloud**

	Input privacy	**Output privacy**
Stochastic guarantee	Anonymization Randomization	Differential privacy
Cryptographic guarantee	Secure multiparty computation Highly functional encryption	

15.2 Anonymization

The definition and requirements for anonymization are discussed in this section. Data anonymity is a state in which data in a personal dataset is not linked to any single person. Instead of using personally identifiable information, using anonymous data allows data analysts to utilize personal data without caring about privacy. It is important that the notion of anonymization ranges from only deleting personal identifiers to converting data to make it difficult to infer its identity.

Encryption for privacy preservation limits an application, but plaintext form anonymization does not. This is the main advantage of anonymization.

If primary data holders wish to store personal data in the cloud to share personal data with another party, anonymization is one good solution.

15.2.1 Personal identifiable information

In this section, personal data is assumed as a collection of records in which each record represents each person's data in a tabulated form. *Identify* means knowing a record in the personal data belongs to a single person. An attribute in the record that identifies the record on its own is called an *identifier*, which includes driver's license numbers, photographs of faces, and fingerprints. Although a legal name does not always identify a record, names are treated as identifiers.

Quasi-identifier is an attribute that does not identify a record on its own but identifies the record along with other attributes in the same record. It includes age, gender, and address. If a combination of quasi-identifiers becomes unique in the dataset, it will have the same function as an identifier.

The remaining attributes are called *non-identifiers*, which are not involved in identifying a record. For example, if someone's illness or thoughts and beliefs are not known by other people, non-identifiers will not lead to them being identified, but providing such attributes with identifiers or quasi-identifiers may cause privacy problems.

The boundaries among identifiers, quasi-identifiers, and non-identifiers are unclear (Figure 15.1). For example, some e-mail addresses contain the names and organizations of the people and are regarded as an identifier, but some are not. If some attributes, such as the number of cavities, are not known by anybody else, they are non-identifiers. However, nobody can guarantee that such an attribute will remain private.

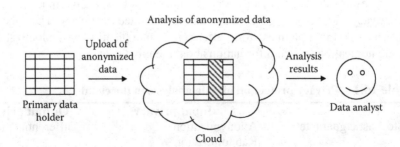

Figure 15.1: Anonymization in the cloud environment.

15.2.2 Personal data anonymity in the cloud

Motivations for outsourcing personal data in the cloud include data storage to reduce data management cost and data utility to analyze or share data. For storage, security is important to prevent information leaks, and privacy is important for utility. In the latter case, data anonymization is one good solution.

Personal data anonymity is a state in which data in a personal dataset is not linked to any single person, and anonymization is the process of making personal-identifiable-data anonymous. To achieve anonymity, identifiers must be deleted from each record (Table 15.2 center), but this is still not sufficient enough. For example, for the record that contains zip code = "232-0011," age = "26," and illness = "hernia" in the complete enumeration survey, if someone has the knowledge about a 26-year-old person in zip "232-0011," his/her illness "hernia" will be known.

To prevent this, quasi-identifiers and non-identifiers should be modified at the expense of accuracy. *Global recoding* means modifying all values of an attribute into several categories, for example, all "asthma" and "tuberculosis" are modified into "lung disease." *Local suppression* means removing some values of an attribute; for example, replace "hernia" with N/A. Other methods include *swap*, to exchange a value of a record with another value in another record, *generalization*, which means abstract numerical values and categorical values.

These methods are not magic tools that should be used to achieve a certain anonymization level. Both *k-anonymity* [14] and *ℓ-diversity* [7] are known anonymity criteria.

k-anonymity guarantees that no records can be distinguished uniquely in a dataset, which means that there are at least $k(>1)$ records that have the same value of quasi-identifiers in combination in the dataset. In Table 15.2 (left), the combination of quasi-identifiers of the first record is (zip code = "232-0011," age = "26"), and this is unique in the dataset. Table 15.2 (center) generalizes zip codes and ages and achieves 3-anonymization.

The *ℓ*-diversity criterion prevents the prediction of non-identifiers, which means that in a set of records that have the same quasi-identifier combination in the k-*anonymized* dataset, there are at least $\ell(1 < \ell \leq k)$ kinds of values of non-identifiers. In Table 15.2 (center), for records #7–9, the combination of quasi-identifiers is zip code = "232-0014," age = "[20–29]," so each record has the same non-identifier value "diabetic." This means that a person in their 20s living in the zip code "232-0014" suffer "diabetes." Table 15.2 (right) achieves 2-diversity by generalizing the value of illness in the record #8.

The level of anonymity is the trade-off of utility. If all values are generalized highly, strong anonymity is achieved, but utility is very low. To keep the data utility high, anonymity will be thrown away. Optimal *k*-anonymity is known as non-deterministic polynomial-time hard (NP-hard), but optimal *k*-anonymity in big data is very difficult. There are some efficient *k*-anonymization algorithms including [8].

15.2.3 Randomization

Randomization is a neighboring idea of anonymization. *Randomization* (*perturbation*) means transforming personal data into synthetic data by adding randomness to the original data, and making identity infer from randomized data difficult. Randomization methods include random value addition (noise injection), value trade within randomly selected records (swap), and replacement with random value. Because randomization is an irreversible operation and original data cannot be rebuilt from randomized data, data privacy can be preserved. In randomized techniques, rebuilding the original data is impossible, but it is possible to infer the statistics behind the data. This statistical inference has been known since the 1960s. Randomization is plaintext processing, such as anonymization, to preserve privacy, and it can calculate reasonable statistics (Figure 15.2).

Table 15.2 (Top) Personal data where identifiers are removed, (Middle) personal data with 3-anonymization, and (Bottom) personal data with 3-anonymization, 2-diversity

Zip code	Age	Illness
232-0011	26	Hernia
232-0015	34	Backache
232-0017	27	Backache
232-0012	45	Rhinitis
232-0013	43	Asthma
232-0014	42	Tuberculosis
232-0014	23	Diabetes
232-0014	24	Diabetes
232-0014	26	Diabetes

Zip code	Age	Illness
232-001x	[20–39]	Hernia
232-001x	[20–39]	Backache
232-001x	[20–39]	Backache
232-001x	[40–49]	Rhinitis
232-001x	[40–49]	Asthma
232-001x	[40–49]	Tuberculosis
232-0014	[20–29]	Diabetes
232-0014	[20–29]	Diabetes
232-0014	[20–29]	Diabetes

Zip code	Age	Illness
232-001x	[20–39]	Hernia
232-001x	[20–39]	Backache
232-001x	[20–39]	Backache
232-001x	[40–49]	Rhinitis
232-001x	[40–49]	Asthma
232-001x	[40–49]	Tuberculosis
232-0014	[20–29]	Diabetes
232-0014	[20–29]	Adult disease
232-0014	[20–29]	Diabetes

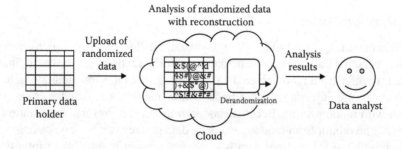

Figure 15.2: Randomization and reconstruction.

The randomization and reconstruction method was proposed in the context of *privacy preserving data mining* (PPDM), introduced by Agrawal in 2000 [17]. It randomizes data and calculates reasonable statistics by applying a specific operation to the randomized data.

This method consists of two steps. The first is randomization, which adds random noise similar to that drawn from a Laplacian distribution to each record, for example. The second is for reconstruction, which estimates statistics from randomized data by using an inverse operation to randomization similar to Bayesian estimation techniques. The reason behind this is, even if information from each piece of data is degraded, statistics can be converged within the range derived from the randomization algorithms.

Quantitative evaluation of privacy level for the reconstruction method has not been studied clearly. One method called *probabilistic* k (*Pk*)*-anonymity* was proposed [2]. This method uses a randomization *retention/replacement perturbation* and enables reconstruction while guaranteeing a privacy level equivalent to k-anonymity. This means that the certainty of any records being identified from randomized data is less than $1/k$. This method leads to new deployment of randomization for personal data anonymization.

15.3 Secure Computation

15.3.1 *Privacy protection with cryptographic technologies*

Suppose a primary data holder outsources private data to a cloud storage. Data on the cloud storage can be secured by encrypting it. Encryption ensures the security of data when the data is leaked from the cloud storage for reasons, such as a theft, an inside job, and disasters. However, when the primary data holder outsources computation with the data to the cloud, the cloud needs to decrypt the ciphertexts and obtain the raw data for computation. As long as the cloud learns the raw data, this process does not guarantee input privacy protection against the cloud.

Secure computation methods (secure computation) are introduced in this section. Secure computation allows the cloud to compute functions with the primary data holder's data and report the results to the primary data holder. At the same time, no private data is leaked to the cloud through the execution of secure computation.

Primarily, three techniques that realize secure computation have been developed in the cryptography community: garbled circuit, homomorphic encryption, and secret sharing.

The idea of secure computation was first introduced by Yao and is known as the garbled circuit [16]. In this method, a function to be evaluated is transformed into a corresponding logic circuit. The garbled circuit provides a way to evaluate the logic circuit without sharing private inputs. Yao's garbled circuit allows us to evaluate arbitral functions as long as the functions are represented by logic circuits. Customizing the circuit can highly improve the computational efficiency for certain types of functions. However, the evaluation time of some functions with an exponentially large circuit representation can be impractically large.

In homomorphic encryption, one can perform a certain operation(s), for example, addition and multiplication, over ciphertexts without decryption. One can obtain the result of the operations by decrypting the resulting ciphertexts. This property of homomorphic encryption naturally realizes secure computation.

Secret sharing was originally designed as a method to hold data among two or more parties securely. In secret sharing, data is partitioned into two or more parts, so that each part seems to be random. The original data can be recovered only when one collects a specified number of shares. Some secret sharing schemes allow arithmetical or logical operations of shares, so that the result of the operations can be recovered by collecting the shares after operations. This type of secret sharing also realizes secure computation.

For secure outsourcing of computation with private data, the cloud is required to process computation specified by the primary data holder without learning the data. In other words, secure outsourcing of computation is a guarantee of input privacy against the cloud. Secure outsourcing of computation is introduced though secure computation, namely, homomorphic encryption and secret sharing in Sections 15.3.1.1 and 15.3.1.2.

15.3.1.1 Homomorphic encryption

Suppose a primary data holder encrypts its data and deposits the ciphertexts to the cloud. If the primary data holder uses the cloud as a storage, encryption guarantees the secrecy of the data against the cloud. However, if the primary data holder wishes to outsource computation with the data to the cloud, the cloud needs to decrypt the data for computation. This scheme does not guarantee the secrecy of the data against the cloud.

Homomorphic encryption can be a solution for secure outsourcing of computation. A regular public-key cryptosystem consists of three algorithms: key generation, encryption, and decryption. Given a security parameter κ, the key generation algorithm outputs a pair of a public key (pk) and a secret key (sk). The public key is used for encryption, and the secret key is used for decryption.

$$\text{KeyGen}(1^{\kappa}) = (\text{pk}, \text{sk}). \tag{15.1}$$

The encryption function generates a ciphertext of a plaintext $x \in \mathbb{Z}_N$ with the public key:

$$\text{Enc}(x, \text{pk}) = \text{ct}, \tag{15.2}$$

where the plaintext space is given by $\mathbb{Z}_N = \{0, 1, ..., N-1\}$, and the size N is associated with the security parameter. The plaintext can be recovered by applying a decryption function to the ciphertext with the secret key as follows:

$$\text{Dec}(\text{ct}, \text{sk}) = x. \tag{15.3}$$

When the cryptosystem is additively homomorphic, the following holds:

$$\text{Dec}(\text{Enc}(m_1 + m_2, \text{pk}), \text{sk}) = \text{Dec}(\text{Enc}(m_1, \text{pk}) \cdot \text{Enc}(m_2, \text{pk}), \text{sk}). \tag{15.4}$$

The ciphertext of the addition of two values can be obtained by applying the dot($\cdot$) operator between two ciphertexts of each value. One can compute scalar multiplication of ciphertexts by repeatedly applying the operation to a ciphertext as

$$\text{Dec}(\text{Enc}(km, \text{pk}), \text{sk}) = \text{Dec}(\text{Enc}(a, \text{pk}) \cdot \text{Enc}(a, \text{pk}) \cdot ... \cdot \text{Enc}(a, \text{pk})), \text{sk}). \tag{15.5}$$

When an encryption function is a one-to-one mapping from plaintexts to ciphertexts, the ciphertext can leak some information about the plaintext if a large number of ciphertexts are collected (statistical attack). We thus require that the distribution of a ciphertext of a value should form a uniform distribution over a specified domain (semantic security). A public-key homomorphic encryption known as Paillier encryption [12] satisfies this requirement.

The cryptosystem is multiplicatively homomorphic if the following holds:

$$\text{Dec}(\text{Enc}(m_1 m_2, \text{pk}), \text{sk}) = \text{Dec}(\text{Enc}(m_1, \text{pk}) \circ \text{Enc}(m_2, \text{pk}), \text{sk}). \tag{15.6}$$

where the circle operator ($\circ$) between two ciphertexts denotes the multiplicative homomorphic operator.

Elgamal encryption [5] is a multiplicatively homomorphic public-key cryptosystem. A public-key encryption with both additive and multiplicative homomorphisms (fully homomoprhic encryption) has been an open problem for a long time. Gentry [6] first described a construction of a fully homomorphic encryption scheme based on a lattice-based cryptosystem in 2009. After this breakthrough, several constructions for fully homomorphic encryption have been reported [1,15].

Recall the problem of outsourcing data analysis. After the primary data holder encrypts data with a homomorphic encryption scheme and provides it to the cloud, the cloud can perform computation with ciphertexts without having to interact with the primary data holder. Unfortunately, the class of functions that can be evaluated within the capability of additively (or multiplicatively) homomorphic encryption is quite limited. For example, the comparison of two numbers is a building block often used for data analysis; however, given two ciphertexts of integers, one cannot learn which one is greater by means of the homomorphic property. In such a case, we can use both the garbled circuit and homomorphic encryption for a solution for privacy-preserving data analysis. Many solutions for privacy-preserving data analysis have been developed, such as decision tree learning [9], clustering [13], and linear regression [10].

In theory, fully homomorphic encryption allows any function of encrypted inputs to be evaluated without interacting with secret key holders. A great amount of effort has been devoted to improving the computational and space efficiency of fully homomorphic encryption. However, it is still impractical to analyze large-scale data encrypted with fully homomorphic encryption. Somewhat (or leveled) homomorphic encryption schemes [1], which allow a limited number of additions and multiplications of ciphertexts, offer practical solutions for preserving privacy in the outsourcing of data analysis.

15.3.1.2 Secret sharing

Secret sharing works based on secure computation. An example of secret-sharing-based secure computation is introduced. This is a three-party computation method in which input data is distributed securely in a 2-out-of-3 manner; computation is completed without ever reconstructing any data in the whole process and only arithmetical/logical operation results are reconstructed. Both addition and multiplication are defined in the following algorithm, and logical operations are defined by using these ADD/MUL algorithms. This type of secure computation has the merit of secret sharing for which practical application has been reported.

Intuitive definition of secret sharing secure computation

Let H denotes primary data holder, $P_i, i = 0, 1, 2$ denotes computation party and A be data analyst.

Distribution H makes x_0, x_1, x_2 from x and allocates share (x_i, x_{i+1}) to P_0, P_1, P_2, where $x_0 = x - x_1 - x_2$, x_1, x_2 are random values.

Reconstruction A collects two shares from P_i and reconstructs the data by $x = x_0 + x_1 + x_2$.

Addition Each P_i calculates the result of addition $(a_i + b_i, a_{i+1} + b_{i+1})$.

Multiplication Each P_i calculates $a_i b_j$ and distributes it again by masking random values in the distrubution to calculate the share of $ab = (a_0 + a_1 + a_2)(b_0 + b_1 + b_2)$.

The protocol achieves arithmetical/logical operation by using the above ADD/MUL with verification step.

15.3.2 Highly functional encryption

In modern elliptic curve cryptography with pairing-based cryptography research, the idea of encryption and decryption keys has been changed. In the context of this study, a functional encryption scheme allows a secret key holder to learn the output of a program on a specific input without learning anything else about the program [11].

One example of functional encryption is searchable encryption [3], which enables a keyword search over encrypted documents. Using searchable encryption, primary data holders deposit encrypted documents on the cloud. Data analysts (searchers), who are allowed to use this database by the primary data holders, search documents with encrypted keywords they wish to search. The cloud matches keywords and documents is an encrypted way and returns matched documents; thus, a database in which no any information is leaked to cloud can be designed (Figure 15.3).

15.4 Output Privacy

The techniques discussed in Section 15.3 are used to conduct data analysis with personal data (input) and release the analysis results (output) without disclosing the information publicly. The major privacy concern of these techniques is the disclosure of the input, whereas what an attacker can infer about the input from the output is not the issue. However, as illustrated by the example in Section 15.4.1, inference of personal data from the output can cause privacy invasion. Differential privacy is a recent definition of privacy motivated by the cryptographic notion of security [4]. Section 15.4.2 introduces the problem of output privacy, namely, differential privacy, which is becoming the *de facto* standard for the output privacy.

15.4.1 Information leakage caused by outputs

Let us consider a database constructed for human resource management at company X. Suppose an employee issues a query to ask "the average salary of employees who joined the company in 201X" and obtains the result "75320 USD." The problem of output privacy involves to what extent does the response help the employee learn about the personal information (in this case, the salary of individual employees).

On the one hand, attackers who do not have any background knowledge about the employees learn nothing more than the query response itself. This thus does not appear to be privacy invasive. On the other hand, what happens if the company hired only three employees (say, A,

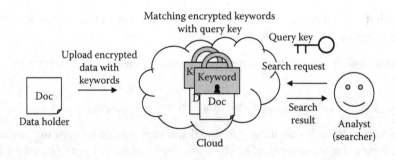

Figure 15.3: Keyword searchable encryption.

B, and C) in 201X, and the attacker is one of them? A is apparently aware of her salary, and she learns that the sum of the salaries of B and C amounts to $x_B + x_c = 38550 \times 3 - x_A$. The disclosure of the response thus causes a certain type of private information leakage. What happens if the company hired 100 employees in 201X? Is the max query more privacy invasive than the average query? Intuitively, the larger the database size is, the less the query response becomes privacy invasive. The max query seems to be more privacy invasive than the average query. Statistical inference caused by disclosure of statistical queries is highly dependent on the size of the database, query type, and background knowledge of attackers. Differential privacy provides a way to quantify how much information leakage is caused by noninteractive query answering.

15.4.2 Differential privacy

For the definition of privacy, we consider the following abstract notion of databases. Suppose a database is a collection of n records, and each record corresponds to information taken from a single individual. The record consists of a single attribute. Let $x_i \in \mathcal{D}$ be the value of the ith individual, where $\mathcal{D}$ is the domain of the attribute. The database is denoted by $X = \{x_1, x_2, ..., x_n\}$. Unless specifically mentioned, we suppose the attribute is numerical $x_i \in \mathbb{R}^m$. In this framework, a user issues a statistical query $f : \mathbb{R}^n \mapsto \mathbb{R}$, for example, average, variance, min, and max, and the database discloses the statistics upon requests.

We can intuitively interpret differential privacy as statistical obfuscation of the membership of individuals. If the outputs of a specific query function are not largely affected by whether the database contains information associated with a single individual, disclosing the outputs would not be that privacy invasive. The formal definition of differential privacy is given as follows:

Definition 15.1 (Differential privacy) Let $\mathcal{M} : \mathbb{R} \mapsto \mathbb{R}$ be a randomization mechanism. For any pair of neighbor databases D_1 and D_2 and any subset of the output domain $S \subseteq \mathbb{R}$, if

$$\frac{Pr[\mathcal{M}(f(D_1)) \in S]}{Pr[\mathcal{M}(f(D_2)) \in S]} \leq \exp(\varepsilon), \tag{15.7}$$

the output of $\mathcal{M}$ is ε-differential privacy with respect to query f.

Let us explain the definition of differential privacy with the example of a human resource database. We say two databases are *neighbors* if all but one record in the two databases are exactly the same. For example, if D_1 contains a record of employee A and D_2 does not, and the other records are all the same, D_1 and D_2 are neighboring. f is a statistical query, say, average. $\mathcal{M}$ works as a function that randomizes the output of query f. The numerator (resp. denominator) of the left-hand side is the probability with which the output of the mechanism taking input as D_1 (resp. D_2) becomes S. Figure 15.4 illustrates the probability density of the mechanism output that takes as input D_1 (resp. D_2). This corresponds to the process described in the left (resp. right) of Figure 15.4). Equation 15.7 indicates that the probability with which the mechanism outputs an arbitrary value is upper bounded by a function of ϵ. The idea behind differential privacy is that the privacy of individual records is protected if the probability (density) ratio of the mechanism is bounded by $\exp(\epsilon)$ for any pair of neighboring databases.

15.4.3 Laplace mechanism

The Laplace mechanism is one of the most fundamental mechanisms that guarantee differential privacy for queries that output numerical values. The behavior of the Laplace mechanism $\mathcal{M}$ is

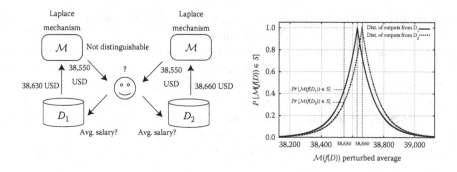

Figure 15.4: (Left) Query responses from database D_1 and D_2 through a Laplace mechanism; (Right) distributions of responses from database D_1 and D_2 through a Laplace mechanism, in which D_1 and D_2 replies 386.3 and 386.6 as the average salary, respectively.

abstractly described for any query f with the *sensitivity* of query f. Given a sensitivity defined for a specific query, the randomization mechanism that guarantees differential privacy for the query is derived. We first introduce the definition of the sensitivity of query f.

Definition 15.2 (Sensitivity) $\Delta_f = \max_{X \sim X'} |f(X) - f(X')|_1$.

Let $X \sim X'$ denote neighboring databases, that is, all the records contained in X and X' are the same except for a single record. Without loss of generality, we can assume that $x_n \neq x'_n$ and $x_i = x'_i$ for $i \neq n$. Then, following Definition 15.2, we derive the sensitivity for typical statistics as follows:

■ Sensitivity of $\text{ave}(X) = \frac{1}{n}\sum_i x_i$

$$\Delta_{\text{ave}} = \max_{X \sim X'} |\text{ave}(X) - \text{ave}(X')|_1 = \max_{X \sim X'} \frac{1}{n}|\sum_i x_i - \sum_i x_i| = \frac{1}{n}|x_n - x'_n|_1 \leq 1/n.$$

(15.8)

■ Sensitivity of $\max(X) = \max_i x_i$

$$\Delta_{\max} = \max_{X \sim X'} |\max(X) - \max(X')|_1 = \max_{X \sim X'} |\max_{i \in \{1,\dots,n\}} x_i - \max_{i \in \{1,\dots,n\}} x'_i| \leq |1 - 0|_1 = 1.$$

(15.9)

When the range of the average query and max query is $[0,1]$, the sensitivity of the average query is $1/n$. Meanwhile, the sensitivity of the max query is 1, and this is relatively larger. For example, let the database be given as $x_1 = x_2 = \cdots = x_n = 0$ and its neighbor be given as $x_n = 1$. Then, the output of the average query changes by $1/n$, while that of the max query changes by 1. As indicated by the example, the sensitivity becomes large when a change in a (not necessarily specific) single record can change the query output significantly. Disclosure of query outputs with a larger sensitivity is not preferable in terms of privacy preservation, because it apparently reveals more information about the membership of an individual. The randomization mechanism should thus give a larger perturbation to the outputs of queries with larger sensitivity. With this observation, given privacy parameters ϵ and Δ_f and the sensitivity of query f, the Laplace mechanism gives a systematic way to guarantee ϵ-differential privacy for query f.

Theorem 15.1 (Laplace mechanism)
If $R = \Delta_f/\varepsilon$, $f(X) + r$ guarantees ε-differential privacy, where $r \sim Lap(|x|/R)$.

Here, the Laplace distribution is defined by

$$\text{Lap}\left(\frac{x}{R}\right) = \frac{1}{2R}e^{-\frac{|x|}{R}}. \tag{15.10}$$

The theorem states that ε differential privacy for query f is achieved by disclosing randomized output $f(X) + r$ instead of $f(X)$ [4]. The mechanism randomizes the outputs with noises with a larger variance when the sensitivity is larger,hich indicates that a larger perturbation is needed forueries withrger sensitivities. At the same time, the mechanism gives a larger perturbation with smaller ε (i.e., stronger privacy protection).

Figure 15.4 (left) is an example of query answering to databases D_1 and D_2 through the Laplace mechanism. Suppose the true response to the average salary query from D_1 is 38630 and that from D_2 is 38660. The figure illustrates that the randomized responses from D_1 and D_2 were both 38550. Figure 15.4 (right) represents the distribution of the randomized outputs from D_1 and D_2. If, for any output value v, the ratio of the probability with which D_1 outputs v to the probability with which D_2 outputs v is less than $\exp(\varepsilon)$, the mechanism guarantees ε differential privacy.

Differential privacy limits what can be inferred from randomized outputs by adversaries with any background knowledge. In this section, we discussed differential privacy for statistical aggregation functions, for example, average and max, which are major applications of differential privacy. In recent studies, differential privacy for more complex statistical aggregation has been extensively investigated, including contingency table [18], certain properties of graphs [19], and predictors of machine learning [20]. For practical applications of privacy preserving information disclosure, differentially private management of dynamically changing data and differentially private interactive query answering are important, and these have remained as areas of future work.

15.5   Personal Data Outsourcing

Personal data outsourcing models have been described in this section. We have discussed privacy-enhancing technologies for two risks in the cloud.

- The cloud and clients do not learn private information contained in the data collection from data provided by the primary data holder (input privacy).

- The client does not learn private information contained in the data collection from data analysis results provided by the cloud (output privacy).

Literally, personal data has been processed privately, and outsourcing/sharing it n the cloud can introduce new problems. However, there are several reasons for personal primary data holders to start using the cloud. One is that as personal data grow bigger, storage and process costs go higher. Another is that expectations for personal data analysis become greater for new businesses for which higher knowledge and technologies are required. Even if personal data outsource has risks, there is business rationality in letting the cloud analyze large personal data.

15.5.1 Single data holder model

First, a simple model for one entity is discussed. The entity is both the primary data holder and the data analyst (Figure 15.5 left). In this model, if the data holder analyzes data in-house, there is no risk of privacy. However, if the cost of storing and processing surpasses the risk of outsourcing for the target personal data, cloud outsourcing is rational. Assumed risks include leak and abuse through improper operation in the cloud. The worth of outsourcing PPDM is to limit the risks of leaks and the abuse of personal data in the cloud.

The second model has two entities, where the data holder and data analyst are non-identical (Figure 15.5, center). This is similar to the first model, but the cloud works as the outsourcer and the mediator. The goals for this model are as follows:

- Data analyst ideally learns only the analysis result of the personal data.

- Cloud ideally learns nothing about the personal data of the data holder.

Note that in Figure 15.5 (left), both the data holder and data analyst are identical that only second goal is required. To achieve this, anonymization, randomization, and encryption are suitable. Anonymization allows the cloud to learn some information of the data, and data can be inaccurate, but they can outsource cloud any analysis. Randomization allows the cloud to learn some information of the data as well, but data accuracy is evaluated statistically, and there can be some analysis limitation. Encryption provides accurate analysis for the cloud, but there is limitation about the analysis and performance.

In the given scenario; see Figure 15.5 (middle), the data analyst is not the data holder; thus, output privacy should be thought about. In this case, differential privacy can play an important role.

15.5.2 Multiple data holder model

The third model is for multiple data holders. In a big data context, data mash-up is believed to bring new values, and it is true for personal data as well, such as the joining or union of data from different sources. If they are open data, there is no problem, but most personal data is private. The issue of sharing these private data over multiple primary data holders has

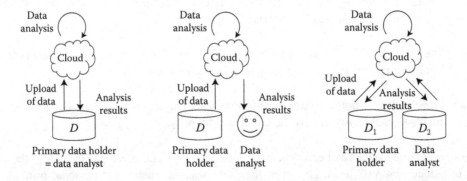

Figure 15.5: (Left) Outsourcing PPDM where a single information owner exists and the information owner is an identical entity to the data analyst, (Center) outsourcing PPDM where a single information owner exists and the information owner is different from the data analyst, and (Right) Outsourcing PPDM where two or more information owners exist.

been studied as a good example of secure computation protocols. Although the multiparty computing model requires each primary data holder to work as a data handler equally, the cloud outsourcing model may give more flexibility. For example, the cloud outsourcing model does not expect each holder to be online. The goals for this model are as follows:

1. Data analyst ideally learns only the analysis result of the personal data and nothing about the other party's information.

2. Cloud ideally learns nothing about the personal data of any data holders.

Anonymization is the most suitable method for outsourcing, but it is difficult to satisfy both analysis accuracy and anonymity, such as joining two pieces of anonymized data precisely. If stronger privacy is required, the encryption approach is preferred. Overcoming the challenges with deploying homomorphic encryption or secure multiparty computation is important with performance and online/offline drawbacks in mind.

15.6 Summary

In this chapter, we discussed the utilization of personal data in the cloud and its risks. The notion of privacy varies depending on the context of utilization; different stakeholders have different opinions about privacy preservation and privacy invasion. Privacy protection is essential in the utilization of big data containing personal data, whereas personal data can be important resources for the sophistication and personalization of services. We introduced technologies for privacy protection to control privacy risks with secure outsourcing of data analysis in terms of input privacy and output privacy. The privacy-enhancing technologies introduced in this chapter were designed assuming simplified utilization models. However, a lot more stakeholders participate in cloud services, so the utilization model would be highly complicated in real online services. Privacy-enhancing technologies that can be employed in more complicated utilization models with more flexibility remain as an area of future work.

References

1. Zvika Brakerski, Craig Gentry, and Vinod Vaikuntanathan. (leveled) fully homomorphic encryption without bootstrapping. In *Proceedings of the 3rd Innovations in Theoretical Computer Science Conference*, pp. 309–325. ACM, Cambridge, MA, 2012.

2. Koji Chida, Katsumi Takahashi, and Dai Ikarashi. A probabilistic extension of k-anonymity. In *Proceedings of Computer Security Symposium 2009*, pp. 1–6. IPSJ, 2011.

3. Giovanni Di Crescenzo, Rafail Ostrovsky, Dan Boneh, and Giuseppe Persiano. Public key encryption with keyword search. In *Proceedings of EUROCRYPT*, pp. 506–522. IACR, Interlaken, Switzerland, 2004.

4. Cynthia Dwork, Frank McSherry, Kobbi Nissim, and Adam Smith. Calibrating noise to sensitivity in private data analysis. In *Theory of Cryptography*, Shai Halevi and Tal Rabin (eds.), pp. 265–284. Springer, Germany, 2006.

5. Taher ElGamal. A public key cryptosystem and a signature scheme based on discrete logarithms. In *Advances in Cryptology*, George Robert Blakley and David Chaum (eds.), pp. 10–18. Springer, Germany, 1985.

6. Craig Gentry et al. Fully homomorphic encryption using ideal lattices. In *STOC*, Bethesda, MD, volume 9, pp. 169–178, 2009.

7. D Kifer, J Gehrke, A Machanavajjhala, and M Venkitasubramaniam. ℓ-diversity: Privacy beyond *k*-anonymity. In *ACM Transactions on Knowledge Discovery from Data*, volume 1, article no.3. ACM, New York, 2007.

8. Kristen LeFevre, David J. DeWitt, and Raghu Ramakrishnan. Incognito: Efficient full-domain k-anonymity In *Proceedings of the 2005 ACM SIGMOD International Conference on Management of Data*, pp. 49–60. ACM, New York, 2005.

9. Yehuda Lindell and Benny Pinkas. Privacy preserving data mining. In *Advances in CryptologyCRYPTO 2000*, pp. 36–54. Springer, 2000.

10. Valeria Nikolaenko, Udi Weinsberg, Sotiris Ioannidis, Marc Joye, Dan Boneh, and Nina Taft. Privacy-preserving ridge regression on hundreds of millions of records. In *2013 IEEE Symposium on Security and Privacy (SP)*, pp. 334–348. IEEE, San Francisco, CA, 2013.

11. Tatsuaki Okamoto and Katsuyuki Takashima. Fully secure functional encryption with general relations. In *Proceedings of CRYPTO*, pp. 191–208. IACR, Santa Barbara, CA, 2010.

12. Pascal Paillier. Public-key cryptosystems based on composite degree residuosity classes. In *Advances in CryptologyEUROCRYPT99*, Jacques Stern (ed.), pp. 223–238. Springer, Germany, 1999.

13. Jun Sakuma and Shigenobu Kobayashi. Large-scale k-means clustering with user-centric privacy-preservation. *Knowledge and Information Systems*, 25(2):253–279, 2010.

14. L Sweeney. *k*-anonymity: A model for protecting privacy. *International Journal of Uncertainty, Fuzziness and Knowledge-Based Systems*, 10(5):557–570, 2002.

15. Marten Van Dijk, Craig Gentry, Shai Halevi, and Vinod Vaikuntanathan. Fully homomorphic encryption over the integers. In *Advances in Cryptology–EUROCRYPT 2010*, Henri Gilbert (ed.), pp. 24–43. Springer, Germany, 2010.

16. Andrew Chi-Chih Yao. How to generate and exchange secrets. In *27th Annual Symposium on Foundations of Computer Science*, pp. 162–167. IEEE, Toronto, Canada, 1986.

17. Rakesh Agrawal and Ramakrishnan Srikant. Privacy preserving data mining. In *Proceedings of SIGMOD 2000*, pp. 439–450. ACM, New York, 2000.

18. B Barak, K Chaudhuri, C Dwork, S Kale, F McSherry, and K Talwar. Privacy, accuracy, and consistency too: A holistic solution to contingency table release. In *Proceedings of the twenty-sixth ACM SIGMOD-SIGACT-SIGART symposium on Principles of database systems*, pp. 273–282. ACM, Beijing, China, 2007.

19. V Karwa, S Raskhodnikova, A Smith, and G Yaroslavtsev. Private analysis of graph structure. In *Proceedings of the VLDB Endowment*, Seattle, WA, volume 4, pp. 1146–1157, 2011.

20. K Chaudhuri, C Monteleoni, and A.D. Sarwate. Differentially private empirical risk minimization. *The Journal of Machine Learning Research*, 12:1069–1109, 2011.

Chapter 16

Composite Big Data Modeling for Security Analytics

Yuh-Jong Hu

Wen-Yu Liu

CONTENTS

16.1 Introduction

A report from McKinsey Global Institute indicates that a success of leveraging big data analytics will be the driver for the next big wave of innovation (Manyika et al. 2011). Two computer science foundations are playing pivotal roles in the success of big data analytics. One is machine learning for robust analytics, and the other is distributed systems for efficient data storage and computing.

Big data analytics has become one of the emerging research frontier areas for innovation, competition, and productivity (Manyika et al. 2011). The key aspects that are needed to establish a big data infrastructure are data, compute infrastructure, storage infrastructure, analytics, visualization, and security and privacy. The issues of storing, computing, security and privacy, and analytics are all magnified by the velocity, volume, and variety of big data (Murthy et al. 2014).

In this chapter, we argue why we use a composite big data modeling technique for security analytics on the emerging high-performance Berkeley Data Analytics Stack (BDAS) Spark platform. This approach is based on a well-established structured machine learning that provides structured modeling and learning analytics for big data (Domingos 2012).

Structured machine learning (or inductive logic programming [ILP]) combines two important fields of artificial intelligence for top-down deductive reasoning and bottom-up inductive reasoning. The field of structured machine learning has been developed over more than three decades and now is marching toward adulthood. It is still making steady progress (Muggleton et al. 2012).

Structured machine learning refers to learning a structured hypothesis from data with rich internal structure. Data might include structured inputs as well as outputs—parts of which may be uncertain, noisy, or missing (Dietterich et al. 2008). Applying structured machine learning for big data analytics is a promising research area but there are still many challenges ahead.

When applying structured machine learning for big data analytics, a learner is provided with a domain ontology that explains possible behaviors of (classification) type-labeled training data. Therefore the desired outputs of a hypothesis should be consistent with both type-labeled training data and domain ontology for unknown function approximations.

Structured knowledge is represented as ontologies, and training data are queried from ontologies by rules to empower the inductive learning analytics. In general, an ontology can be encoded by a description logic (DL)-based ontology language such as OWL2 and a rule can be encoded by a logic program (LP)-based rule language, such as RuleML (Eiter et al. 2008). Both ontology and rule languages are machine readable with no semantic ambiguity.

In this chapter, we use RDF(S) graph-based ontology language and SPARQL ontology query language instead of using OWL2 and RuleML for ontology modeling and query (Brickley and Guha 2014; Harris and Seaborne 2014). Although the expressive power of graph-based ontologies and queries is weaker than logic-based ones, the RDF(S) linking open data

(LOD) cloud* can provide big LOD datasets in a lightweight JSON-LD format for learning analytical processing (Spomy et al. 2013).

In a pure inductive learning with no prior domain knowledge, we formulate a hypothesis by finding the empirical regularities over the training dataset, which verifies the robustness of a hypothesis by computing its testing dataset error. On the other hand, in a pure structured modeling with few or no dataset available, we use prior domain knowledge to construct an ontology and enable a query deductively by checking the instance satisfiable. Composite big data modeling and analytics tightly integrate inductive reasoning of machine learning with deductive reasoning of structured modeling to complement the inferencing capabilities of each other.

When prior domain knowledge is available at the initial stage of data analytics, we can boost feature selection of inductive learning and possibly speed up the hypothesis search in vast hypothesis space. Moreover, if we have large labeled dataset available to compute its training and testing errors, this overcomes the shortcomings of applying incomplete and uncertain prior knowledge needed to build an ontology.

Ontology learning improves a graph ontology by deriving implicit new and discarding unnecessary old relationships. Therefore, the revised ontology is more appropriate reflecting features association or cause–effect relationships discovered in the big data (Lehmann et al. 2014). Rule learning is closely related to ontology learning because a rule enforces approximate ontology query by using predicate vocabularies declared in the ontology with filtering conditions (Fürnkranz et al. 2012). This deductive reasoning unfolds the positive and negative training data and prepares for a leaner's inductive reasoning.

Composite big data analytics solves a cyberspace security intrusion detection use case. Based on a previous big data analytics pipeline (Labrinidis et al. 2012), we propose a revised data analytics and modeling process with seven consecutive steps including acquisition and recording, cleaning and extracting, aggregation and semantic annotation, query and analytics, modeling and verification, reactive and proactive actions, and intrusions detection and intrusion alerts (see Figure 16.1).

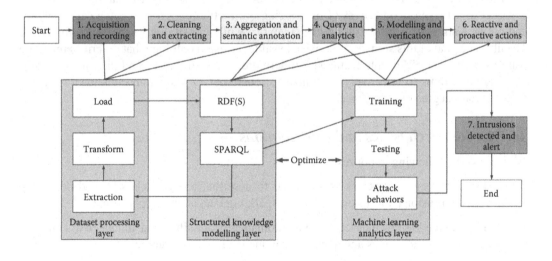

Figure 16.1: A composite big data analytics and modeling consists of seven consecutive steps or processes to detect intelligent security intrusion.

---

*LOD: The Linking Open Data cloud diagram.

We have established a three-layer architecture corresponding to the above composite big data modeling and analytics pipeline: (1) dataset processing (DAP) layer, (2) structured knowledge modeling (SKM) layer, and (3) machine learning and analytics (MLA) layer. First, we provide feature extraction, transform, and load (ETL) services on the DAP layer to enable intrusion patterns discovery on the SKM layer. Then, the SKM layer is shown as an ontology and rule knowledge integration system, where the RDF(S) ontology language is used to construct the intrusion ontologies with machine-understandable syntax.

RDF(S) ontologies first provide a simple intrusion behavior taxonomy with feature inheritance to discover additional complex intrusion behaviors. Moreover, when combined with RDF(S) ontology learning, SPARQL queries of RDF(S) ontologies with rule learning provide training datasets for the MLA layer, which verifies the robustness of the SKM layer and discovers possibly evolving unknown intrusion patterns.

Third, in the MLA layer, we first apply the decision tree algorithm for supervised learning using complete prior domain knowledge with correct classification type-labeled data assumption. This supervised learning in the decision tree provides an optimal performance benchmark for other semi-supervised learning algorithms to justify their robustness in terms of training and testing errors. We apply two well-known machine learning algorithms, naïve Bayes and Bayesian network, for semi-supervised learning with incomplete prior domain knowledge to classify type-unlabeled data in the search for a structured hypothesis.

We are aware that the semi-supervised learning is more appropriate to classify a large number of type-unlabeled data with only incomplete domain knowledge and noisy datasets available. A selected machine learning algorithm first computes a training error by using a type-labeled data queried from current ontologies. The same machine learning algorithm computes a testing dataset error by using a set of type-unlabeled data to verify the robustness of current ontologies. The revised ontologies provide a new set of type-labeled training data for classifying type-unlabeled testing data through rule learning. Eventually, this structured machine learning iterative process is halted if the training and testing errors reach a predefined threshold.

In the cyberspace intrusion detection scenario, a hypothesis is constructed and tested for a learner to achieve the objectives of minimizing offline training error in batch data analytics and online testing error in streaming data analytics while classifying various intrusion types. The ultimate goal of this study is to derive a feasible intrusion ontology that describes intrusion patterns with a set of feasible features. Furthermore, our goal was to find plausible machine learning algorithms to effectively recognize various types of known and unknown intrusion with a certain degree of confidence.

16.1.1 Research issues and contributions

We address the following major research issues:

1. How can security features be extracted to describe, analyze, classify, and detect the misuse intrusion patterns on the BDAS Spark big data analytics platform?

2. How can the big raw datasets be semiautomatically classified and annotated to empower the evolving ontologies and queries for inductive learning analytics?

3. How can structured machine learning, as semantics-enabled (semi-)supervised learning, be established-leverage the power of top-down deductive reasoning and bottom-up inductive reasoning for perfect and imperfect domain knowledge?

4. How can the semantics-enabled (semi-)supervised learning be enforced as a combination of ontologies and queries (or rules) with a specific machine learning algorithm to detect the known and unknown intrusion detection patterns?

This chapter makes the following contributions. We show how structured machine learning is useful for solving one of the important big data analytics and modeling problems. More specifically: (1) Under perfect and imperfect domain knowledge assumptions, we show how the composite big data modeling, as a combination of RDF(S) ontologies and SPARQL queries, is useful for learning analytics. (2) Semantics-enabled (semi-)supervised learning is proposed on solving a misuse intrusion detection problem. (3) Based on solving an intrusion detection problem experiences, we point out how the semantics-enabled (semi-)supervised learning can be extended further to other general big data analytics and modeling problems.

This chapter is organized as follows. In Section 16.1 we give an introduction. In Section 16.2 we provide background knowledge. In Section 16.3 we address related work. Then top-down SKM concepts are presented in Section 16.4. In Section 16.5, we present bottom-up machine learning analytics concepts. We propose the semantics-enabled (semi-)supervised learning for big data analytics in Section 16.6. In Section 16.7, we present the composite big data analytics in a three-layer architecture. An intelligent security use-case scenario is demonstrated in Section 16.8. Finally, in Section 16.9, we conclude this chapter and point out possible future work.

16.2 Background

16.2.1 Structured machine learning

Machine learning has been developing for at least three decades before big data analytics emerged. Machine learning sometimes is referred to as data mining or statistical learning. From a computer science viewpoint, machine learning focuses on data clustering, classification, and prediction accuracy, while data mining focuses on the discovery of meaningful patterns. As statistical learning, it focuses on interpretable parameters (Hastie et al. 2013; James et al. 2013).

Structured machine learning has made steady progress since the first ILP workshop in 1991 (Muggleton et al. 2012). Since then, structured machine learning has been an active research field (Dietterich et al. 2008). The other extended fields of ILP are Markov logic networks, statistical relational learning (SRL), and (probabilistic) logical learning (Domingos and Lowd 2009; Getoor and Taskar 2007; Raedt 2008).

16.2.2 Big data analytics platforms

The first pioneer big data analytics platform was Apache Hadoop ecosystem with MapReduce programming paradigm (Agneeswaran 2014; Singh and Reddy 2014). Hadoop was originally designed for batch but not streaming data analytics. In fact, MapReduce is more suitable for lightweight but not heavyweight data analytics. However, lightweight data analytics cannot derive entity implicit correlations or cause–effect relationships in the big datasets. We need machine learning algorithms to provide heavyweight data analytics. BDAS Spark provides an integrated platform for batch and streaming heavyweight data analytics with high-performance in-memory resilient distributed datasets (RDD) processing (see Figure 16.2).

An RDD is created from/to the Hadoop Distributed File System (HDFS) that can be retained in main memory across the machine learning iterations, leading to significant performance boosting. In addition, heavyweight data analytics can be achieved by calling machine learning libraries MLlib or the emerging SparkR.* The BDAS Spark platform has been used in the composite big data learning analytics to enable heavyweight data analytics for intelligent security.

---

* SparkR: R on Spark.

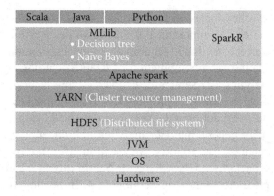

Figure 16.2: A Spark platform for the composite big data analytics model.

16.2.3 A hybrid intrusion detection model

A *misuse* intrusion detection model discovers attacks based on the features extracted from known intrusion patterns (Anderson et al. 1995). The misuse intrusion detection operation is operated in near real time with a low false-positive intrusion detection error rate. If this misuse detection system lacks the new intrusion signatures in its database, unless it uses the machine learning algorithms, it is almost impossible to detect any new or novel attacks. On the contrary, in an *anomaly* detection model, it assumes the normal activities are modeled and identified in the datasets. Whenever an activity deviates from pre-defined, normal activities, it will be recognized as an anomaly behavior. But an anomaly intrusion detection model has a high false-positive error rate (Chandola et al. 2009).

Supervised learning methods for anomaly detection depend on attack-free training datasets to model normal activities. However, it is difficult to obtain the attack-free training datasets to describe normal activities. On the contrary, an unsupervised anomaly detection model does not need attack-free training datasets. The majority of activities are normal and only a few of outliers are considered anomaly behaviors that deviate from the majority of behaviors in the datasets. A *hybrid* intrusion detection model is a better option because it combines the advantages of misuse and anomaly detection models.

16.3 Related Work

The deductive or inductive learning systems with perfect domain knowledge, including PROLOG-EBG explanation-based learning algorithm and inverted deduction ILP, were proposed two decades ago (Mitchell 1997). PROLOG-EBG is a deductive learning system (DLS), which assumes that the domain knowledge is correct and complete. PROLOG-EBG uses its domain theory to *reduce* the acceptable hypotheses. In contrast, inverted deduction ILP uses its background knowledge to *enlarge* the set of hypotheses so that it becomes an inductive learning system.

However, if training datasets and domain knowledge contain errors, and thus we only have imperfect domain knowledge and datasets available. Finding the best structured hypothesis to classify new types of unlabeled data involves solving an optimization problem. Thus, a structured hypothesis that best fits the training dataset and domain knowledge should minimize some combined measure errors when using a dataset and domain ontology.

First-order inductive logic (FOIL) learns first-order rules similar to Horn clauses. FOIL performs a *general-to-specific* search. At each step, it adds a single new literal to the rule preconditions. The extended first-order combined learner (FOCL) allows rule learning in an imperfect domain knowledge and dataset assumption. Single rule and ruleset learning attempts to cover as many types of positive data as possible by introducing new preconditions for each rule (Fürnkranz et al. 2012). Appropriate rule preconditions are discovered by querying a revised ontology that provides an updated training dataset for types of intrusion patterns. This enhances the learning analytics.

Markov logic is a language that combines first-order logic and Markov networks' undirected graphs (Domingos and Lowd 2009). A knowledge-based (KB) in Markov logic is a set of first-order rules with weights. In first-order logic, rules are hard constraints such that not even a single rule can be violated. In Markov logic, rules are soft constraints. A world that violates a rule is less probable than the one that satisfies it. The features and weight of each rule can be manually assigned or be learned from the data. Therefore, the features and a weight of a rule represents its strength as a constraint for a world.

Comprehensive intrusion detection models were first proposed in 1987 (Denning 1987). Applying machine learning methods for intrusion detection has become an important research area for computer security analysis (Joseph et al. 2012; Maloof 2006; Sommer and Paxson 2010). Moreover, using big data analytics to enable intelligent security is another emerging research area to explore for detecting novel new attacks (Cárdenas et al. 2013).

For example, Beehive provides large-scale log analysis for detecting suspicious activity in enterprise network (Yen et al. 2013). In François et al. (2011), BotTrack can track botnets using NetFlow and PageRank techniques. Advanced persistent threat (APT) attacks are identified using unfolded attack pyramid techniques in large-scale distributed computing (Giur and Wang 2012). In fact, APT attack is a type of zero-day attack (Bilge and Dumitras 2012). We are the first to use structured machine learning from the Semantic Web perspective for intrusion detection whereas others only use signature-based or pure machine learning techniques (Zhang et al. 2008).

16.4 Top-Down SKM

16.4.1 *Deductive reasoning*

Reasoning with large datasets was extensively studied in the field of deductive databases. The deductive reasoning techniques are classified as two parts for query answering in the structured knowledge systems. One is from the DL-based OWL ontology viewpoint. This approach provides reasoning tasks for knowledge base and concept satisfiability. This computes the subsumption hierarchy and answers conjunctive queries.

The other is from the LP (DL) Datalog rule (or query) viewpoint. This provides sound and complete query services in the ontology, which enhances the reasoning capability of the ontology reasoner. For example, the Semantic Web Rule Language (SWRL), which combines OWL and RuleML, allows users to write Horn-like rules expressed in terms of OWL concepts to reason about OWL individuals. Thus, the deductive reasoning of the rules can be used to infer new knowledge from existing OWL knowledge bases (Eiter et al. 2008).

We used RDF(S) graph-based ontologies to describe the concepts of intrusion patterns and use SPARQL query services similar to the rule's deductive reasoning to discover the *class* and *property* subsumption hierarchy of the RDF(S) schema and *subject–predicate–object* triples in the RDF(S) triple stores. The deductive reasoning of an RDF(S) ontology query provides a type-labeled training dataset for a machine learning analytics algorithm to enforce its inductive reasoning.

16.4.2   Ontology learning

Ontology learning is a structured knowledge concept refinement process to create, revise, and update established ontology schema for its evolution to catch existing and newly included data relationships in a training dataset. If we can import other experts' domain knowledge, then ontology learning becomes a knowledge fusion and integration process through ontology matching, alignment, and merging (Lehmann et al. 2014). Ideally, we need (semi-)automated ontology learning to revise and update ontologies.

We can discover new features and properties after new training and testing of dataset errors are generated from previous analytical learning stages. This ontology learning approach is different from the traditional approaches that provide (semi-)automatic concepts and property taxonomy via a natural language processing (NLP) technique (Lehmann et al. 2014).

16.4.3   Rule learning

Rule learning in an instance-type classification is defined as finding a set of rules used for classifying a type of testing data given a set of training data. In fact, the rule learning concept corresponds to data classification by a decision tree. Each leaf in a decision tree denotes an instance-type classification and each internal node indicates a feature to be evaluated for branching out of the subtree. The selection of a feature order to be evaluated is based on information gain by computing entropy.

A classification rule is represented as a Datalog, e.g., *If body then head*, where a *body* contains a conjunction of conditions, and each condition is a feature satisfaction constraint. A *head* contains a prediction with a classification-type or implicit concepts and properties verified by learning analytics. Moreover, Datalog can be used for expressing and enforcing security and privacy policies (Bonatti 2011).

A rule *covers* a positive (or negative) instance if the set of features satisfy conditions of a rule. A rule's head is a type of data classification label or prediction value if a rule covers this data or prediction value. In a pure LP rule, a rule's head only covers type-positive data, so a negation-as-failure (NAF) for closed world assumption (CWA) is assumed.

When rule learning is probabilistic ILP, then a rule might cover both positive- and negative-labeled data. This avoids overfitting that only uses training dataset learning. It might also reduce testing dataset error. Single rule learning is enforced as the principle *from general to specific* to reduce the number of negative-labeled data. In contrast, a rule set learning is enforced as the principle *from specific to general* to increase the number of positive-labeled data (Fürnkranz et al. 2012).

16.4.4   From ontology learning to rule learning and vice versa

The type-labeled training dataset in an RDF(S) ontology module is imported to a rule module via the SPARQL approximate ontology queries. The query syntax can be shown as follows:

Select [an instance classification type]
from ontologies
where [a set of features condition check]

SPARQL queries provide an equivalent capability of the rules (Ceri et al. 1989). The expressions for *a set of features condition check* corresponds to a rule's *body*, and *an instance classification type* corresponds to a rule's *head*.

During rule learning, the SPARQL approximate ontology query is similar to the BlinkDB query on the Spark platform, which provides a type-labeled training dataset (Singh and Reddy 2014). The features, selection ensures that the true-positive instance classification error rates verified by the analytics module for training and testing datasets are below an acceptable threshold level in the (semi-)supervised learning. Otherwise, a new stage of ontology and rule learning is initiated to further reduce the above error rates.

The type-labeled training data with a set of selected features is sent to the learning analytics layer to compute the testing error of type-unlabeled dataset. The training and testing error rates with a set of revised features are further forwarded to the SKM layer to revise its ontology schema. In addition, the type-unlabeled data are classified through a machine learning algorithm such as the Bayesian network to enhance the target function learning accuracy.

More specifically, structured knowledge K_p is shown as ontologies O and queried via the SPARQL R for delivering type-labeled training data D_l to a learning analytic module. The desired output is a hypothesis $h_p \in H$ consistent with the domain knowledge K_p and type-labeled training data $\forall X^i \in D_l$. For each new type-unlabeled instance, $X^j \in D_u$, it is classified by the naïve Bayes and Bayesian network algorithms to increase the number of type-labeled data in structured modeling for ontology learning and rule learning.

In a semantics-enabled (semi-)supervised learning scenario, we first enforce a decision tree algorithm in a perfect domain knowledge K_p for supervised learning to obtain the training and testing errors with optimal performance benchmarks to evaluate a hypothesis $h_{ip} \in H$ with accuracy of imperfect domain knowledge via unlabeled data.

16.5 Bottom-Up Machine Learning Analytics

16.5.1 Inductive reasoning

In a supervised machine learning, inductive reasoning induces a classifier for type-labeled training data. The classifier generalizes the training data so that it can assign a class label to each new data in the testing dataset. An inductive learning algorithm searches for a member of a given family of function approximation hypotheses, including decision trees, the naïve Bayes and the Bayesian network. They optimize the given quality criteria such as the minimum of false positives and false negatives as well as maximum of true positives and true negatives.

16.5.2 Supervised versus semi-supervised learning

Let $D = (X^1, \cdots, X^n)$ be a set of n inputs. The goal of *supervised learning* is to learn a function f from $\forall X^i \in D$ to $\forall Y^i \in Y$, where $\forall Y^i$ are called the types or classes of the data $\forall X^i \in D$. Given a training set n instances, where each instance $X^i \in D$ has d features indicated as $(X_1^i, \cdots, X_d^i)$. Typically, it is assumed that the $\forall X^i$ instances are drawn *i.i.d.* (independently and identically distributed) from a common distribution on D.

This *i.i.d.* assumption sometimes fails for big datasets because various data sources are integrated from a combination of unstructured, semi-structured, and structured formats without the existence of a statistical distribution. However, if we use a generative learning expectation maximization (EM) algorithm to maximize the log likelihood of a model with hidden variables, we still must assume a probability distribution for each classification type variable such as Gaussian or multinomial distribution. Otherwise, we cannot compute each unlabeled data classification type. Moreover, it is almost impossible to pre-classify all of the big data. Therefore, in practicality, a supervised learning approach is not plausible for big data analytics.

The goal of *unsupervised learning* is only to find an interesting structure such as density, clustering, outliers, and dimensionality reduction for the unclassified data $\forall X^i \in D$. However, this approach cannot provide a specific intrusion type alert. Thus, it cannot help trigger a suitable countermeasure action. Furthermore, applying unsupervised learning methods in an anomaly detection model usually has a high false-positive rate on intrusion detection.

16.5.3 Feature selection and pattern recognition

Security features can be extracted to describe, analyze, classify, and detect the intrusion for a hybrid model on the BDAS Spark big data analytics platform. These security features can be specified in the RDF(S) ontology to describe the concepts of denial-of-service (DoS) attacks (see Figure 16.3).

First, we specify the most important security features in d dimension; for instance, $X^i := (X^i_1, ..., X^i_d)$ that should be extracted from the datasets $D = D_l \cup D_u$, where D_l is a set of

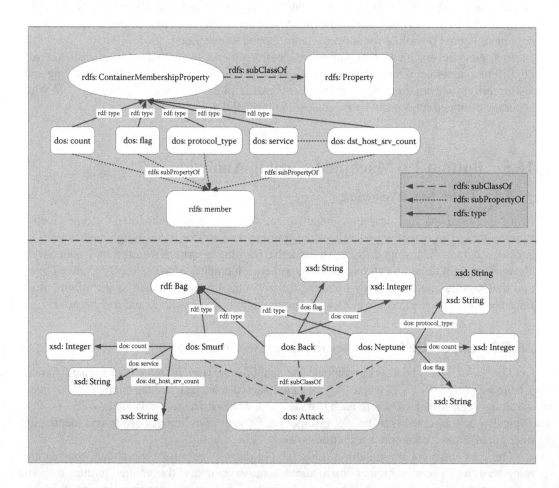

Figure 16.3: In RDF(S) ontology, we use a set of five security features for describing the concepts of different types of denial-of-service (DoS) intrusion attacks.

(classification) type-labeled data, and D_u is a set of type-unlabeled data. To model the initial security expert domain knowledge K_l, ontologies are represented in RDF(S) with nodes as basic features selected by a security expert. Each feature node is a Boolean variable, X_j^i, and a sink node is also a Boolean variable Y^i for two possible classification types: *attacked* and *not-attacked.*

Referring to a framework for constructing features and models for intrusion detection (Lee and Stolfo 2000), possible network security features for intrusion pattern recognition are protocol type, flag, packet size, source and destination IP addresses, ports, header fields, time stamps, packet inter-arrival time, session duration, session volume, and so on (Sommer and Paxson 2010). In addition, possible data types for each Boolean variable X_j^i are the selector, order, hierarchical, relational, and set-value (Fürnkranz et al. 2012). If possible, we might consider additional contextual features including *time, space,* and *sequence order* with their possible cause–effect relationships between those Boolean feature variables to describe a complete intrusion behavior that effectively recognizes intrusion types.

16.6   Semantics-Enabled Semi-Supervised Learning

Semi-supervised learning is halfway between supervised and unsupervised learning (Seeger et al. 2010). The data-set D can be divided into two parts: data $D_l := (X^1, ..., X^l)$, for which a classification type labels $Y := (Y^1, ..., Y^l)$ are provided. Data $D_u := (X^{l+1}, ..., X^{l+u})$, and the classification types are unknown. These are classification type-unlabeled data. For the past few years, the most active research area in semi-supervised learning has been on the investigating of graph-based methods for effective data classification to obtain a plausible and accurate model for data type classification, meaningful relationship recognition, and prediction. We propose supervised and semi-supervised learning methods for intrusion detection with RDF(S) semantic annotation. In this semantics-enabled (semi-)supervised learning, we address several research challenges and propose possible solutions as follows.

First, a deterministic decision tree learning algorithms was used to learn the security target function $f(X)$ from a classification type-labeled instances $X^i \in D_l$. Initially, a small set of type-labeled data D_l is created by a security expert by querying the semantic annotated RDF(S) ontologies. In the supervised learning, the security expert has a perfect domain knowledge because (s)he can examine complete classification type-labeled data.

Second, if the learning process converges, then we need to determine how many instances of the training data D_l are sufficient to learn and approximate with a certain confidence for a true target function $f(X)$. This target function is a model to describe the true intrusion behavior B_t. This implies that we should at least obtain the minimum amount of type-labeled data $X^l \in D_l$, when we stop and finalize the function approximation learning processes.

The semantics-enabled semi-supervised learning processes for feedback loop operations are shown as Figure 16.4. In the learning processes, we first apply rule learning to facilitate the approximated queries. Next, we obtain a set of type-labeled samples as a training data for the naïve Bayes and Bayesian network algorithms to classify extra type-unlabeled data. Finally, the newly type-labeled data are fed into RDF(S) ontology for another round of approximate queries and learning processes. In the meantime, the ontology based on the machine learning analytics layer feedback might be revised and evolved. A security expert might select another set of security feasible features that responds to the results of learning analytics.

The prior security domain knowledge K_p guides and generalizes from the incrementally type-classified data $x^i \in D_l$ to confidently predict a new type-unlabeled data $x^j \in D_u$ on its approximate classification type.

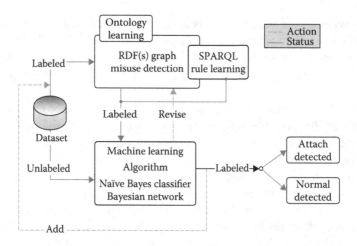

Figure 16.4: Semantics-enabled semi-supervised learning facilitates ontology learning and rule learning for a hybrid intrusion detection on the structured knowledge modeling for helping the naïve Bayes and Bayes network algorithms on the learning analytics, and vice versa.

In supervised-learning, we first assume that a learner is provided with a perfect domain knowledge K_p and a perfect data D_p. If a learner is provided with an imperfect domain knowledge K_{ip}, and an imperfect data D_{ip}, then we are dealing with an optimization problem to minimize the error rates of imperfect ontology K_{ip} and data D_{ip} on the target function approximation with a hypothesis testing in the hypothesis space (Mitchell 1997).

16.6.1 Perfect domain knowledge learning

Semantics-enabled supervised learning is first proposed for perfect domain knowledge learning. Here we assume a security expert has perfect domain knowledge on using structured features to identify intrusion patterns. A security expert can observe all learning data with their correct data type classification. We assume that the set of security features to identify intrusion patterns are fixed. Moreover, type-labeled data are clean and correct when describing the intrusion behavior for learning analytics.

The object of learning analytics is to find a hypothesis compliance with structured domain knowledge and data. This approach provides a baseline optimal performance benchmark to evaluate the accuracy of imperfect domain knowledge learning. We refer to the DLS for learning a target function shown in the book (Mitchell 1997).

16.6.1.1 Deductive learning system

$$DLS = \begin{cases} (\forall < X^i, f(X^i) > D_l)(h_p \wedge X^i \vdash f(x_i)) \\ D_l \wedge K_p \vdash h_p \\ (\forall < X^i, f(X^i >) \in D_l)(K_p \wedge X^i) \vdash f(X^i) \end{cases}$$

where:
    K_p is the perfect domain knowledge in RDF(S) ontology
    D_l is the classification type-labeled data queried from RDF(S) ontologies
    $h_p \in H$ is the structured feature of a hypothesis shown as a decision tree

f is an unknown target function
$f(X^i)$ is the target value in positive and negative data classification types
X^i is the ith type-labeled training data

A hypothesis h_p for machine learning inductive reasoning is shown as a decision tree algorithm, and the security domain knowledge is represented as RDF(S) ontologies. By using SPARQL approximate queries as rules for sampling positive/negative type that are fully observed data from the ontologies, we established a decision tree that has the minimum training and testing errors. In fact, a hypothesis $h_p \in H$ as a decision tree learning corresponds to a set of rule learning. Each rule learning applies the *from general to specific* principle to unfold the type observed positive/negative data for a path traversing in a tree. Moreover, a set of rule learning applies the *from specific to general* principle to unfold the type observed positive/negative data from a tree's total paths traversing with minimum training and testing errors.

16.6.2 *Imperfect domain knowledge learning*

Semantics-enabled semi-supervised learning is also proposed for imperfect domain knowledge learning. A domain expert only can see (classification) types that are partly observed and partly unobserved data while establishing a structured domain ontology. Imperfect ontologies provide a set of type-labeled data for learning analytics to estimate each type-unlabeled data. This is an optimization problem for minimizing the error rates of type-unlabeled data. More details are shown as follows:

$$arg \min_{h_{ip} \in H} \alpha_{D_{ip}} error_{D_{ip}}(h_{ip}) + \beta_{K_{ip}} error_{K_{ip}}(h_{ip})$$

where:
K_{ip} is the imperfect domain knowledge in RDF(S)
D_{ip} is the partly type-labeled and type-unlabeled data for hypothesis h_{ip} learning
$h_{ip} \in H$ is the hypothesis to learn through a set of features combination

The weight parameters $\alpha_{D_{ip}}$ and $\beta_{K_{ip}}$ in the objective function are tunable to indicate the importance of using data and domain expert knowledge while minimizing the error rates of semi-supervised learning analytics. The term $error_{D_{ip}}(h_{ip})$ is a ratio of data misclassified by a hypothesis h_{ip} for type-unlabeled data.

In the real experiment (see Section 16.8), we hide each instance's classification type for an unobserved data so that we can compute the testing error rate, $error_{D_{ip}}(h_{ip})$, after a specific learning algorithm computes a maximum likelihood estimate (MLE) for each type-unlabeled data. In addition, $error_{K_{ip}}(h_{ip})$ denotes the ratio of type classification discrepancies between a static ontology taxonomy and a learning algorithm's MLE classification.

In the imperfect domain knowledge learning, we first assume security features proposed by a domain expert for intrusion detection in $h_{ip} \in H$ are mutually independent. The naïve Bayes learning algorithm is used to classify a new unobserved data for its MLE value. Then we assume security features are not all mutually independent. The Bayesian network learning algorithm is used. Initially we assume the structured domain knowledge is known. The data for learning includes partly type-labeled and partly type-unlabeled variable data. By using the EM technique, we can compute the MLE value from type-unlabeled to become type-labeled data for afterward learning analytics. We are exploiting whether increasing the number of type-labeled data can help make learning analytics more robust.

16.7 Composite Data Analytics and Modeling

In the composite analytics and modeling, a structured knowledge system model is firstly established through a RDF(S)-based graph ontology. This RDF(S) ontology is represented as a set of Subject, Predicate, and Object, triples. A new knowledge base for a new data analytics and modeling domain is initiated by writing down plausible RDF(S) ontologies for learning. This is followed by induction reasoning of a machine learning algorithm that can adjust and revise this RDF(S) ontology. The rules, represented as the SPARQL query language, is to access the RDF(S) ontology, and further discover anomalous behaviors that exist in the ontology to train a learning analytics module.

The structured knowledge system provides bootstrapping capabilities to aid the learning analytics on the deficiency of initial feature recognition and extraction for anomalous pattern clustering and classification. In this composite big data modeling and analytics theory, we examine the research issues of establishing the optimal hybrid modeling and analytics processes to satisfy the accurate and effective intrusion detection criteria (see Figure 16.5).

16.7.1 DAP layer

In the DAP layer, we provide ETL processing services from diversified raw datasets. The raw dataset might include structured, semi-structured, and unstructured data formats. In principle, we can use the JSON-LD lightweight format for RDF(S) ontology-linked data. This key-value linked data format fits nicely into the big data platform for effective distributed processing (see Figure 16.6).

16.7.2 SKM layer

The SKM layer is shown as the integration of ontology and rule knowledge systems. We used RDF(S) ontology language to construct graph-based ontologies for the taxonomy of intrusion threats. In fact, RDF(S) ontologies can detect the intrusion threat patterns' similarity through security features inheritance. In addition, the SPARQL approximate query with rule learning when combined with the RDF(S) ontology learning offers additional type-labeled data for learning analytics to discover several well-known existing and evolving intrusion threats (see Figure 16.7).

16.7.3 MLA layer

In the MLA layer, while classifying feature-based anomalous patterns for possible intrusion detection, we used three appropriate machine learning algorithms, for example, a decision tree, naïve Bayes, and Bayesian network, to compute offline training errors and online testing errors. The training and testing errors from the MLA layer can help the SKM layer to revise its ontology schema through ontology learning. Following the rule learning, there is a new set of positive/negative type-labeled data through the revised ontology querying for improving learning analytics accuracy in the MLA layer. This forms a loop to enhance the structured modeling of security analytics for effective intrusion detection (see Figure 16.8).

16.7.3.1 Decision tree learning

Supervised decision tree learning provides an optimal performance benchmark of training and testing errors by using a perfect domain knowledge learning. A set of features is first selected by a security expert. In a top-down induction of a decision tree, selecting the best decision attribute

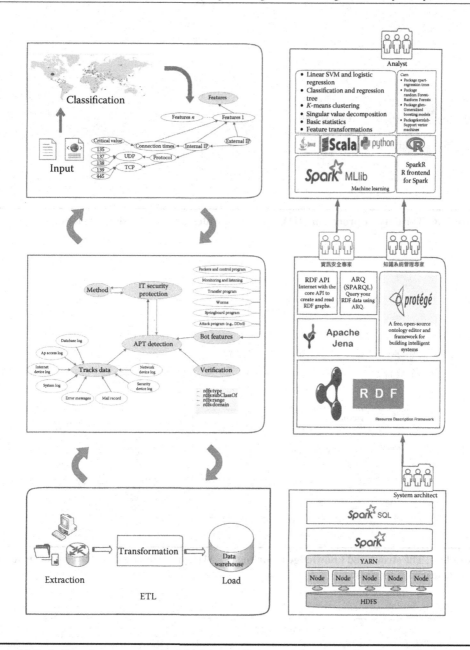

Figure 16.5: A three-layer composite big data analytics architecture including dataset processing (DAP) layer, structured knowledge modeling (SKM) layer, and a machine learning and analytics (MLA) layer.

order for each node is based on information gain. Information gain is the mutual information between input feature X^i and target variable Y. We picked the most expected reduction in entropy of target variable Y, for example, the largest information gain by sorting on feature variable vector X^i for $i = 1, ..., n$ (see Figure 16.9).

In fact, a supervised decision tree learning on the MLA layer correspond to a rule learning on the SKM layer (see Figure 16.10). A path from a root node to a leaf node is just a single rule. For each tree, an interior node is a Boolean feature variable and a leaf node is a type of

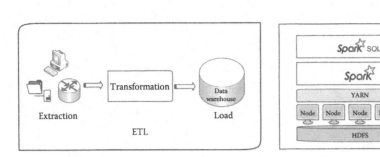

Figure 16.6: The dataset processing (DAP) layer provides ETL processing services.

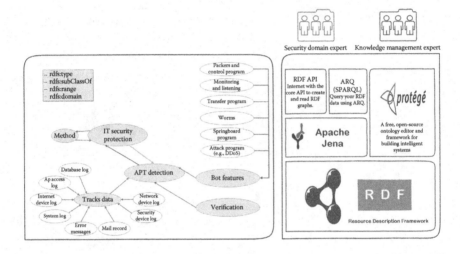

Figure 16.7: The structured knowledge modeling (SKM) layer provides deductive ontology learning and rule learning services.

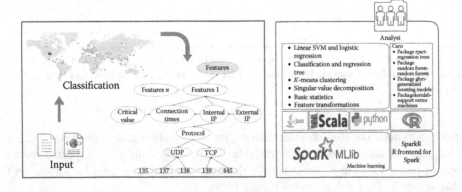

Figure 16.8: The machine learning and analytics (MLA) layer provides the inductive learning analytics services.

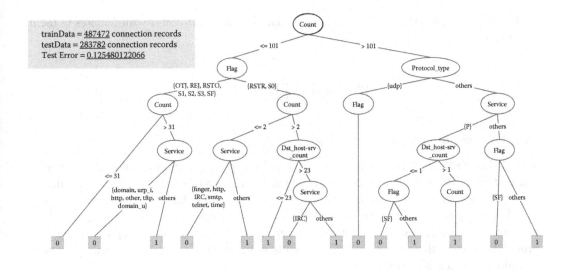

Figure 16.9: The supervised decision tree learning identified the order of five Boolean feature attributes for training and testing KDD Cup 1999 datasets.

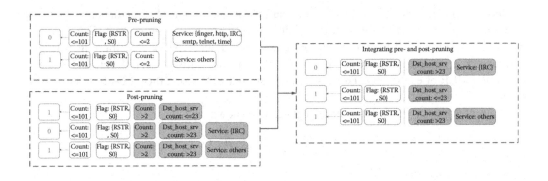

Figure 16.10: The rule learning corresponds to the supervised decision tree learning and provides a set of type-labeled data for a machine learning analytics.

Boolean classification. For example, C4.5 is the most frequently used rule learning method for converting a decision tree to a set of rules. Moreover, it prunes each rule independently and sorts final rules into desired sequence to compute its respective training and testing errors (Quinlan 1993).

16.7.3.2 Naïve Bayes learning

In the semi-supervised learning, a set of RDF(S) classified seed data are first used as prior background information for the Bayes rule to estimate the posterior probability of a new type-unlabeled instance X^u. This MLE computed value corresponds to a classification type, Y^u. In the naïve Bayes algorithm, we assume that any two security features $X_i \neq X_j$ are conditionally independent given $Y = y^k$, so

$$P(X^k) = P(X_1^k, ..., X_d^k | Y) = \prod_{i=1}^{d} P(X_i | Y)$$

A classification rule for a type-unlabeled new data is a Boolean vector with features in the d dimension,

$$X^u = < X_1^u, ..., X_d^u >$$

$$Y^u \leftarrow arg \max_{y^k} \prod_{i=1}^{d} P(X_i^u | Y = y^k)$$

16.7.3.3 Bayesian network learning

A Bayesian (or Bayes) network is a directed acyclic graph and a set of conditional probability distributions (CPDs). Each node denotes a random variable and each edge denotes dependence. The CPD for each node X_i defines $P(X_i | Pa(X_i))$, where $Pa(X_i)$ denotes immediate parents of the X_i in the graph. In fact, the naïve Bayes is a special case of the Bayes network where any two security features are conditionally independent. A Bayes network represents the joint probability distribution over a collection of random variables for a vector of security features regarding intrusion detection.

$$P(X^k) = P(X_1^k, ..., X_d^k) = \prod_{i=1}^{d} P(X_i^k | Pa(X_i^k))$$

There are four categories of Bayes network learning problems. Graph structures may be known or unknown and variable values may be fully observed or party unobserved. The easier one is a given graph with fully observed variable values. Each type-labeled training data gives a value for a random variable in the graph. We can use fully observed variable values to obtain the MLE of each CPD. This approach is similar to training the CPDs of a naïve Bayes classifier.

MLE of $\theta_{I|i,j,l,m,n}$ for fully observed data can be shown as:

$$P(X^k | \theta) = P(I^k, d^k, c^k, f^k, s^k, p^k)$$

$$P(X^k | \theta) = P(I^k | d^k, c^k, f^k, s^k, p^k) P(d^k) P(c^k) P(f^k) P(s^k | d^k c^k f^k) P(p^k | s^k)$$

$$\log P(X^k | \theta) = \log P(I^k | d^k, c^k, f^k, s^k, p^k) + \log P(d^k) + \log P(c^k)$$
$$+ \log P(f^k) + \log P(s^k | d^k c^k f^k) + \log P(p^k | s^k)$$

$$\frac{\partial \log P(X^k | \theta)}{\partial \theta_{I|i,j,l,m,n}} = \frac{\partial \log P(I^k | d^k, c^k, f^k, s^k, p^k)}{\partial \theta_{I|i,j,l,m,n}}$$

$$\theta_{I|i,j,l,m,n} = \frac{\sum_{k=1}^{K} \delta(I_k = r, d_k = i, c_k = j, f_k = l, s_k = m, p_k = n)}{\sum_{k=1}^{K} \delta(d_k = i, c_k = j, f_k = l, s_k = m, p_k = n)}$$

However, when the graph is given, the variable values are mostly observed except the type of classification variable. This becomes the semi-supervised learning problem, where we use partly observed variable data to classify the type-unlabeled variable by using its expectation value (Seeger et al. 2010).

In Figure 16.11, we considered intrusion detection problem with partly type-labeled and partly type-unlabeled data with a hidden classification type variable Y. Five known

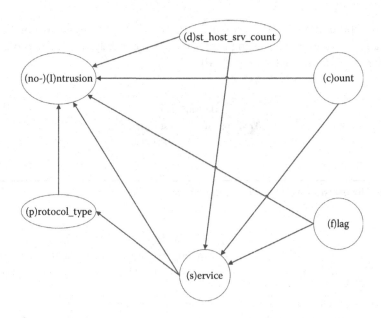

Figure 16.11: The intrusion discovery is based on the security feature dependencies in the Bayes network.

security feature variable values are observed including *(p)rotocol_typ, (s)ervice, (f)lag, (c)ount,* and *(d)st_host_srv_count.* There is one classification type-hidden (or -unobserved) variable *(no-) (I)ntrusion.*

An EM generative algorithm is used to maximize the MLE parameters θ with hidden variable Y of the type-labeled and type-unlabeled data. This problem is related to self-learning in the MLA layer.

- E-step estimates the labels for type-unlabeled hidden variable Y according to the current decision function by computing $q^u(Y) = P(Y|X^u, \theta)$.

- M-step: Maximize over θ estimates the decision function with the current type-labeled data $\log P(X^l|Y^l, \theta)P(Y^l|\theta)$ and the type-unlabeled data $q^u(Y)\log(P(X^u|Y, \theta)P(Y|\theta))$

$$\sum_l \log(P(X^l|Y^l, \theta)P(Y^l|\theta)) + \sum_u \sum_Y q^u(Y)\log(P(X^u|Y, \theta)P(Y|\theta))$$

We can compute the expected value of each unobserved Boolean variable $(no-)(I)ntrusion$ in the *E step*. Then we can compute the MLE value of type-unlabeled Boolean variable for $(no-)(I)ntrusion$ in the *M step* with replacing each count by its expected count. This EM process iterates until it converges.

16.8 An Experiment of Intrusion Detection Scenario

The experiment for the intelligent security scenario is based on the KDD Cup 1999 dataset.* Each instance has about 40+ specific security features. Based on Figure 16.11, we use R library of self-learning via the EM optimization. We compute the optimal results based on decision tree

---

*http://kdd.ics.uci.edu/databases/kddcup99/kddcup99.html.

algorithm with 500 training data (see Figure 16.12). After eight rounds of the EM processes, the final results on training and testing errors for the naïve Bayes and the Bayesian network algorithms are very promising (see Figure 16.13).

Decision tree	
Training error	Testing error
0	0.213230

Figure 16.12: The decision tree experimental results.

		Naïve Bayes		Bayesian network	
		Training error	Testing error	Training error	Testing error
Round 1	Training data: 500 (SPARQL) Testing data: 214 (sample)	0.186	0.17149533	0.152	0.2630841
Labeled class error (10000 records): 0.0574					
Round 2	Training data: 10500 (SPARQL + EM) Testing data: 4500 (sample)	0.07161905	0.17873334	0.03492381	0.1748444
Labeled class error (20000 records): 0.04625					
Round 3	Training data: 30500 (SPARQL + EM) Testing data: 8785 (sample)	0.054	0.17810421	0.03596721	0.17465382
Labeled class error (50000 records): 0.04192					
Round 4	Training data: 80500 (SPARQL + EM) Testing data: 34500 (sample)	0.04541615	0.17859709	0.03110559	0.17513042
Labeled class error (100000 records): 0.026					
Round 5	Training data: 180500 (SPARQL + EM) Testing data: 77375 (sample)	0.0213518	0.1784467	0.1301939	0.1763331
Labeled class error (200000 records): 0.024855					
Round 6	Training data: 380500 (SPARQL + EM) Testing data: 163071 (sample)	0.0146544	0.17920414	0.006467806	0.17701798
Labeled class error (300000 records): 0.02386					
Round 7	Training data: 680500 (SPARQL + EM) Testing data: 291642 (sample)	0.03037913	0.17895981	0.003776635	0.17688398
Labeled class error (450000 records): 0.02386222					
Round 8	Training data: 1130500 (SPARQL + EM) Testing data: 311025 (All)	0.02898364	0.178976	0.002442282	0.1768732

Figure 16.13: The experimental results after eight rounds of EM optimization processes for the naïve Bayes and Bayesian network. The final training errors and testing errors converge.

16.9   Conclusion and Future Works

We have shown that the structured machine learning techniques can be useful for big data analytics and modeling. Structured domain knowledge is modeled as RDF(S) ontologies, and SPARQL queries provide a labeled training dataset for an inductive learning to evaluate its learning robustness. Semantics-enabled (semi-)supervised learning is the technique to empower the structure machine learning concept for security analytics.

In perfect domain knowledge learning, a security expert has complete knowledge about the intrusion behaviors. All data are correctly classified and available in the learning process. Moreover, the dataset is clean with no hidden and type-unlabeled variables. This structured machine learning provides an optimal performance benchmark for imperfect domain knowledge learning. We first apply a decision tree algorithm by enforcing the supervised learning to find a learning hypothesis to describe the intrusion behaviors.

In imperfect domain knowledge learning, a security expert does not have complete and correct knowledge about the intrusion behaviors. Furthermore, some data might not be correctly classified and some data have a hidden variable as a type-unlabeled classification. We use the naïve Bayes and Bayesian network algorithms to classify each type-unlabeled data by using semi-supervised learning. The type-unlabeled data are newly classified by using the EM generative method through maximizing the log likelihood of type-labeled and type-unlabeled data. By using the newly classified type-labeled data, structured knowledge module can facilitate its ontology and rule learning to enhance the accuracy of type-unlabeled data classification.

The KDD Cup 1999 dataset has been used to evaluate the semantics-enabled (semi-)supervised learning technique. It still needs further empirical study to determine how it can improve the ontology learning and rule learning and be helpful for machine learning algorithms, and vice versa regarding training and testing error rates reduction for big data analytics and modeling.

References

Agneeswaran, S. V. (2014). *Big Data Analytics Beyond Hadoop*. Pearson, Upper Saddle River, NJ.

Anderson, D., T. Frivold, and A. Valdes (1995). Next-generation intrusion detection expert system NIDES a summary. Technical report, Computer Science Laboratory, SRI International.

Bilge, L. and T. Dumitras (2012). Before we knew it: An empirical study of zero-day attacks in the real world. In *CCS'12*, pp. 833–844. ACM, Raleigh, NC.

Bonatti, A. P. (2011). Datalog for security, privacy and trust. In *Datalog 2010*, LNCS 6702, pp. 21–36. Springer, Oxford, UK.

Brickley, D. and V. R. Guha (2014). RDF schema 1.1. Technical report, W3C Recommendation.

Cárdenas, A. A. et al. (2013). Big data analytics for security intelligence. Technical report, Cloud Security Alliance (CSA).

Ceri, S. et al. (1989). What you always wanted to know about Datalog (and never dared to ask). *IEEE Trasactions on Knowledge and Data Engineering 1*(1), 146–166.

Chandola, V., A. Banerjee, and V. Kumar (2009). Anomaly detection: A survey. *ACM Computing Surevey 41*(3), 15–58.

Denning, E. D. (1987). An intrusion detection model. *IEEE Trasactions on Software Engineering 13*(2), 222–232.

Dietterich, G. T. et al. (2008). Structured machine learning: The next ten years. *Machine Learning 73*, 3–23.

Domingos, P. (2012). A few useful things to know about machine learning. *Communications of the ACM 55*(10), 78–87.

Domingos, P. and D. Lowd (2009). *Markov Logic: An Interface Layer for Artificial Intelligence.* Morgan & Claypool, San Rafael, CA.

Eiter, T. et al. (2008). *Rules and Ontologies for the Semantic Web.* Springer, Berlin, Germany.

François, J. et al. (2011). Bottrack: Tracking using netflow and pagerank. In *IFIP Networking,* Valencia, Spain.

Fürnkranz, J. et al. (2012). *Foundations of Rule Learning.* Springer, Berlin, Germany.

Getoor, L. and B. Taskar (2007). *Introduction to Statistical Relational Learning.* The MIT Press, Cambridge, MA.

Giur, P. and W. Wang (2012). Using large scale distributed computing to unveil advanced persistent threats. Technical report, AT & T Security Research Center.

Harris, S. and A. Seaborne (2014). SPARQL 1.1 query language. Technical report, W3C Recommendation.

Hastie, T., R. Tibshirani, and J. Friedman (2013). *The Elements of Statistical Learning: Data Mining, Inference, and Prediction,* 2nd Edition. Springer, New York.

James, G. et al. (2013). *An Introdution to Statistical Learning with Applications in R.* Springer, New York.

Joseph, A. D. et al. (2012). Machine learning methods for computer security. Technical Report 1, Manifesto from Dagstuhl Perspective Workshop.

Labrinidis, A. et al. (2012). Challenges and opportunities with big data. Technical report, Computing Research Consortium (CSR).

Lee, W. and J. S. Stolfo (2000). A framework for constructing features and models for intrusion detection systems. *ACM Transactions on Information and System Security 3*(4), 227–261.

Lehmann, J. et al. (2014). Concept learning. In J. Lehmann and J. Vöelker (Eds.), *Perspectives on Ontology Learning,* pp. 71–91. IOS Press, Burke, VA.

Maloof, A. M. (2006). *Machine Learning and Data Mining for Computer Security: Methods and Applications.* Springer, London.

Manyika, J. et al. (2011). Big data the next frontier for innovation, competition, and productivity. Technical report, McKinsey Global Institute.

Mitchell, M. T. (1997). *Machine Learning.* McGraw-Hill, New York.

Muggleton, S. et al. (2012). ILP turns 20: Biography and future challenges. *Machine Learning 86*(1), 3–23.

Murthy, P. et al. (2014). Big data taxonomy. Technical report, Cloud Security Alliance (CSA).

Quinlan, J. R. (1993). *C4.5: Programs for Machine Learning*. Morgan Kaufmann, San Mateo, CA.

Raedt, D. L. (2008). *Logical and Relational Learning*. Springer, Berlin, Germany.

Seeger, M. et al. (2010). A taxonomy for semi-supervised learning methods. In O. Chapelle et al. (Eds.), *Semi-Supervised Learning*, pp. 15–31. MIT Press, Cambridge, MA.

Singh, D. and K. C. Reddy (2014). A survey on platforms for big data analytics. *Journal of Big Data 1*(8), 1–20.

Sommer, R. and V. Paxson (2010). Outside the closed world: On using machine learning for network intrusion detection. In *IEEE Symposium on Security and Privacy*, pp. 305–316. IEEE, Oakland, CA.

Spomy, M. et al. (2013). JSON-LD 1.0. Technical report, W3C Proposed Recommendation.

Yen, T. F. et al. (2013). Beehive: Large-scale log analysis for detecting suspicious activity in enterprise networks. In *ACSAC'13*. ACM, New Orleans, LA.

Zhang, J., M. Zulkernine, and A. Hague (2008). Random-forest-based network intrusion detection systems. *IEEE Transactions on Systems, Man, and Cybernetics 38*(5), 649–659.

Chapter 17

Exploring the Potential of Big Data for Malware Detection and Mitigation Techniques in the Android Environment

Rasheed Hussain

Donghyun Kim

Michele Nogueira

Junggab Son

Heekuck Oh

CONTENTS

Smartphone technology has seen an unprecedented growth in production because of its services to consumers, ranging from telephony to mobile Internet. Among the major market players in the smartphone operating systems, Android claims a major portion of the market share. However, malicious code writers have penetrated into the Android environment. This necessitates malware detection and mitigation mechanisms in the Android environment. To date, various approaches have been made to detect malware in the Android environment. This chapter presents a comprehensive review on malicious software (malware) detection and mitigation in the Android environment—one of the important security threats prevailing in the environment. It begins with a brief taxonomy of the currently available mechanisms to detect and/or thwart the malware in the mobile environment in general, and in the Android environment in particular. Furthermore, the chapter outlines current solutions for malware detection and mitigation mechanisms in the Android environment in detail. It covers important dimensions of these solutions that include generic malware detection approaches, signature-based approaches, feature extraction-based approach, machine learning (ML)-based schemes, permission-based techniques, and data- and text mining-based malware detection techniques. Because malware families are constantly growing, new mechanisms are needed to meet the challenges of malware detection and mitigation in the Android environment. New directions for malware detection techniques in the Android environment have also been summarized. We argue that, thanks to the growing potential of big data analysis-based techniques, malware detection based on big data and cloud computing for Android will likely outperform the existing techniques because of its capability of handling a large volume of data, efficiency, and robustness.

17.1 Introduction

Over the past few years, the smartphone industry has skyrocketed with advancements in smartphone technologies. Unlike in the past, smartphones are capable of replacing old-fashioned personal computers because of their rich onboard computation, communication, and storage capabilities. Among the various functionalities offered by smartphones, mobile Internet is most popular, which has enabled consumers to use the Internet virtually anywhere and anytime.

Surprisingly, smartphones are used not only for personal necessity but also for business purposes. The aforementioned statement has been confirmed by the report from DigiTimes Research, which states that global shipment of smartphones reached around 1.2. billion in 2014, which is 30% more than that in 2013 [1]. Usually, smartphones come with a built-in operating system (OS) and a number of preinstalled applications by the manufacturer and third-party service providers. Big market players in the smartphone industry include Google, Apple, Blackberry, and Microsoft. These companies provide the consumers with their own application markets where users can download applications of their choice. Moreover, cellular service providers also provide their customers with their own application markets or stores, such as SK Telecom and Korea Telecom of South Korea. It is to be noted that Google claims a major portion of the smartphone industry by employing the Android platform in smartphones.

Despite the sudden boom in the smartphone industry over the past decade, security issues in these devices have also become challenging. Cyberspace is a double-edged sword. On the one hand, it provides enormous services and applications to the users; on the other hand, it puts the privacy and security of the users at stake. For instance, smartphones host a number of sophisticated sensors that could leak highly sensitive data about users' location, their physical activities, their likes and dislikes, and so forth. Moreover, sophisticated attackers could also invade pictures and videos stored in the memory of the phone. Recently, smartphones have been used for online shopping; therefore, sensitive personal information such as credit card numbers and other credentials may also fall into wrong hands. The most serious threat is injecting malicious software (malware) programs and malicious code to perform all aforementioned functionalities without the user's knowledge and mandate.

The aforementioned issue has compelled the vendors to provide consumers with adequate protection against such attacks. As mentioned earlier, Android is the market leader in the smartphone industry; therefore, attackers tend to target Android more than other platforms. Hence, serious security measures (both defensive and offensive) are essential. The statistics of the malware in Android markets are overwhelming. A report in [2] showed that 2000 new malware samples were discovered almost every day, and by the end of May 2014, around 650,000 malware samples were detected in the Android environment. Such staggering amount mandates the research community to take necessary preventive measures. If the system is able to detect the malware well before it fully infiltrates the system, then the personal data and other sensitive information can be saved from the prying eyes.

There are a number of reasons for the increasing number of malware in the Android environment, for instance, the volume of applications, the open-source nature, and malicious users, to name a few. However, the most dominating reason is that the Android environment hosts applications not only from Google but also from a huge number of third-party providers. This capability to incorporate third-party applications into the Android market (currently known as Google Play) makes it more vulnerable to security threats such as malware. To this end, the market operators have employed revision mechanisms in which the apps are analyzed before uploading them to the store. However, in a broader perspective and with the increase in the volume of applications, such strategy is not sufficient to rule out the inclusion of sophisticated malwares. More precisely, malware writers have made their code even more sophisticated, so that it can bypass the pre-uploading analysis and other precautionary measures taken by the application markets.

This chapter covers various aspects of malware detection and mitigation techniques, and we cover the major *de facto* standards established so far. We particularly discuss the static and dynamic analyses to detect malware in Android apps. Moreover, other techniques such as signatures, application program interface (API) calls-based techniques, application behavior analysis, permission analysis, malware detection in the Internet of things (IoT) environment through a linear support vector machine (SVM), ML-based approaches, semantic-based approaches, text mining- and information retrieval-based approaches, big data- and cloud computing-based approaches, and kernel- and user-level monitoring approaches are outlined. A thorough taxonomy of the currently available solutions to detect malware in the Android environment is also presented. It is to be noted that historically the volume of Android apps in the markets is drastically increasing, and so is the percentage of malware apps. Therefore, traditional approaches of detecting malware in Android will be either inefficient or insufficient due to the advent of new malware families and the surprisingly growing volume of Android malware. Hence, other means of an extensive analysis on a large volume of data are essential. To this end, big data analysis techniques are promising methods for analyzing a huge amount of data. Therefore, from this comprehensive survey, the currently available approaches—a large portion of which use ML, classifiers, text and data mining, and big data- and cloud-based techniques—can easily outperform the existing techniques due to increased capability of data analysis in an efficient manner. This chapter also summarizes that in the future, big data and cloud-based malware detection in the Android environment will address the performance and robustness issues of the currently available techniques.

We first outline the taxonomy of the available solutions to detect malware in the Android environment. Then, we briefly describe in detail the generic approaches for malware detection and mitigation, which include characterization and sampling of malware and its types followed in packet classification-based techniques. There are a number of generic approaches for malware detection including extraction of data flow features, remote monitoring of applications, behavior-based anomaly detection, pattern mining, contrasting permission patterns, and static and dynamic feature extraction techniques. We will cover most of the generic approaches followed by ML- and big data-based malware detection and mitigation mechanisms in the Android environment. Furthermore, we cover semantic-based approaches, text mining approaches, permission analysis-based approaches, and kernel- and user-level analysis approaches before concluding this chapter.

17.2 Taxonomy of Malware Detection and Mitigation in the Android Environment

Figure 17.1 illustrates the comprehensive taxonomy of malware detection and mitigation techniques available in the literature. We divide these techniques abstractly into five categories: generic approaches, ML-based schemes, feature extraction techniques, semantic-based approaches, and text mining schemes. These categories are further divided into subcategories as shown in Figure 17.1. The generic approaches are mainly subdivided into signature-based, feature extraction-based, and pattern analysis-based approaches. It is to be noted that some baseline techniques are common in the subcategories of the major categories. For instance, signature-based approach or feature extraction-based approach can be applied both in generic methods and in ML-based methods; however, their characteristics are different in either case, and their target is to detect and/or mitigate malware in the Android environment. These techniques are explained in Sections 17.3 through 17.8 in a thorough detail.

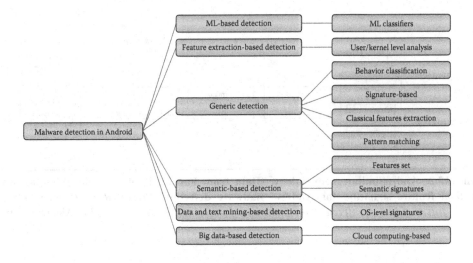

Figure 17.1: Taxonomy of malware detection mechanisms in the Android environment.

17.3   Generic Malware Detection and Mitigation Techniques

This section outlines generic strategies available in the literature to detect malware in the Android environment. Zhou et al. aimed to characterize the existing Android malware [3]. They present a large collection of 1260 Android malware samples and cover 49 different Android malware families. Moreover, they perform a timeline analysis to argue on the discovery of these malware samples. Furthermore, the collection of malware is broken down into their installation, activation, and consistency in payloads. The most daunting challenge posed by malware is that they are constantly evolving with the advances in programming practices.

17.3.1   *Signature-based approach*

One of the most common methods to detect malware is to check the signatures of the suspected application against the known malware signature family.

17.3.1.1   *AndroSimilar*

AndroSimilar is a robust signature-based mechanism that singles out malware apps through extracting comprehensive features of the application and generating signatures pertaining to those features [4]. It detects various unknown, obfuscated, repackaged apps* of the known malware families. The detection process in AndroSimilar is carried out through a statistical similarity digest hashing mechanism, which is based on the byte stream. A 35% of threshold in cases of a signature match flags the sample as a threat. The high-level methodology of AndroSimilar is given in Figure 17.2.

The important characteristic of AndroSimilar is the similarity index among the signatures of the known malware and the app under observation. In the figure, the detection process proceeds as follows.

First of all, apps are collected from Google Play and other third parties in the market, and AndroSimilar takes these apps as input where entropy features are generated by the byte blocks.

---

*The terms *application* and *app* are used interchangeably in this chapter.

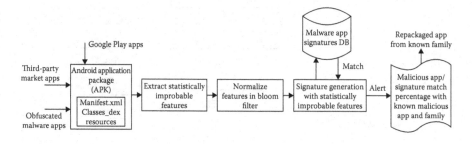

Figure 17.2: AndroSimilar methodology. (Data from Faruki, P. et al., AndroSimilar: Robust statistical feature signature for Android malware detection, *Presented at the Proceedings of the 6th International Conference on Security of Information and Networks*, Aksaray, Turkey, 2013.)

Afterward, these entropy features are normalized to represent the respective features. The normalized features are also used to select strong attributes that will aid in malware detection. To narrow down further, the most popular features are selected according to the similarity digest scheme. These popular features will serve as representative attributes for those particular blocks. AndroSimilar stores 128 most popular features in the bloom filter for signature generation, and the signature database is generated for malicious apps. Once the generation is completed, the collected apps are compared with the signatures to detect any match with a known signature beyond a given threshold (for AndroSimilar, the statistical threshold is kept at 35%). Finally, the percentage similarity is identified with the known malicious sample to detect code obfuscation and repackaging of the malicious sample.

With the help of AndroSimilar, a total of 6779 apps were analyzed from Google Play store, of which 6739 were correctly detected, whereas 40 were found suspicious, with a 99.4% accuracy. However, among 545 apps from third-party markets, 539 were correctly detected and 6 of them were suspicious, with a 98.89% accuracy.

17.3.1.2 DroidLegacy

DroidLegacy is an automatic analysis method that deals with the piggybacked malware and presents a method for creating obfuscation resilient signatures [5]. It also contains a method for scanning an Android application package (APK) file of the target malware. Piggybacked malware is equivalent to a hidden parasitic Windows virus that infects other legitimate applications. The aforementioned resiliency is achieved through constructing signatures by using Android API calls. The application behavior is reflected in the use of a different set of API calls; therefore, the parameter related to API calls remains an invariant throughout the app life cycle. However, piggybacked malware is another challenge that separates the legitimate application from the injected malware code. It is also to be noted that piggybacked malware is usually loosely coupled with the core of the host application. DroidLegacy takes advantage of this fact and partitions an APK into loosely coupled modules, and the rider code modules are identified. After an unlabeled APK is partitioned into loosely coupled modules, the Android API calls made by each module are compared with the signature of each malware family. DroidLegacy is the work in continuation with the findings of the piggybacked applications [6]. From an efficiency standpoint, DroidLegacy is compatible with parallel processing and vantage point trees [7] because once the signatures are generated for the known malware families, the new untested APK can be processed and analyzed in parallel. It can then be compared with the signatures through vantage point trees. The method to extract a signature in order to identify a repackaged malware family consists of the following steps.

Partitioning: The classes of each APK of the prescreened collection are partitioned into modules. It is to be noted that there must be a high cohesion between the classes in the same module and low coupling between classes in different modules. This is the key phenomenon based on which signatures are compared. First, a class dependence graph (CDG) is generated, and then the classes are partitioned based on the clustering mechanism of the scheme of Zhou et al. [6]. The CDG of DroidLegacy is significantly different from [6] in the sense that the level of granularity is different in both CDGs, and thus the level at which obfuscations are ignored. In other words, the package-level granularity implies that the method in [6] would work well when the benign code and the malware code (rider code) stay in distinct classes. However, the class-level granularity in DroidLegacy would yield good results when the classes remain distinct. The partitioning process of the APK in DroidLegacy is shown in Figure 17.3.

Locating malware modules: Then from each APK, a combination of modules are identified that have the highest intermodule similarity. An efficient algorithm is executed to identify which modules of an APK tend to have a malware functionality. The algorithm takes as input a collection of APKs and outputs the possible malware modules as a vector consisting of one module from each APK. The similarity between two modules may be measured using the Jaccard similarity for their set of API calls.

Extracting malware signatures: After the corresponding malware modules are identified in each repackaged malware APK, it is time to extract a signature to represent that particular family. The malware family is characterized by a set of API calls. In other words, any APK that contains a module with API calls that are similar to the signatures could be suspected as a member of that family.

17.3.1.3 AntiMalDroid

AntiMalDroid is a software behavior signature-based malware detection framework [8]. It uses the SVM algorithm, and it can detect malicious software and its variants in the runtime. Moreover, it extends malware characteristics database dynamically.

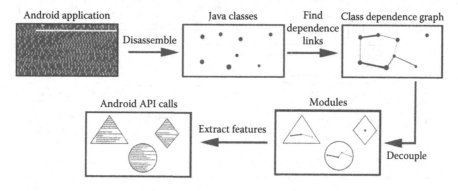

Figure 17.3: Partitioning of Android APK. (Data from Deshotels, L. et al., DroidLegacy: Automated familial classification of Android malware, *Presented at the Proceedings of ACM SIGPLAN on Program Protection and Reverse Engineering Workshop 2014*, San Diego, CA, 2014.)

The basic structure of the AntiMalDroid framework consists of two modules: characteristic learning and malware detection. As illustrated in Figure 17.4, the characteristic learning module consists of submodules that include characteristic monitoring module, characteristic learning module, behavior characteristics signature module, and signature database. The other important module is the malware detection module that includes the runtime behavior monitoring module, behavior signature module, decision module, and response module. The behavior signatures are compared with the signatures in the signature database, and based on the comparison, the response module gives a response if the signature is matched with the malware signature in the database.

The AntiMalDroid framework consists of three stages. The first stage is the software behavior signature generation. Behavior-based malicious code detection is ideal for resource-constrained mobile devices because, compared to feature-based malware detection, the signature database of the behavior signature is much smaller. It is also to be noted that the new malwares usually include new behavior signatures that are inconsistent with the normally known behavior; therefore, behavior-based signatures can aid in detecting new malware. AntiMalDroid defines software behavior as intent issued and system resources that are accessed by the application in the Android OS.

The second stage is the learning process for the malware signatures. The currently available approaches for malware detection are mostly rule based [9], which are able to detect only a predefined rule database of malware and unable to detect new variants of malware. Therefore, AntiMalDroid uses active learning, which was originally proposed by Gale et al. [10]. In the learning process, the SVM-based active learning algorithm is used to detect malware. The SVM-based algorithm needs a smaller set of training samples, and the classifier achieves a higher classification accuracy. This results in improved detection of malware with a reduced construction cost of the training samples.

The final stage is malware detection. The main feature of the SVM-based algorithm is the principle of structural risk optimization. Therefore, the SVM-based algorithm results in better and effective detection of malware. At the detection stage, a risk factor is also introduced.

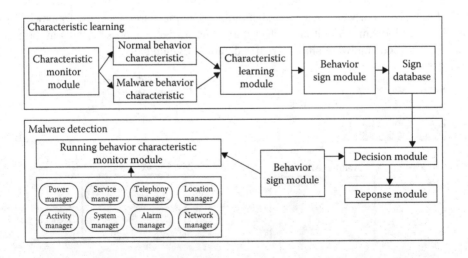

Figure 17.4: AntiMalDroid framework. (Data from Zhao, M. et al., AntiMalDroid: An efficient SVM-based malware detection framework for Android, in *Information Computing and Applications*, vol. 243, C. Liu, J. Chang, and A. Yang, Eds., Springer, Berlin, Germany, pp. 158–166, 2011.)

The risk factor is used for every short sequence of a malicious act to give a weight (the base value is set to 1), and accordingly the value of the risk factor is increased if the behavior of the system or the user security threat is greater. This allows the system to trigger the countermeasure for the possibly detected malware.

The aforementioned three schemes are based on the behavior signatures. Another framework called RiskRanker has also been developed that does not rely on the signatures [11]. It is a proactive scheme that spots zero-day Android malware* by analyzing a sheer number of Android apps in the markets. The main characteristic of RiskRanker is that it does not rely on the signature specimens that were used in the previous schemes. Rather, it is a risk-driven scheme in which potential risks are classified as high, medium, and low. The level of the risk refers to the severity of the situation; for instance, high risk means exploitation of the platform-level software vulnerabilities to compromise the integrity of the entire phone. However, medium-risk apps can cause financial losses or exposure of personal information, such as subscribing for a premium service unbeknownst to the user and cause him/her financial troubles. Low risk is identical to medium risk, but the effect is milder. Based on the risk, RiskRanker performs a two-order risk analysis. In the first-order risk analysis, RiskRanker identifies the capabilities of the app and then categorizes it into either high risk or medium risk, whereas in the second-order risk analysis, it investigates to uncover the suspicious behavior of the apps. For instance, some apps may be designed to evade and/or bypass the first-order analysis. Therefore, RiskRanker locates such apps and maps them onto the corresponding risk categories. This way, RiskRanker substantially reduces the number of suspicious apps that require verification. The generic architecture of the RiskRanker is given in Figure 17.5.

17.3.2   Feature extraction-based approach

Signature-based approaches are able to detect the existing malware through a signature comparison, and the list of the signature requires to be updated frequently. Unlike the previous approach, in feature extraction-based approach, different app features are extracted and analyzed through anomaly detection algorithms. The common feature extraction-based approaches for malware detection are outlined below.

17.3.2.1   SmartMal

SmartMal is a service-oriented architecture (SOA) that is integrated into malware detection [12]. Applying SOA concepts for malware detection has two basic advantages: It greatly reduces the workload of the detection algorithm, and the user behavior analyses are stored on the central and/or distributed servers. This phenomenon helps in the thin-client concept because all the processing threads will run on the servers. SmartMal also provides a behavior analysis algorithm with SOA concepts with the help of integrating distributed components into a hierarchical kernel model. It is an integrated client–server model, and the general framework is given in Figure 17.6.

The application runs on the mobile devices that are responsible for logging all the abnormal information, which is then sent to the server. The abnormal information is represented as vectors and extensible markup language (XML), and it is sent to servers over general packet

---

*Zero-day malware is a malicious code present in the application and is unknown to the vendor. The attackers exploit it before the vendor comes into play.

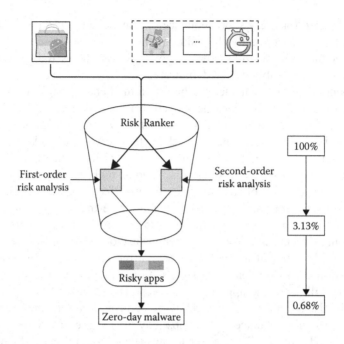

Figure 17.5: Architecture of RiskRanker. (Data from Grace, M. et al., RiskRanker: Scalable and accurate zero-day Android malware detection, *Presented at the Proceedings of the 10th International Conference on Mobile Systems, Applications, and Services,* **Lake District, 2012.)**

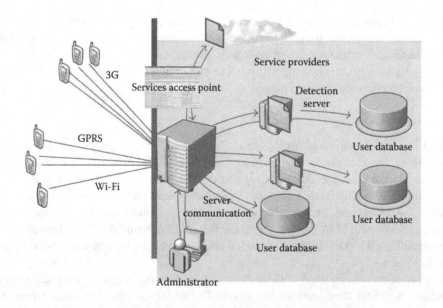

Figure 17.6: SmartMal architecture. (Data from Wang, C. et al., *The Scientific World Journal,* **2014, 101986, 2014.)**

radio services (GPRS), 3/4G, and/or Wi-Fi networks. The detection algorithm in the distributed servers will identify the possible threat and send the result to the major servers, and then the servers will notify the concerned terminals.

The basic framework of SmartMal is divided into two architectures: client and server.

1. *Client architecture*: The main function of the client is to extract abnormal features. Feature extractor is the main module of the client, and the features are extracted through APIs that are provided by the Android application or from the Linux kernel. The extracted features are then clustered into the kernel level, application level, and user behavior level. Among them, the user behavior level includes the features that reflect the user behavior. It is to be noted that the feature extraction frequency is controlled by the user. After that, the communication module sends the features to the remote server and receives anomaly alerts, if any. The client can also configure the parameters through a graphical user interface.

2. *Anomaly detection server*: The major functionality of this server is to decide whether the feature vector is normal or abnormal. It consists of a database, a detecting module, a communication module, a client module, and a graphical interface. The database is used to store the extracted features, and then complex detection algorithms are applied to the features. The communication module of the anomaly detection server communicates with the client back and forth for feature communication and results.

3. *Service-oriented hierarchical model*: The hierarchical structure of the distributed servers consists of services, a service scheduler, and a transmitter. The service provider presents the clients with a service access point (SAP), and each SAP controls one specific kind of service. The service scheduler is responsible for the service scheduling and mapping. It also ensures that the system's load balancing is intact. After that, the transmitter dispatches the tasks to different servers for execution. When the task is executed, the results are collected by the transmitter.

4. *Remarked features*: During abnormal detection, all collected behaviors do not necessarily mean that they pose any threat; therefore, it is challenging to figure out malware among the extracted set of behaviors. SmartMal achieves this by modeling a weight for each feature as a synthesized combination of subjective and objective weights to identify the behavior. After the weight for each target feature has been calculated, these features are moved forward to another step in which different weight vectors are combined. This way through the efficient algorithm in place, the malware is detected.

17.3.2.2   DroidAPIMiner

In DroidAPIMiner, the relevant features are extracted to capture the behavior of malware at the API level [13]. DroidAPIMiner mainly addresses the shortcomings of the permission-based warning mechanism. For instance, a large number of requested permissions by an Android app are not used within the application's code, but rather are required by advertisement packages. Therefore, DroidAPIMiner builds a robust and lightweight classifier for Android apps to detect malware. It relies on the API-level information within the bytecode to select the best features in order to distinguish between benign apps and malware apps. The API-level information conveys enough semantics about the apps behavior. Furthermore, DroidAPIMiner performs frequency analysis on the set of APIs to list out the malware from the benign ones.

The approach of DroidAPIMiner can be divided into three phases: feature extraction, feature refinement, and model learning and generation as shown in Figure 17.7.

In the feature extraction phase, similar to the previous schemes, benign and malware APK samples are collected to determine and extract the necessary features for malware to function. Then in the feature refinement phase, the API calls that are exclusively invoked by third-party packages, such as advertisement packages, are removed. Hence, the remaining API calls are only those APIs whose support in the malware set is significantly higher than in the benign set. Finally, in the model learning and generation phase, the representative vectors are fed to the standard classification algorithms that build the models.

As mentioned earlier, DroidAPIMiner extracts the necessary features for malware functioning. The Android app's bytecode has significant information that reflects its behavior. Therefore, comprehensive information can be retrieved from the bytecode, ranging from coarse-grained (package-level) to fine-grained (instruction-level) information. It is to be noted that DroidAPIMiner only focuses on the extracting package and API-level information because they are enough to capture the behavior of the application. The extraction process consists of extracting dangerous APIs through analyzing the samples. Such APIs are invoked by malware. Moreover, it consists of extraction of package-level information and API parameters.

Based on the API-level analysis, DroidAPIMiner identified the top APIs invoked by Android malware. These APIs are shown in Figure 17.8. To understand the malware behavior, the APIs are classified on the basis of the type of requested resources and utilities such as what resources are accessed, and what actions are performed. These classes include application-specific resource APIs, Android framework resource APIs, Dalvik virtual machine (DVM)-related resource APIs, system resource APIs, and utility APIs.

17.3.2.3   User-trigger dependence

User-trigger dependence (UTD) is a quantitative program analysis and data flow feature-based approach for detecting malware in Android apps [14]. It presents a high-precision Android app classification method based on a complex feature that exploits the app behavior to decide on the maliciousness of the app. One important characteristic of the UTD is that the features it exracts from the apps reflect the casual relations in the execution of the app. The UTD classification recognizes legitimate and desirable behavioral patterns, which is different from identifying

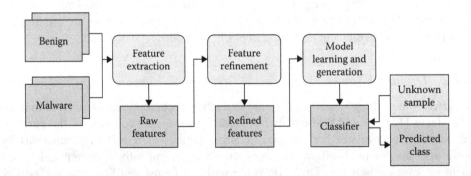

Figure 17.7: DroidAPIMiner malware detection approach. (Data from Aafer, Y. et al., DroidAPIMiner: Mining API-level features for robust malware detection in Android, in *Security and Privacy in Communication Networks*, vol. 127, T. Zia, A. Zomaya, V. Varadharajan, and M. Mao, Eds., Springer, Berlin, Germany, 2013, pp. 86–103.)

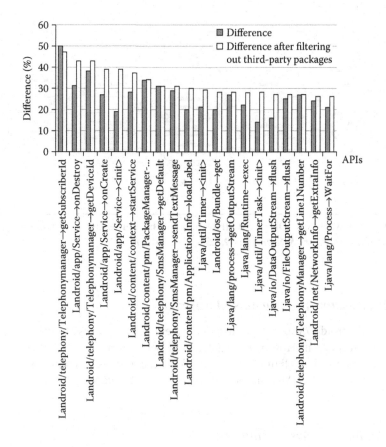

Figure 17.8: Top 20 APIs with the highest differences between malware and benign apps. (Data from Aafer, Y. et al., DroidAPIMiner: Mining API-level features for robust malware detection in Android, in *Security and Privacy in Communication Networks*, vol. 127, T. Zia, A. Zomaya, V. Varadharajan, and M. Mao, Eds., Springer, Berlin, Germany, 2013, pp. 86–103.)

malicious patterns. In other words, UTD classification is based on whether a program possesses benign features or not. A def-use graph is analyzed to extract a TriggerMetric feature for every encountered API call. This TriggerMetric feature gives approximation on the decision whether the user triggers the occurrence of the call. One unique characteristic of the UTD is that it does not advocate the use of fewer features in program classification; rather, it aims to advocate the enforcement of the benign property (Figure 17.9).

To extract TriggerMetric features from Android apps, a data dependence graph is generated and analyzed, which includes general data flow dependences, event-specific data dependences, reachability analysis for unused code, and backward depth-first search for finding dependence paths. On the basis of the UTD classification, an application is classified as benign or malicious. The values based on which the decision is taken are extracted from the TriggerMetric features of the app. Given the TriggerMetric values retrieved from the analysis before, the classification is done based on certain rules. For instance, let a score of an Android app be V and an assurance threshold be $T \in (0, 100\%)$; if $V < T$, then the app is classified as malware, otherwise benign. Therefore, it can be seen that the choice of T decides the accuracy of the classification. Similarly, other rules specific to similarity functions are also applied before taking a final decision on an app.

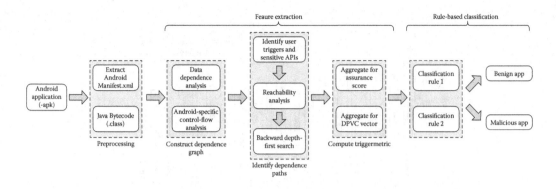

Figure 17.9: UTD workflow. DPVC, distribution of the percentages of valid call. (Data from Elish, K.O. et al., *Computers & Security*, 49, 255–273, 2015.)

17.3.3 Permission mining-based approach

Generally, the Android OS uses a permission system to gain access to the system and to the user's private information. However, to date, used permission and its utilization in malware detection have received negligible attention. With the permission mining-based approach, the used permission in the Android environment is employed to detect malware in Android systems.

17.3.3.1 Mining permission patterns

Moonsamy et al. proposed a novel pattern mining algorithm to identify a set of permissions in Android applications, and these permissions are leveraged to detect the difference between benign and malicious applications [15]. As mentioned earlier, to make applications secure, Android uses a permission system to restrict applications from accessing users' private resources, and this system provides an oversight for the security of Android [16]. In other words, an application must acquire the user's consent for the requested permissions to access the resource(s). Thus, the permission system provides security to the user from the invasive behavior of malicious applications. However, in reality, most users do not bother to read all those required permissions, and usually these permissions are asked at the time of installation and a majority of the users approve it. However, a problem still exists: whether the patterns in a permission combination can provide better performance in cases of malware detection. Therefore, the aim is to propose an efficient pattern mining mechanism to identify a set of contrast permission patterns. These permission patterns will aid in identifying malware applications.

Statistical solutions such as frequency counting and probabilistic models are common methods to analyze Android permissions [3,17]. In this framework, biclustering algorithm is used to group both applications and permissions. At the final stage, a novel contrast permission pattern mining algorithm is used to identify a specific permission pattern.

Statistical analysis on Android permission: In this step, a statistical analysis is carried out to study the required and used permissions for both benign and malicious applications. For this purpose, four datasets are extracted—required permissions and used permissions, each for both benign applications and malicious applications. Table 17.1 lists the details of top 20 required permissions for benign and malicious applications. It was found that

Table 17.1 Permissions required by benign and malicious applications

Clean applications		Malicious applications	
Required permission	*Frequency*	*Required permission*	*Frequency*
INTERNET	1121 (91.36%)	INTERNET	1199 (97.72%)
ACCESS_NETWORK_STATE	663 (54.03%)	ACCESS_COARSE_LOCATION	1146 (93.40%)
READ_PHONE_STATE	391 (31.87%)	VIBRATE	994 (81.01%)
WRITE_EXTERNAL_STORAGE	362 (29.50%)	WRITE_EXTERNAL_STORAGE	823 (67.07%)
ACCESS_COARSE_LOCATION	236 (19.23%)	READ_SMS	779 (63.49%)
VIBRATE	210 (17.11%)	WRITE_SMS	762 (62.10%)
WAKE_LOCK	188 (15.32%)	READ_CONTACTS	680 (55.42%)
ACCESS_FINE_LOCATION	162 (13.20%)	BLUETOOTH	633 (51.59%)
GET_TASKS	125 (10.19%)	WRITE_CONTACTS	542 (44.17%)
SET_WALLPAPER	102 (8.31%)	DISABLE_KEYGUARD	491 (40.02%)
ACCESS_WIFI_STATE	64 (5.22%)	WAKE_LOCK	471 (38.39%)
RECEIVE_BOOT_COMPLETED	60 (4.89%)	RECORD_AUDIO	461 (37.57%)
READ_CONTACTS	58 (4.73%)	ACCESS_FINE_LOCATION	446 (36.35%)
WRITE_SETTINGS	45 (3.67%)	ACCESS_NETWORK_STATE	416 (33.90%)
CAMERA	43 (3.50%)	READ_PHONE_STATE	414 (33.74%)
CALL_PHONE	42 (3.42%)	SET_ORIENTATION	413 (33.66%)
SEND_SMS	34 (2.77%)	CHANGE_WIFI_STATE	384 (31.30%)
RESTART_PACKAGES	32 (2.61%)	READ_LOGS	361 (29.42%)
RECEIVE_SMS	31 (2.53%)	BLUETOOTH_ADMIN	342 (27.87%)
RECORD_AUDIO	27 (2.20%)	RECEIVE_BOOT_COMPLETED	325 (26.49%)

Source: Moonsamy, V. et al., *Future Generation Computer Systems*, 36, 122–132, 2014.

malicious applications requested a total of 14,758 permissions, which are more than benign applications that requested 4,470 permissions. Another important behavior is that some of the permissions are only requested by the benign applications and not by the malicious ones, and vice versa. Furthermore, common permissions and unique permissions are also analyzed.

However, Table 17.2 lists the top 20 used permissions for benign and malicious applications. In the table, it can be seen that there is a small difference in which 16 out of 20 popular used permissions are common in both datasets. Now there is a need for an efficient visualization algorithm to find out the relationship between the permissions and the applications.

Visualizations using biclustering: Biclustering is a cluster analysis method [18]. The unique characteristic of the biclustering is that it applies the common clustering approach to both rows and columns at the same time. In the malware detection case, biclustering can help a group of applications that request or use different permissions.

Biclustering is achieved by performing agglomerative hierarchical clustering (AHC) [19,20] on both rows and columns. First, it is applied to the columns of the data followed by the rows. After running the AHC algorithm, the question is: How can we find out the permission combinations and use them as the patterns for malware detection?

To answer the above question, this framework uses contrast permission pattern mining (CPPM). The output permission patterns were expected to identify the clean and malicious applications. The two major processes involved in CPPM are candidate permission itemset generation and contrast permission pattern selection.

17.3.3.2 Contrasting permission patterns

Ping et al. introduced the contrasting permission patterns to differentiate between malware applications and benign applications from the permissions standpoint [21]. A framework based on contrasting permission patterns for Android malware detection is also proposed. Based on the proposed framework, this system also has an ensemble classified, namely, Enclamald, which is used to detect whether an application is potentially malicious. As previously mentioned, there is a sophisticated permission mechanism in the Android platform that restricts applications' access to the sensitive private resources. Each application has a list of particular permissions through which it can interact with the system APIs, databases, and the message-passing system. It is to be noted that, in essence, this framework is similar to the scheme of Moonsamy et al. [15]. However, this work particularly focuses on the contrasting permission patterns.

Android not only allows applications from specific vendors in the market but also exclusively allows apps from third parties that pose a threat to the users. To reduce the damage, Android restricts the access to privileged system resources and provides a mandatory access control that is based on the permission system. There are 130 official permissions defined by Android under four threat levels: Normal, Dangerous, Signature, and SignatureOrSystem. Normal permissions are put into the category of low risk, and they do not request users' explicit approval. However, dangerous permissions give access to the private user data and also control over the device. This access can possibly have a negative impact on the user; therefore, user consent is necessary. A signature permission is granted only if the application is signed with manufacturer's certificate, and for such permission, user consent is not necessary. SigntureOrSystem permission is granted only to applications that are in the Android system image or signed with the same certificate as the application that declared the permission.

Table 17.2 Well-known used permissions by benign and malicious applications

Clean applications		Malicious applications	
Used permission	*Frequency*	*Used permission*	*Frequency*
INTERNET	1029 (83.86%)	INTERNET	1161 (94.62%)
WAKE_LOCK	816 (66.50%)	ACCESS_COARSE_LOCATION	1125 (91.69%)
ACCESS_NETWORK_STATE	738 (60.15%)	VIBRATE	954 (77.75%)
VIBRATE	608 (49.55%)	WAKE_LOCK	826 (67.32%)
READ_PHONE_STATE	457 (37.25%)	ACCESS_WIFI_STATE	584 (47.60%)
ACCESS_COARSE_LOCATION	372 (30.32%)	ACCESS_NETWORK_STATE	519 (42.30%)
SET_WALLPAPER	126 (10.27%)	READ_SMS	473 (38.55%)
ACCESS_FINE_LOCATION	116 (9.45%)	WRITE_CONTACTS	426 (34.72%)
GET_ACCOUNTS	98 (7.99%)	READ_PHONE_STATE	354 (28.85%)
ACCESS_WIFI_STATE	85 (6.93%)	RECORD_AUDIO	319 (26.00%)
READ_SMS	82 (6.68%)	SET_WALLPAPER	297 (24.21%)
RESTART_PACKAGES	65 (5.30%)	ACCESS_FINE_LOCATION	199 (16.22%)
GET_TASKS	61 (4.97%)	GET_ACCOUNTS	178 (14.51%)
CHANGE_CONFIGURATION	55 (4.48%)	GET_TASKS	124 (10.11%)
RECEIVE_SMS	37 (3.02%)	RECEIVE_BOOT_COMPLETED	111 (9.05%)
FLASHLIGHT	37 (3.02%)	ACCESS_CACHE_FILESYSTEM	101 (8.23%)
WRITE_CONTACTS	34 (2.77%)	WRTIE_OWNER_DATA	59 (4.81%)
RECEIVE_BOOT_COMPLETED	23 (1.87%)	CHANGE_CONFIGURATION	52 (4.24%)
WRTIE_OWNER_DATA	12 (0.98%)	READ_HISTORY_BOOKMARKS	49 (3.99%)
WRITE_SETTINGS	10 (0.81%)	EXPAND_STATUS_BAR	41 (3.34%)

Source: Moonsamy, V. et al., *Future Generation Computer Systems*, 36, 122–132, 2014.

Consequently, Android apps have also a number of required permissions that are asked during installation, and usually users do not care for such permissions and allow them unknowingly.

Detection framework based on hybrid profile: Basically, malware detection can be categorized into two classes: anomaly detection and misuse detection. In anomaly detection, knowledge of normal behavior is used to decide on the maliciousness of the application, whereas in misuse detection, the characterization of what is known to be malicious is used to decide on the maliciousness. Both detection techniques use a profile that mimics the normal/abnormal behavior of the application. More precisely, the normal profile is used in anomaly detection, whereas an abnormal profile is used in misuse detection.

The proposed framework uses a hybrid profile that combines both normal and malware profiles, as well as the common profile. The key idea is that the contrasting permission patterns are effective representations of the fine line between benign and malware apps. The patterns are used as a hybrid profile that consists of three subprofiles: malware profile, clean profile, and common profile. The overall detection framework is shown in Figure 17.10.

The final stage of the malware detection is the construction of Enclamald using contrasting permission patterns achieved in the previous steps. Afterward, an efficient algorithm is used to construct the classifier based on which the decision is made whether the application is malicious or benign.

17.4 ML-Based Approach to Detect Malware

From a bird's-eye view, there are two kinds of malware detection techniques: static analysis and dynamic analysis. In both analyses, applications profiles are constructed. The behavior of the applications in question (both benign and malicious) is profiled with a set of feature vectors. These feature vectors are the snapshot of system information, for instance, memory usage, network utilization, and power consumption. ML techniques are also effective to detect malware and have been widely used [22–31]. It is an area from artificial intelligence in which new algorithms are designed to generalize the behaviors using datasets. ML algorithms are trained with the already known feature vectors to predict the classification of the unknown features. However, in ML-based approaches, it is important that the learning should be enough and intense. Amos et al. presented the evaluation of a number of existing ML classifiers

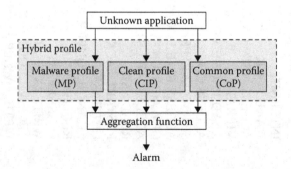

Figure 17.10: Malware detection through contrasting permissions. (Data from Xiong, P. et al., *China Communications*, **11, 1–14, 2014.)**

using a dataset containing a huge number of real applications [22]. The next section outlines a few of the many ML-based techniques to detect and mitigate malware in the Android environment.

17.4.1 Manifest analysis for malware detection in Android

Manifest analysis for malware detection in Android (MAMA) is a new technique for the detection of malicious Android executables that employes features after analyzing the manifest file of the Android app [32]. It uses permissions and the feature tags in the manifest file, and these features are later on used to build an ML algorithm to detect a malicious application.

MAMA uses a different feature set in the whole malware detection process. These features are gathered from the Android Manifest.xml file. The manifest file has a predefined structure. Moreover, the feature set to be gathered is of two types: the permissions required for the application under the uses-permission tag and the features under the uses-features group in the Android manifest file.

In order to extract these features, permissions used by each application are extracted through the android asset packaging tool. The main reasons for selecting these features are that the process incurs low computing overhead and that they contain different behaviors that may be present within them.

Permissions of the manifest file: At this stage, the permissions requested by the applications in the dataset are analyzed. The structure for declaring a uses-permission in the Manifest.xml file is given below:

```
<uses -- permission androin:name = ''string''/>
```

There are several strings that are used to declare permission usage of different Android applications such as "Android.permission. CAMERA" or "Android.permission. SEND_SMS". MAMA analyzes a number of permissions and their frequencies to determine their distribution within the collected dataset. The most used extracted permissions from the application are shown in Figure 17.11.

It can be seen that INTERNET is one of the most required permissions in both benign apps and malware apps. However, malware apps present an even more intensive use of the permissions related to the sending and receiving of text messages than benign apps. This way other permissions can also be analyzed. In order to input these permissions to the ML algorithm, the manifest file is analyzed and searched for the uses-permission tag. The string is retrieved and an input vector is generated for each of the possible permissions. MAMA uses supervised algorithms that employ a labeled dataset. The ML algorithms used by MAMA include k-nearest neighbors, decision trees, Bayesian networks, and SVM.

17.4.2 Application's network behavior

Shabtai et al. proposed a system to protect smartphones from malicious applications [31]. The proposed system identifies malicious attacks or masquerading apps installed on the mobile device and the republished popular applications injected with a malicious code (i.e., repacked apps). This system also detects a new form of malware that is able to update themselves. The detection process is based on an application's network traffic patterns. A learning process is carried out for each application's network traffic locally (on the device). Then semi-supervised learning strategies are used to carry out the learning process for the normal behavioral patterns and for detecting any potential malware.

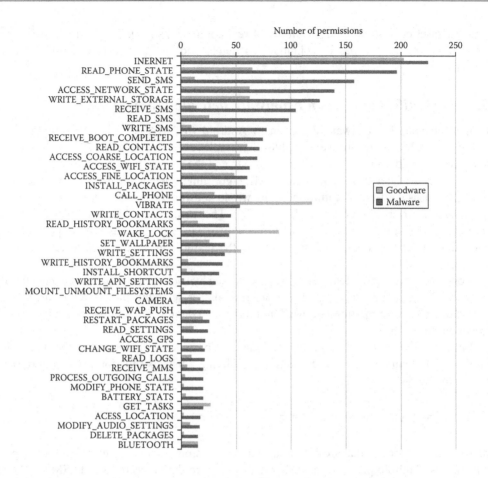

Figure 17.11: Most used extracted permissions. (Data from Sanz, B. et al., *Cybernetics Systems***, 44, 469–488, 2013.)**

A survey shows that 70% of the known malware apps steal user's private information and/or credentials [17]. Therefore, Shabtai et al. took the lead to work on self-updating malware apps that have even more dire consequences than the normal malware. A behavior anomaly detection system is introduced to monitor and detect meaningful deviation in an application's network behavior.

The main process flow of the system is to monitor the applications running on the phone, learn the models that represent their normal network behavior, and detect any deviations from the learned patterns. Such continuous analysis would not only detect the existing malware but can also offer suggestions about the unexpected behavioral change of the applications. This system supports two cases: In the first case, the application is already installed on the phone, and in the second case, a new application is downloaded and installed. It is to be noted that in the first case, the network traffic pattern is subject to change for three reasons: change in user's behavior, application update, and malicious attack. In such a case, the system has to detect the deviation from the normal traffic pattern and classify it into one of the aforementioned reasons. In the second case, the system must identify whether the newly downloaded application is benign, or a modified version of the previous one with possibly malicious behavior.

Shabtai et al.'s system is based on the client–server architecture in which the client-side software is responsible for actively monitoring the application, learning its behavior, and detecting any deviation, whereas the server-side software is responsible for comparing the behavior of the new application against the application's known traffic patterns in order to identify whether the application is acting benignly or it is a malware modification of the previous benign application. Local learning is carried out on the smartphone, and the local-side software is implemented as a full-fledged Android OS running on the device. An overview of the client-side software is shown in Figure 17.12.

The system architecture (at the client side) consists of a client app that is installed on the phone. The responsibility of the client-side module is to learn the user-specific local models and detect any deviations from the observed patterns and behaviors. The functions performed by this app include feature extraction, feature aggregation, local learning, and anomaly detection.

In the feature extractor module, a defined number of features are extracted for a predefined period. APIs exist to carry out this operation. The common features include sent/received data in bytes and percentage of the total transmitted data, network state, and time since application's last sent/received data. After the features are extracted, the feature aggregator module provides a true representation of the extracted features in an aggregated form. The aggregated features include average, standard deviation, minimum and maximum of sent/received data, the total amount of the transmitted data, and time intervals between sent/received events. As the main goal of the system is the local learning carried out for user-specific patterns for each application, it is related to the semi-supervised learning algorithms family. However, this family of the anomaly detection algorithms assumes that the training data includes samples only for normal data. For the current system, semi-supervised learning problem is converted to a set of supervised problems for which fast algorithms already exist.

Shabtai et al. used a cross-feature analysis approach [33] that is further analyzed by Noto et al. [34]. The cross-feature analysis approach assumes that in a normal behavior pattern, there exists a strong correlation between the features, and it can be exploited to detect deviations caused by abnormal activities. At the end of the algorithm, anomalies that decide on whether the application is malware or benign are detected.

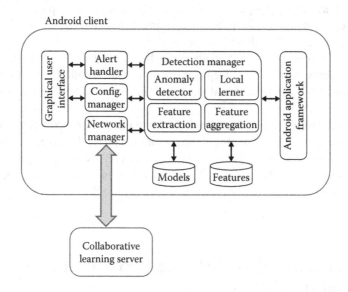

Figure 17.12: Architecture of the client-side software. (Data from Shabtai, A. et al., *Computers & Security*, 43, 1–18, 2014.)

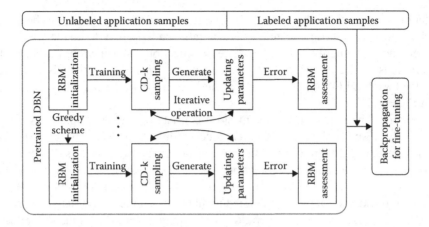

Figure 17.13: Framework of deep learning. DBN, deep belief network; RBM, restricted Boltzmann machines; CD, contrastive divergence. (Data from Yuan, Z. et al., *SIGCOMM Computer Communication Review*, 44, 371–372, 2014.)

17.4.3 Droid-Sec

Droid-Sec is an ML-based scheme that utilizes comprehensive features extracted from both static and dynamic analyses of an Android app to detect malware [35]. The motivation for Droid-Sec, like others before, is the line of defense that other malware detection schemes follow, which is the warning to the users about the permission required by an app. This approach is indeed not effective because it presents the permissions of an app in a stand-alone fashion. Besides, it requires sophisticated knowledge about the malware and the app, which may not be the field of expertise of a common user. To make things worse, benign and malware apps may need the same set of permissions. Such a phenomenon makes it difficult to distinguish between benign and malicious apps. In contrast, deep learning, a new ML-based technique, has gained much attention in artificial intelligence. Droid-Sec also leverages deep learning in which more than 200 features are extracted from both static and dynamic analyses of each Android app, and deep learning is applied to identify malware apps.

Traditional ML techniques are considered to be shallow in the architecture and can be trained in a particular way; however, deep learning is another paradigm in which the system can be trained in many ways with different algorithms. The Droid-Sec framework for deep learning is shown in Figure 17.13.

It is to be noted that deep learning consists of two phases: unsupervised pretraining phase and supervised back-propagation phase. In the pretraining phase, Droid-Sec uses the deep belief network paradigm for pretraining [36], whereas in the back-propagation phase, the pretrained neural network needs to be fine-tuned with labeled values in a supervised manner. After these two phases, the deep learning is completed and it is able to distinguish benign apps from the malicious ones.

17.5 Semantic-Based Text Mining-Based Approach for Malware Detection

With the development of the apps, malware authors have also increased their capability to evade the countermeasure and/or restrictions applied in app markets. Therefore, more sophisticated mechanisms are essential to deal with the growing number of malware apps in the Android environment. It is also true that the open-source nature of Android demands extra care against such malware apps. Another family of the malware detection mechanism is based on semantic- and text mining-based approaches. Semantic- and text mining-based approaches are closely related to the ML-based mechanisms, in which the former extracts semantics from the behavior of the app, whereas the latter makes a decision based on the reports from the behavior, patterns, and/or features of the app [37–39].

17.5.1 Apposcopy

Apposcopy is a semantic-based approach to identify a class of Android malware that steals users' private information [37]. It combines the advantages of the pattern-based malware detection mechanisms and taint analyzers. It incorporates a high-level specification language for describing semantic characteristics of malware families and a powerful statistical analysis for deciding if a given application matches with the signature of the malware family. It is resistant to low-level code transformation because the semantics and the high-level signature specification allow the analysts to point out the key characteristics of the malware family. It provides two types of semantic properties by leveraging the signature-based specification language: control flow and data flow.

The signatures specified in the aforementioned language, Apposcopy's static analysis, must contain two methods: construction of a new high-level representation of an Android app referred to as intercomponent call graph (ICCG) and a static taint analysis. ICCG is used to decide whether the Android app in question matches the control flow properties specified in the signature, whereas taint analysis is used to check for the consistency of the given application with a specified data flow property.

As previously mentioned, Apposcopy incorporates a malware specification language, which is a datalog program with built-in predicates. Users first specify signatures for the specific malware family by specifying a unique predicate. The user has a choice to add any other helper predicate to the same signature. It is to be noted that datalog is a program that consists of a set of rules and a set of facts. The format is the same as the one that is used in the formal logic; for instance, parent(''Bill'', ''Mary'') means that *Bill* is the *parent of Mary*. Now Apposcopy incorporates built-in predicates as well as component-type predicates, which represent different components provided by the Android framework. Android application generally has four components: activity, service, broadcast receiver, and content provider. Other built-in predicates include predicate ICC for intercomponent communication, predicate calls that are a control flow predicate, and predicate flows that are a data flow predicate.

As mentioned earlier, Apposcopy performs a static analysis to decide whether an application under consideration matches with the signature of the malware family. The steps involved in the static analysis include pointer analysis and call graph construction, and intercomponent control flow analysis. After that, to answer the data flow queries, Apposcopy performs a taint analysis.

17.5.2 DroidSIFT

DroidSIFT is a semantic-based mechanism that classifies malware in Android through dependency graphs [38]. It stops the transformation attack based on a mechanism in which the weighted contextual API dependency graph is extracted and used as a program semantic in order to construct a feature set. By using graph similarity metrics to uncover homogeneous apps behaviors while keeping implementation similarities, DroidSIFT is also effective against different variants of malware and zero-day malware.

DroidSIFT also builds the database of the behavior graphs for Android apps. The graphs represent the API semantics and the program semantics of those apps. When a new app is encountered, a query is made to the database to find the behavior similarity. Upon success, the corresponding element in the feature vector is set. It is to be noted that each element of the feature vector is associated with an individual graph in the database. It builds two graph databases for malicious and benign behaviors. Then the feature vectors extracted from these behaviors are used to train two new classifiers that are used for anomaly detection and signature detection. Anomaly detection is capable of detecting zero-day malware, whereas signature detection is used to detect variants of the malware.

DroidSIFT addresses the shortcomings of the vetting systems that were previously proposed. The basic architecture of DroidSIFT is given in Figure 17.14.

DroidSIFT performs two kinds of classifications: anomaly detection and signature detection. When a new application is being analyzed by DroidSIFT, the vetting process is conducted first to detect any deviation from the behavior of the benign application behavior in the Droid-SIFT database. Then the signature detection process is conducted to determine if the new app falls into any malware family within the signature database.

If the application passes the aforementioned tasks, there is still a possibility that new malware species are there. In such a case, DroidSIFT sends the app back to the developer with a report of the suspicious behavior. The general architecture of DroidSIFT is shown in Figure 17.15.

Figure 17.14: Architecture of DroidSIFT. (Data from Zhang, M. et al., Semantics-aware Android malware classification using weighted contextual API dependency graphs, *Presented at the Proceedings of the ACM SIGSAC Conference on Computer and Communications Security***, Scottsdale, AZ, 2014.)**

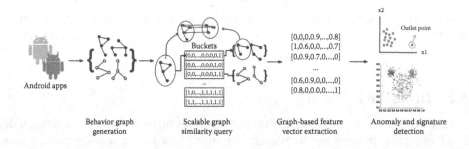

Figure 17.15: Workflow of DroidSIFT. (Data from Zhang, M. et al., Semantics-aware Android malware classification using weighted contextual API dependency graphs, *Presented at the Proceedings of the ACM SIGSAC Conference on Computer and Communications Security***, Scottsdale, AZ, 2014.)**

The workflow of DroidSIFT consists of four major steps. The first step is behavior graph generation. DroidSIFT considers graph similarity as a feature vector. Thus, the Android byte-code is transformed to its graph representation through a static program analysis. The graph representation includes entry point discovery and a call graph analysis to understand the context of the API calls. The outcome of the analysis is expressed through weighted contextual API dependency graphs. These graphs give us the security-related behavior of the application and is used in further processing of the application.

After the graphs are generated for both benign and malicious apps, the graph databases are queried to find a high similarity between the graphs. It is to be noted that finding a best match rather than a perfect match in the graph similarity is essential to identify polymorphic malware. To this end, we have a similarity feature vector through a similarity check analysis. Each element in the vector is associated with a graph already there in the database. The final step is to perform anomaly and signature detection. The feature vectors produced in the previous steps are used to train the classifier for signature detection. However, the anomaly detection discovers zero-day Android malware, and the signature detector reveals the family of the malware.

17.5.3 Dendroid

Dendroid is a malware detection system based on text mining and information retrieval techniques [39]. It employs the code structure analysis of the Android OS malware families. After that, it adopts the standard vector space model in the text mining applications. This enables Dendroid to measure similarities between malware samples and then automatically assign them to a certain malware family.

The high-level overview of Dendroid is shown in Figure 17.16. It can be observed from the figure that in the modeling phase, all different code structures are extracted from the malware samples. After that, a vector space model is used to associate a unique feature with every malware sample. This vector is henceforth used to illustrate two aspects: automatic classification of unknown malware into malware families based on their code structure and a revolutionary analysis of malware families based on the hierarchical clustering.

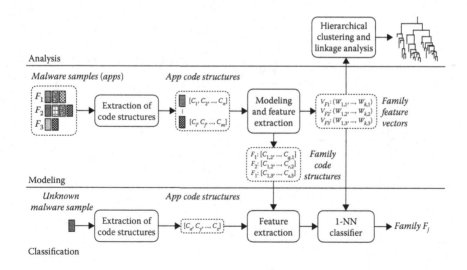

Figure 17.16: Overview of Dendroid. NN, nearest neighbor. (Data from Suarez-Tangil, G. et al., *Expert Systems with Applications*, 41, 1104–1117, 2014.)

Dendroid is novel from the standpoint of code structures. Another important point to note is that Dendroid focuses on the internal structure of the code (methods) rather than the specific sequence of instructions.

17.6 Cloud Computing- and Data Mining-Based Malware Detection

17.6.1 MobSafe

MobSafe is a system through which mobile apps are evaluated based on a cloud computing platform and data mining [40]. It combines both static and dynamic analyses to comprehensively evaluate Android apps. Moreover, it leverages a home-brewed cloud computing platform and data mining approach to evaluate Android apps. Android security evaluation framework (ASEF) and static Android analysis framework (SAAF) are used in the implementation of MobSafe.

The cloud infrastructure for MobSafe is shown in Figure 17.17. MobSafe uses about 40 servers and a 40 TB storage based on HDFS. The working procedure of MobSafe is shown in Figure 17.18.

When an APK file is uploaded to the system, it checks whether the result already exists in the system by submitting the hash value corresponding to the APK. If the value already exists, then the result is returned to the user. Otherwise, the APK is stored in Hadoop. Afterward, a demon is created to invoke ASEF and SAAF to collect the logs and then store them in the Hadoop-specified directory. Moreover, the hash and key values are already inserted into the system. The front end of MobSafe is web based and the APK can be uploaded there for analysis. ASEF provides an emulation of the human behavior for the app under consideration. The app is first installed on an Android virtual environment and then ASEF takes control of the analysis. At the end of the analysis, the results are compiled. However, SAAF is the static analyzer that analyzes the APK file directly for its contents. The code of the APK file is decoded into the smali code to analyze the permissions of apps and other different patterns. This way the decision is taken on the health of the Android application.

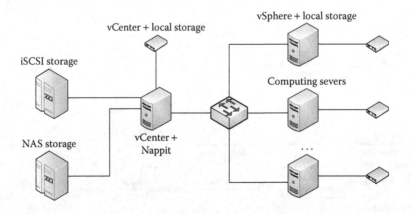

Figure 17.17: Infrastructure of MobSafe. iSCSI, Internet small computer system interface; NAS, network-attached storage. (Data from Xu, J. et al., *Tsinghua Science and Technology*, 18, 2013.)

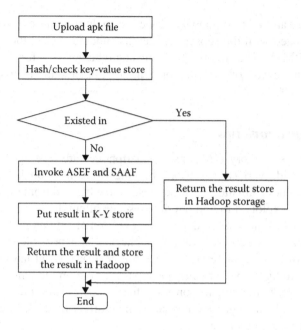

Figure 17.18: Workflow of MobSafe. (Data from Xu, J. et al., *Tsinghua Science and Technology*, 18, 2013.)

17.7   Anti-Analysis for Android Malware Detection

Petsas et al. proposed an anti-analysis with the help of which Android malware can evade the analysis mechanisms in an emulated environment [41]. The evasion properties are incorporated into the apps through static properties, dynamic sensor information, and virtual machine-related detailed characteristics of the Android emulator. When these properties were put to real Android apps (more precisely the evasion mechanisms), the preventive tools and services were found vulnerable. This scheme investigates how an Android app can know whether it is running on an emulator or real hardware. To find the vulnerabilities, the first and most important step is to identify the features of the execution environment through heuristics. This system includes a wide range of heuristics that include both simple heuristics and sophisticated ones. In order to assess the findings of this system, a set of existing malware was repackaged by incorporating the developed heuristics and submitted to the online analysis tool. The findings of such system can be very alarming for the security of the malware detection mechanisms.

At the abstract level, the anti-analysis tools employed by the malicious user(s) to evade the malware detection can be classified into three categories: (1) static heuristics based on simple or static information, (2) dynamic heuristics based on the sensor behavior observation, and (3) hypervisor heuristics based on the incomplete emulation of the actual hardware.

17.7.1   Static heuristics

Static heuristics can be used for emulated environments by checking for the static contents or the values that are unique, for instance, device identifiers and serial numbers. For instance, every smartphone has an international mobile station equipment identity, a unique number in the global system for mobile communication network. This value has already been exploited in order to hinder an analysis by malware detection tools running on emulators.* Similarly, another value associated with the subscriber identity module card is the international mobile subscriber identity (IMSI). The simplest heuristic could be to check for the IMSI value, which is historically null in the case of an emulator. Similarly, other static heuristics include current build and routing tables.

17.7.2   Dynamic heuristics

For dynamic heuristics, sensory functions of smartphones are leveraged. Smartphones have become home to a wide variety of sensors, ranging from the global positioning system to accelerometer. The values from most of these sensors are obtained from the environment. These sensory values can be exploited to find out the presence of emulators. By default, the Android emulator does not support mobility, and the current version of emulators does not support many other sensors. Therefore, if the sensor values are checked, it is possible statistically to find out whether the app is running in the emulated environment or on real hardware. Moreover, the straightforward way to figure out whether the app is running on the emulator is by registering a sensor in the system. If the registration fails, then it means that the app is running in the emulated environment. If the registration is done, then further methods could be used to check if the app is running on the emulator.

17.7.3   Hypervisor heuristics

In the hypervisor heuristics, incomplete emulations are exploited. For instance, it can be done by analyzing the behavior of the program at a low level. In this system, the hypervisor heuristic includes indentifying quick emulator (QEMU) scheduling.† It is to be noted that QEMU does not update the virtual program counter at every instruction execution. The reason for not doing such practice is performance because increasing/incrementing virtual program counter would need an additional instruction, and thus additional resources for the emulator. In the current scheme, such practice is exploited to figure out the presence of the emulator and thus evade it.

17.8   Permission-Based Analysis for Malware Detection

Android security mechanism includes the permission control in which an app is restricted from the access to the core facilities of the device or critical resources. Therefore, responsibilities are defined for both app developers to accurately define the permission requests and users to carefully grant permissions without endangering the security of their device. Wang et al. developed a system to explore the permission-induced risk in the Android environment [42]. This system works at three levels: At the first level, the risk of individual permission is analyzed and then the risk of a group of permissions is analyzed. At the second level, the usefulness of risky permissions to detect a malware app is evaluated. At the third level, the detection results are analyzed in depth.

An application's behavior is characterized by the requested permissions because it will execute the methods, use the phone resources based on the requested permission, and henceforth constitute the app's behavior. Alongside other important questions, the most important one is

---

*http://vrt-blog.snort.org/2013/04/changing-imei-provider-model-and-phone.html.
†http://www.dexlabs.org/blog/btdetect.

that whether there exists any permission rule that can be used to identify unknown malware applications, also referred to as zero-day malware applications.

In order to carry out the permission-induced risk analysis, first three feature-ranking techniques are employed to evaluate the risk of granting each permission. Based on these evaluations, the permissions are ranked in descending order. After that, the permission sets are evaluated by selecting the subset of features to check for the collaborative risk of several permissions. Then the problem of classification for detecting malware apps is solved through building ML classifiers. In the final stage, detection rules are extracted so that, based on the permission requests, the malware applications can be reported. The whole process is shown in Figure 17.19.

In order to rank permissions, three methods are employed: mutual information, Pearson correlation coefficient (CorrCoef), and T-test. It is to be noted that permissions are used as features. The ranking results from the aforementioned methods aid in selecting the most relevant permission for distinguishing between benign and malware apps. In the second step, the subset of risky permissions based on either their combinations or their cooperation with each other is identified. For the subset of feature selection, two methods are employed: sequential forward selection and principal component analysis.

After identifying risky permissions and their subsets, it is time to distinguish between benign and malware apps. For this reason, three classification algorithms are employed: SVM, decision tree, and random forest. In order to see the effect of this system, Figure 17.20 shows the occurrence rate of each top-ranked permission with CorrCoef in both benign and malware apps. The results show that permissions that are ranked top distinguish malware apps from benign apps based on their frequency. In addition, the difference of occurrence rate between malware and benign apps for permissions READ_SMS, RECEIVE_SMS, and SEND_SMS is above 50%; moreover, it is more than 15% for the permission WRITE_SMS. It can be confirmed from these statistics that the usage pattern of permissions related to SMS is different in benign and malware apps. In other words, many malware apps attempt to request SMS-related permissions.

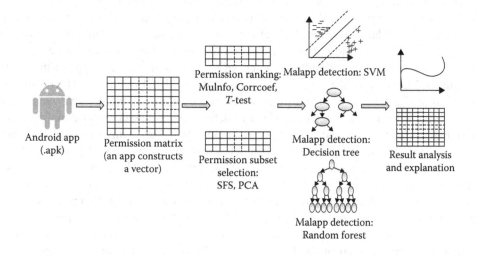

Figure 17.19: Process of permission-induced risk analysis. PCA, principal component analysis; SFS, sequential forward selection. (Data from Wei, W. et al., *Proceedings of the IEEE Transactions on Information Forensics and Security,* **vol. 9, pp. 1869–1882, 2014.)**

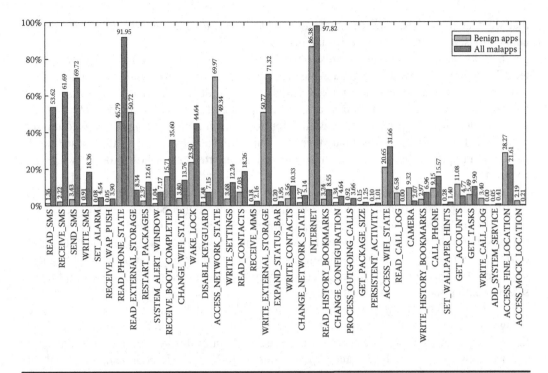

Figure 17.20: Occurrence percentage of top 40 risky permissions with CorrCoef. (Data from Wei, W. et al., *Proceedings of the IEEE Transactions on Information Forensics and Security*, vol. 9, pp. 1869–1882, 2014.)

17.9 Conclusions

The malicious software threat for the Android environment has raised an alarm for users because their private data are at stake. The unique feature of the malware is that it is very difficult to know the kind of malware in advance, although heuristics algorithms that help predict malware exist. Malware writers have also become sophisticated in their coding. Therefore, malware detection in the Android environment should be handled with care. To date, a number of mechanisms exist to detect and/or counter various kinds of malware apps in the Android environment. Nonetheless, the types and families of Android malware are still increasing. This chapter outlined a taxonomy of malware detection techniques in the Android environment. Some of these schemes have been explained in detail, which include generic malware detection, signature-based malware detection, big data- and cloud computing-based malware detection, permissions-based malware detection, and data- and text mining-based malware detection. To deal with malware detection in the Android environment, a single scheme will not be sufficient to cover every dimension of the malware. Therefore, a combination of existing schemes would suffice the application security in a better way. Moreover, the challenging aspect of the malware detection is the variant families of malware in the Android environment. To date, the exact number of the existing malware families is unknown, and it is drastically increasing in both number and sophistication. Therefore, traditional methods of malware detection may be inadequate due to poor performance and robustness. This chapter provides a comprehensive and in-depth survey of the existing schemes for detecting malware in the Android environment; however, the increasing volume of malware in the app markets and the analysis of current techniques mandate the need for big data- and cloud-based malware detection techniques. The current techniques also use text mining, data mining, and ML techniques, which in essence are processing-intensive techniques. Therefore, using big data- and cloud-based malware detection techniques will prove to be efficient based on the established fact of the performance offered by big data analytic strategies. In future, with the success of big data analytics, malware detection in the Android environment will be easier, more effective, and robust than the currently available mechanisms.

References

1. DigiTimes Research, *Global Smartphone Shipments to Reach 1.24 Billion in 2014*. http://www.digitimes.com/news/a20131125PD218.html, 2013.

2. V. Svajcer, *Sophos Mobile Security Threat Report*, Mobile World Congress, Sophoslabs, Barcelona, Spain, 2014.

3. Z. Yajin and J. Xuxian, Dissecting Android malware: Characterization and evolution, in *Proceedings of the IEEE Symposium on Security and Privacy (SP)*, Washington, DC, 2012, pp. 95–109.

4. P. Faruki, V. Ganmoor, V. Laxmi, M. S. Gaur, and A. Bharmal, AndroSimilar: Robust statistical feature signature for Android malware detection, *Presented at the Proceedings of the 6th International Conference on Security of Information and Networks*, Aksaray, Turkey, 2013.

5. L. Deshotels, V. Notani, and A. Lakhotia, DroidLegacy: Automated familial classification of Android malware, *Presented at the Proceedings of ACM SIGPLAN on Program Protection and Reverse Engineering Workshop 2014*, San Diego, CA, 2014.

6. W. Zhou, Y. Zhou, M. Grace, X. Jiang, and S. Zou, Fast, scalable detection of "Piggy-backed" mobile applications, *Presented at the Proceedings of the 3rd ACM Conference on Data and Application Security and Privacy*, San Antonio, TX, 2013.

7. P. N. Yianilos, Data structures and algorithms for nearest neighbor search in general metric spaces, *Presented at the Proceedings of the 4th Annual ACM-SIAM Symposium on Discrete Algorithms*, Austin, TX, 1993.

8. M. Zhao, F. Ge, T. Zhang, and Z. Yuan, AntiMalDroid: An efficient SVM-based malware detection framework for Android, in *Information Computing and Applications*. vol. 243, C. Liu, J. Chang, and A. Yang, Eds., Springer, Berlin, Germany, 2011, pp. 158–166.

9. W. Enck, M. Ongtang, and P. McDaniel, On lightweight mobile phone application certification, *Presented at the Proceedings of the 16th ACM Conference on Computer and Communications Security*, Chicago, IL, 2009.

10. D. D. Lewis and W. A. Gale, A sequential algorithm for training text classifiers, *Presented at the Proceedings of the 17th Annual International ACM SIGIR Conference on Research and Development in Information Retrieval*, Dublin, Ireland, 1994.

11. M. Grace, Y. Zhou, Q. Zhang, S. Zou, and X. Jiang, RiskRanker: Scalable and accurate zero-day Android malware detection, *Presented at the Proceedings of the 10th International Conference on Mobile Systems, Applications, and Services*, Lake District, 2012.

12. C. Wang, Z. Wu, X. Li, X. Zhou, A. Wang, and P. C. K. Hung, SmartMal: A service-oriented behavioral malware detection framework for mobile devices, *The Scientific World Journal*, vol. 2014, p. 101986, 2014.

13. Y. Aafer, W. Du, and H. Yin, DroidAPIMiner: Mining API-level features for robust malware detection in Android, in *Security and Privacy in Communication Networks*, vol. 127, T. Zia, A. Zomaya, V. Varadharajan, and M. Mao, Eds., Springer, Berlin, Germany, 2013, pp. 86–103.

14. K. O. Elish, X. Shu, D. Yao, B. G. Ryder, and X. Jiang, Profiling user-trigger dependence for Android malware detection, *Computers & Security*, vol. 49, pp. 255–273, 2015.

15. V. Moonsamy, J. Rong, and S. Liu, Mining permission patterns for contrasting clean and malicious Android applications, *Future Generation Computer Systems*, vol. 36, pp. 122–132, 2014.

16. M. Frank, D. Ben, A. P. Felt, and D. Song, Mining permission request patterns from android and facebook applications, in *Proceedings of the IEEE 12th International Conference on Data Mining (ICDM)*, 2012, pp. 870–875.

17. A. P. Felt, M. Finifter, E. Chin, S. Hanna, and D. Wagner, A survey of mobile malware in the wild, *Presented at the Proceedings of the 1st ACM Workshop on Security and Privacy in Smartphones and Mobile Devices*, Chicago, IL, 2011.

18. S. C. Madeira and A. L. Oliveira, Biclustering algorithms for biological data analysis: A survey, *IEEE/ACM Transactions on Computational Biology and. Bioinformatics*, vol. 1, pp. 24–45, 2004.

19. G. J. Szekely and M. L. Rizzo, Hierarchical clustering via joint between-within distances: Extending ward's minimum variance method, *Journal of Classification*, vol. 22, pp. 151–183, 2005.

20. A. Fernández and S. Gómez, Solving non-uniqueness in agglomerative hierarchical clustering using multidendrograms, *Journal of Classification*, vol. 25, pp. 43–65, 2008.

21. P. Xiong, X. Wang, W. Niu, T. Zhu, and G. Li, Android malware detection with contrasting permission patterns, *China Communications*, vol. 11, pp. 1–14, 2014.

22. B. Amos, H. Turner, and J. White, Applying machine learning classifiers to dynamic Android malware detection at scale, in *Proceedings of the 9th International Wireless Communications and Mobile Computing Conference*, 2013, pp. 1666–1671.

23. B. Sanz, I. Santos, C. Laorden, X. Ugarte-Pedrero, P. Bringas, and G. Álvarez, PUMA: Permission usage to detect malware in Android, in *International Joint Conference CISIS'12-ICEUTE'12-SOCO'12 Special Sessions*, vol. 189, Á. Herrero, V. Snášel, A. Abraham, I. Zelinka, B. Baruque, H. Quintián et al., Eds., Springer, Berlin, Germany, 2013, pp. 289–298.

24. G. Dini, F. Martinelli, A. Saracino, and D. Sgandurra, MADAM: A multi-level anomaly detector for Android malware, in *Computer Network Security*, vol. 7531, I. Kotenko and V. Skormin, Eds., Springer, Berlin, Germany, 2012, pp. 240–253.

25. A. Shabtai and Y. Elovici, Applying behavioral detection on Android-based devices, in *Mobile Wireless Middleware, Operating Systems, and Applications*, vol. 48, Y. Cai, T. Magedanz, M. Li, J. Xia, and C. Giannelli, Eds., Springer, Berlin, Germany, 2010, pp. 235–249.

26. A. Shabtai, U. Kanonov, Y. Elovici, C. Glezer, and Y. Weiss, "Andromaly": A behavioral malware detection framework for Android devices, *Journal of Intelligent Information Systems*, vol. 38, pp. 161–190, 2012.

27. V. Rastogi, C. Yan, and J. Xuxian, Catch me if you can: Evaluating android anti-malware against transformation attacks, *IEEE Transactions on Information Forensics and Security*, vol. 9, pp. 99–108, 2014.

28. K. Allix, T. F. Bissyand, Q. Jerome, J. Klein, R. State, Y. L. Traon, Large-scale machine learning-based malware detection: Confronting the "10-fold cross validation" scheme with reality, *Presented at the Proceedings of the 4th ACM Conference on Data and Application Security and Privacy*, San Antonio, TX, 2014.

29. H. Gascon, F. Yamaguchi, D. Arp, and K. Rieck, Structural detection of Android malware using embedded call graphs, *Presented at the Proceedings of the 2013 ACM Workshop on Artificial Intelligence and Security*, Berlin, Germany, 2013.

30. W. Dong-Jie, M. Ching-Hao, W. Te-En, L. Hahn-Ming, and W. Kuo-Ping, DroidMat: Android malware detection through manifest and API calls tracing, in *Proceedings of the 7th Asia Joint Conference on Information Security*, Tokyo, Japan, 2012, pp. 62–69.

31. A. Shabtai, L. Tenenboim-Chekina, D. Mimran, L. Rokach, B. Shapira, and Y. Elovici, Mobile malware detection through analysis of deviations in application network behavior, *Computers & Security*, vol. 43, pp. 1–18, 2014.

32. B. Sanz, I. Santos, C. Laorden, X. Ugarte-Pedrero, J. Nieves, P. G. Bringas et al., MAMA: Manifest analysis for malware detection in Android, *Cybernetics Systems*, vol. 44, pp. 469–488, 2013.

33. H. Yi-an, F. Wei, L. Wenke, and P. S. Yu, Cross-feature analysis for detecting ad-hoc routing anomalies, in *Proceedings of the 23rd International Conference on Distributed Computing Systems*, Providence, RI, 2003, pp. 478–487.

34. K. Noto, C. Brodley, and D. Slonim, Anomaly detection using an ensemble of feature models, in *Proceedings of the IEEE 10th International Conference on Data Mining*, 2010, pp. 953–958.

35. Z. Yuan, Y. Lu, Z. Wang, and Y. Xue, Droid-Sec: Deep learning in Android malware detection, *SIGCOMM Computer Communication Review*, vol. 44, pp. 371–372, 2014.

36. Y. Bengio, Learning deep architectures for AI, *Foundation and Trends in Machine Learning*, vol. 2, pp. 1–127, 2009.

37. Y. Feng, S. Anand, I. Dillig, and A. Aiken, Apposcopy: Semantics-based detection of Android malware through static analysis, *Presented at the Proceedings of the 22nd ACM SIGSOFT International Symposium on Foundations of Software Engineering*, Hong Kong, China, 2014.

38. M. Zhang, Y. Duan, H. Yin, and Z. Zhao, Semantics-aware Android malware classification using weighted contextual api dependency graphs, *Presented at the Proceedings of the 2014 ACM SIGSAC Conference on Computer and Communications Security*, Scottsdale, AZ, 2014.

39. G. Suarez-Tangil, J. E. Tapiador, P. Peris-Lopez, and J. Blasco, Dendroid: A text mining approach to analyzing and classifying code structures in Android malware families, *Expert Systems with Applications*, vol. 41, pp. 1104–1117, 2014.

40. J. Xu, Y. Yu, Z. Chen, B. Cao, W. Dong, Y. Guo et al., MobSafe: Cloud computing based forensic analysis for massive mobile applications using data mining, *Tsinghua Science and Technology*, vol. 18, pp. 418–427, 2013.

41. T. Petsas, G. Voyatzis, E. Athanasopoulos, M. Polychronakis, and S. Ioannidis, Rage against the virtual machine: Hindering dynamic analysis of Android malware, *Presented at the Proceedings of the 7th European Workshop on System Security*, Amsterdam, the Netherlands, 2014.

42. W. Wei, W. Xing, F. Dawei, L. Jiqiang, H. Zhen, and Z. Xiangliang, Exploring permission-induced risk in android applications for malicious application detection, *Proceedings of the IEEE Transactions on Information Forensics and Security*, vol. 9, pp. 1869–1882, 2014.

Index

Note: Page numbers followed by '*f*' and '*t*' refer to figures and tables, respectively.

Printed in the United States
By Bookmasters